Modern Web Cartography

Open Source Map Solutions with OpenStreetMap, Overpass, Nominatim, Leaflet and Folium

Patrick Marie

Apress®

Modern Web Cartography: Open Source Map Solutions with OpenStreetMap, Overpass, Nominatim, Leaflet and Folium

Patrick Marie
Paris, France

ISBN-13 (pbk): 979-8-8688-2334-3 ISBN-13 (electronic): 979-8-8688-2335-0
https://doi.org/10.1007/979-8-8688-2335-0

Managing Director, Apress Media LLC: Welmoed Spahr
Acquisitions Editor: James Robinson-Prior
Editorial Assistant: Gryffin Winkler

Cover designed by eStudioCalamar

Cover image designed by TungArt7 on pixabay

Distributed to the book trade worldwide by Springer Science+Business Media New York, 1 New York Plaza, New York, NY 10004. Phone 1-800-SPRINGER, fax (201) 348-4505, e-mail orders-ny@springer-sbm.com, or visit www.springeronline.com. Apress Media, LLC is a Delaware LLC and the sole member (owner) is Springer Science + Business Media Finance Inc (SSBM Finance Inc). SSBM Finance Inc is a **Delaware** corporation.

For information on translations, please e-mail booktranslations@springernature.com; for reprint, paperback, or audio rights, please e-mail bookpermissions@springernature.com.

Apress titles may be purchased in bulk for academic, corporate, or promotional use. eBook versions and licenses are also available for most titles. For more information, reference our Print and eBook Bulk Sales web page at http://www.apress.com/bulk-sales.

Any source code or other supplementary material referenced by the author in this book is available to readers on GitHub. For more detailed information, please visit https://www.apress.com/gp/services/source-code.

If disposing of this product, please recycle the paper.

To my father, who was an enthusiast of geography and passed his passion on to me.

Table of Contents

About the Author

Patrick Marie is a software engineer with over 38 years' experience working for IBM in various technical roles, covering a wide range of topics including technical publishing, image archiving, workflow, application integration, web application servers, enterprise service bus, and business activity monitoring. Recently retired, Patrick is an enthusiast of mapping and cartography and enjoys combining his love of maps and his passion for programming and open source software.

Patrick enjoys traveling, especially by bike and on foot. He often creates customized maps for his trips using open source tools like those described in this book, which he then uses on his mobile phone while traveling.

He is the author of *Le Guide des PAO* (published by DUNOD).

About the Technical Reviewer

Nikhil Kapoor brings over 16 years of experience in artificial intelligence and machine learning, specializing in building scalable models, intelligent platforms, and enterprise-grade solutions. He has delivered end-to-end AI/ML systems at a major cloud provider, integrating cloud-native technologies to drive innovation. His work spans finance, investment management, audit, and telecom R&D, and he has led teams across the full software development lifecycle.

Acknowledgments

I would like to thank the people who helped me complete this book:

- **My friend and former colleague at IBM, Carl Farkas**, who spent a great deal of time and did an excellent job reviewing the technical content and refining the style of this book. Carl is a native English speaker (which I am not), a great IT professional, and an accomplished cyclist. We went on many cycle touring trips together.
- **My friend and former engineering schoolmate, Stéphane Faroult**, who encouraged me to write the book in English and guided me in approaching American publishers. Stéphane is a renowned database expert and the author of *The Art of SQL*, a highly respected reference on relational databases. He has written and published many advanced computing books in the United States.
- **My publisher, James Robinson-Prior**, who believed in my book project and made it possible for me to bring it to life, and **all the people at Apress** who helped me turn my manuscript into a real book.
- **My family**, for their unwavering support during this project.

Many thanks to all of them!

Introduction

From time immemorial, people have sought to represent the space in which they live through maps. These maps have served a variety of purposes:

- For travelers, whether merchants, soldiers, migrants, or, more recently, tourists, they enabled more efficient, faster, and safer movement.
- For land owners and managers, they facilitated the management of agricultural plots, estates, seigneuries, kingdoms, and states.
- For residents, they offered a better understanding of their region, their country, and neighboring countries. This knowledge led to a greater grasp of the world's physical, social, and economic realities and helped them to better understand events covered by the media.
- For emigrants, they could serve as a means of remembering the country they had left behind.
- For dreamers, they enabled the imagination of faraway places they might never visit.

Maps are flat representations of territories (with a few exceptions, such as globes, relief maps, and the 3D maps now offered by computer technology). These representations must be faithful to the geometry of the territory. However, they are not intended to render natural reality in its entirety (otherwise, maps based on satellite photos would represent the pinnacle of cartography). Instead, maps aim to provide readers with information about a given territory according to a specific theme:

- Topographic maps, for example, show relief (with contour lines), watercourses, coastlines, and natural terrain features in detail. They are particularly useful for hikers or bikers.
- Human geography maps highlight urban agglomerations, population densities, and economic activities (industry, agriculture, mining, etc.).

- Road maps show routes and distances.
- Tourist maps highlight places to visit and show how to reach them.
- Cadastral maps show the division of land into parcels.
- Etc.

For centuries, maps relied on written media: first clay tablets, then parchments, wall frescoes, globes, and then paper. Although paper maps still exist, they have now been largely supplanted by digital maps accessible on electronic devices. This transformation has been underpinned by several major technological developments: the rise of the personal computer, the spread of the Internet, and the democratization of GPS and smartphones. Today, mapping and navigation applications are ubiquitous: they are integrated into all smartphones, and most new cars are equipped with GPS. These devices can even be found on bicycles and tractors.

This explosion in digital maps has been accompanied by a significant development in software for designing, displaying, and using them. While the first tools were mainly proprietary solutions developed by companies such as Esri, Google, or Microsoft, the open source movement has grown considerably. It now enables users to free themselves from the constraints imposed by commercial solutions while fostering innovation and collaboration.

It was in this context that the OpenStreetMap project emerged, accompanied by a rich ecosystem of open source tools, offering a free, collaborative, and powerful alternative for creating and exploiting digital maps.

The aim of this book is to facilitate the understanding and rapid acquisition of digital map creation skills within an open source web environment. It focuses on OpenStreetMap and a set of complementary tools, including Overpass, Nominatim, Leaflet, and Folium. These tools are introduced concisely and illustrated with practical examples of map production, integrated into interactive web applications.

Readers are encouraged to replicate these examples in their own development environments. Some are simple to reproduce and require no more than a web browser and an HTML file, while others – representative of real-world web applications – require additional tools for implementation and testing. Typically, web applications in this book will be developed and executed using the following technologies:

PART I

Exploring the Open Source Mapping Ecosystem and Core Tools

In this part, we will present an overview of the mapping landscape, focusing particularly on OpenStreetMap and its related projects.

Each tool will then be explored individually through simple examples demonstrating its major functionalities:

- **Overpass**: To retrieve OpenStreetMap data using general queries
- **Nominatim**: For geocoding and reverse geocoding requests
- **Leaflet**: To create maps using JavaScript code executed in the browser
- **Folium**: To build maps using Python code executed on the server

Part 2 will then explain how to use these tools together to build more advanced maps.

CHAPTER 1

Introduction to Open Source Mapping Tools and the Digital Mapping Landscape

Digital cartography is typically associated with interactive maps displayed on the screen of a computer, tablet, smartphone, or GPS device. However, this represents only the tip of the iceberg. Behind it lie:

- Large volumes of data stored in databases: objects to be represented on maps, their geographical positions, and the information describing them.
- Software to store and update this data, as well as to access and process it, i.e., to search for and extract data, particularly for the purpose of building maps.

The mapping tools described and used in this book are open source and part of the broader OpenStreetMap ecosystem.

Open source software is generally available in source-code form and can be freely used, modified, and redistributed under its license terms. Such software is often the fruit of the collaboration of numerous developers, although some projects are also maintained by individuals or small teams.

Open source software differs from **proprietary** software, which is produced and marketed by software publishers. Proprietary software is subject to licenses (with the exception of freeware, such as Skype and VS Code), which customers must purchase

P. Marie, *Modern Web Cartography*, https://doi.org/10.1007/979-8-8688-2335-0_1

in order to use it. Proprietary software is developed by the internal teams of software publishing companies, with its source code kept confidential and remaining the exclusive property of the publisher.

The open source movement, as we know it today, emerged in the late 1990s and has grown steadily ever since. According to *The Business Research Company*, the open source market was estimated at around **USD 41.8 billion** in 2024, corresponding to approximately **5–6%** of the global software market (valued at **USD 730–740 billion**). This revenue comes primarily from services, as the software itself is typically free to use, although its distribution or commercial packaging may sometimes incur costs.

According to *GitHub's Octoverse 2022* report, about **90%** of companies report using open source software in some way. More precisely, in the information systems of these companies, open source software generally coexists with proprietary software.

OpenStreetMap

OpenStreetMap (often abbreviated to **OSM**) is a collaborative online mapping project that aims to build a geographic database in the open source world. The project was launched in 2004 by British entrepreneur Steve Coast. University College London subsequently supported the project, notably by hosting some of its servers.

The data in the OSM database is updated by a large number of volunteer participants worldwide (several million contributors are registered). Contributors can base their updates on GPS tracks they have produced, on field observation, on aerial and satellite photos, or even on their personal knowledge of a local territory. Automatic updating of OSM from artificial intelligence tools analyzing aerial images is not currently permitted: data produced by AI tools, or supplied by government institutions, must be reviewed and validated by a human before being introduced into OSM.

OSM data can be queried and used freely via APIs, including Overpass and Nominatim, which will be used in the remainder of this book. More specifically, OSM data is available under the Open Database License (ODbL), guaranteeing its free use subject to attribution and identical sharing.

OSM is used by technology companies, government institutions, and humanitarian projects such as the Humanitarian OpenStreetMap Team (HOTOSM). Companies such as Meta, Microsoft, and Tesla also use it for their maps and artificial intelligence algorithms. Finally, a large number of websites and cell phone applications rely on OSM data, such as OsmAnd, Maps.me, OSRM, Komoot, GraphHopper, OpenRouteService, Strava, and Amazon Logistics.

OpenStreetMap also offers a website (`https://www.openstreetmap.org`) for displaying maps and searching for places and routes in a way comparable to other online mapping sites (such as Google Maps, Bing Maps, Mappy, etc.).

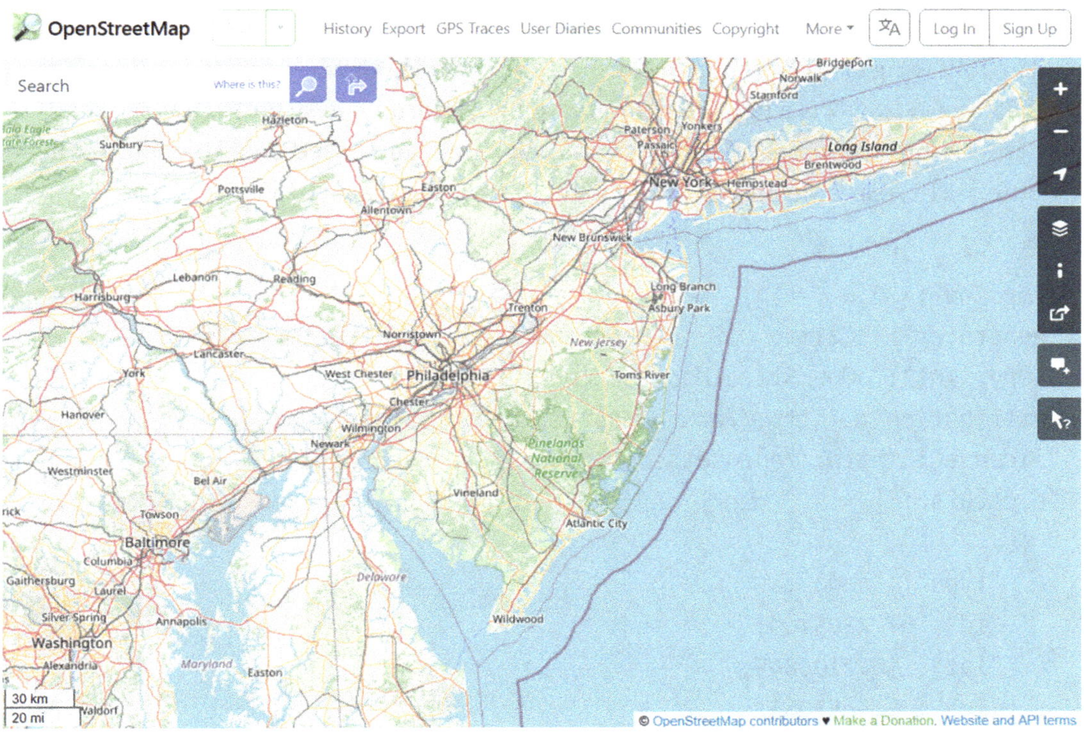

Figure 1-1. *Screenshot of the* `www.openstreetmap.org` *website using map data© OpenStreetMap contributors, under ODbL*

The heart of OpenStreetMap is a database containing a vast amount of geographic data on every country in the world. From an external perspective, this data falls into three main categories:

- **Nodes**: Geographical points identified by latitude and longitude, to which a certain amount of information can be added via tags.

- **Ways**: Polylines (lines made up of contiguous segments) or polygons, defined by the nodes constituting their vertices and extremities (i.e., corners and ends). Ways can be completed with information added via tags.
- **Relations**: More complex objects that can be made up of nodes, ways, and even other relations. Additional information can be associated with relations via tags.

The following example, in JSON format, gives an idea of what data is stored in the OSM database (JSON is a lightweight data exchange format, easy for humans to read and write, simple for machines to parse and generate, and based on a key/value pair structure). This example only includes a node, a way, and a relation:

```
{
  "version": "0.6",
  "generator": "CGImap 0.8.1 (6263 thorn-01.openstreetmap.org)",
  "copyright": "OpenStreetMap and contributors",
  "attribution": "http://www.openstreetmap.org/copyright",
  "license": "http://opendatacommons.org/licenses/odbl/1-0/",
  "elements": [
    {
      "type": "node",
      "id": 6772305769,
      "lat": 50.2106895,
      "lon": 8.5857061,
      "timestamp": "2019-09-04T15:05:00Z",
      "version": 1,
      "changeset": 74092935,
      "user": "Jonaes",
      "uid": 4433367,
      "tags": {
        "bicycle": "yes",
        "information": "guidepost",
        "ref": "OU.425.1",
        "tourism": "information"
      }
```

```
},
{
  "type": "way",
  "id": 5123563,
  "timestamp": "2019-09-04T15:05:00Z",
  "version": 7,
  "changeset": 74092935,
  "user": "Jonaes",
  "uid": 4433367,
  "nodes": [
    35352060,
    35352061,
    35352062,
    35352063,
    2056819178,
    35352001
  ],
  "tags": {
    "bicycle": "yes",
    "foot": "yes",
    "highway": "track",
    "horse": "yes",
    "lcn": "yes",
    "motor_vehicle": "agricultural;forestry",
    "surface": "asphalt",
    "tracktype": "grade1",
    "traffic_sign": "DE:260,DE:1026-36"
  }
},
{
  "type": "relation",
  "id": 10000000,
  "timestamp": "2019-09-18T18:21:44Z",
  "version": 2,
  "changeset": 74641141,
```

```
      "user": "HoloDuke",
      "uid": 75317,
      "members": [
        {
          "type": "node",
          "ref": 6772305769,
          "role": "sign"
        },
        {
          "type": "node",
          "ref": 35352001,
          "role": "intersection"
        },
        {
          "type": "way",
          "ref": 5123563,
          "role": "to"
        }
      ],
      "tags": {
        "bicycle": "yes",
        "color:arrow": "green",
        "color:back": "white",
        "color:text": "green",
        "destination": "Friedrichsdorf;Bad Homburg",
        "destination:symbol": ";train_station",
        "distance": "8.0;4.6",
        "type": "destination_sign"
      }
    }
  ]
}
```

Overpass, Nominatim, and Other OSM APIs

OpenStreetMap offers developers several types of API for accessing and updating its data.

Overpass

Overpass is a general-purpose API that provides a simple and efficient way of querying and extracting data from the OpenStreetMap geographic database. Unlike OpenStreetMap's Main API (explained later), which is designed for managing map objects (creation, modification, deletion), Overpass is dedicated solely to data search and retrieval and is optimized for this purpose.

Overpass exposes an API that enables OSM data searches via HTTP requests. The syntax of these requests can be based on:

- A dedicated query language (Overpass Query Language), comparable to SQL, but specifically designed for geographic data retrieval. This language allows users to specify the types of objects to be retrieved, the selection criteria for their attributes, and the geographic area in which to search for them. This query language is documented on several sites, including `https://wiki.openstreetmap.org/wiki/Overpass_API/Language_Guide` and `https://dev.overpass-api.de/overpass-doc/en/index.html`. It will be described in the next chapter.
- XML can also be used to build requests, using the Overpass XML format. However, Overpass QL is generally preferred over Overpass XML because it is more compact and human-readable. Still, some applications and tools may expect queries in XML format.

Although Overpass is the officially recommended OpenStreetMap API, it does not directly query the OSM database. OSM data is regularly extracted to feed and maintain a database dedicated to Overpass, which is optimized for fast searches.

Overpass is complemented by Overpass Turbo (`https://overpass-turbo.eu/`), a website for testing and developing queries using the Overpass Query Language (or Overpass XML). Overpass queries are executed and their results displayed either on a map or in text form. The main page of Overpass Turbo is shown here under.

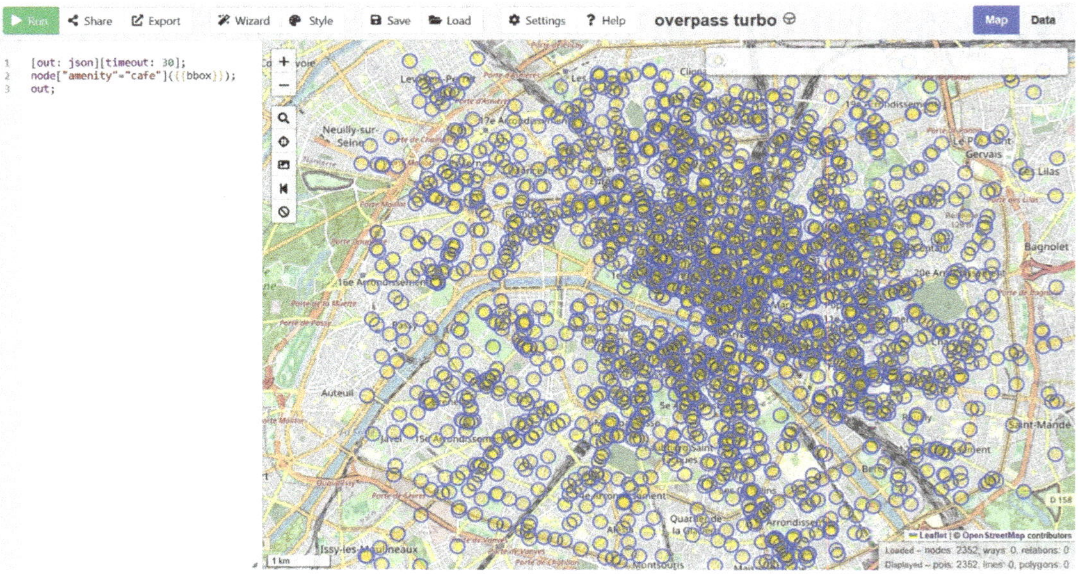

Figure 1-2. *Screenshot of Overpass Turbo (overpass-turbo.eu) after having executed a query, interface licensed under MIT. Map data © OpenStreetMap contributors (ODbL)*

In this example, the Overpass request on the left searches for all cafés located within the map area displayed on the right (Paris, France). Once the search is executed, the cafés found are highlighted with blue circles on the map.

Overpass was designed with public access in mind, but self-hosting is also supported: it is possible to download the Overpass code and install it on private servers. This approach is recommended for heavy or frequent querying, sensitive applications, or when offline access or full control over the data is required.

The Overpass API will be used to build several prototypes in the remainder of this book.

Nominatim

Nominatim is a geocoding engine based on OpenStreetMap. It enables:

- **Geocoding (or direct geocoding)**: Determine the geographic coordinates (latitude and longitude) of a location from an address or name.

- **Reverse geocoding**: Determine a location's address from its geographic coordinates.

Nominatim does not directly query the OpenStreetMap database. Its data is regularly extracted from the OSM database and stored in another database, in a form indexed and optimized for geocoding and reverse geocoding searches. It is this database that is used to respond to queries received by Nominatim.

Nominatim exposes an API enabling geocoding searches via HTTP requests. The query syntax is documented on the following site: `https://nominatim.org/release-docs/latest/api/Overview/`, and will be summarized in the chapter dedicated to this component.

Nominatim also provides a web user interface (`https://nominatim.openstreetmap.org/ui/`) for testing geocoding and reverse geocoding queries.

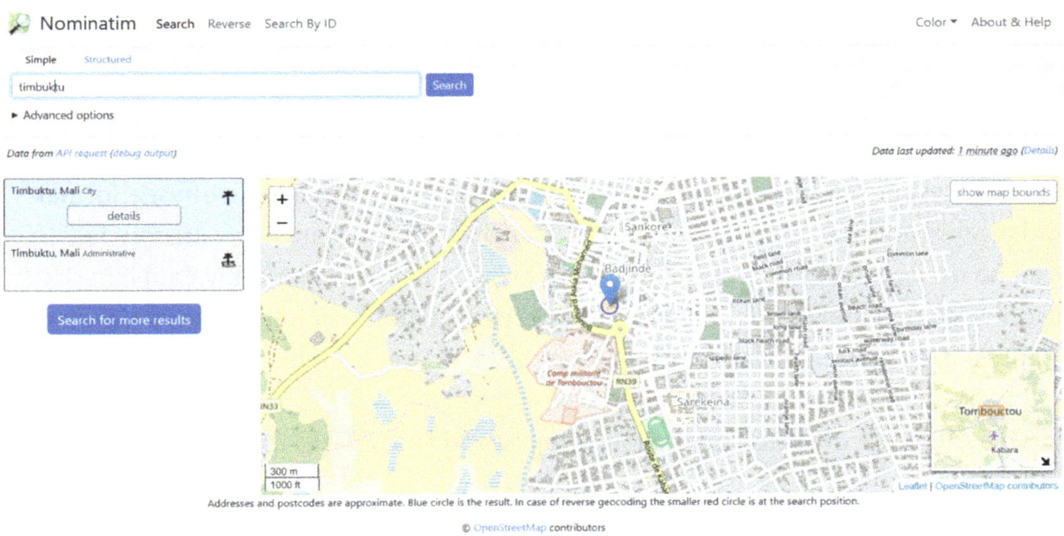

Figure 1-3. *Screenshot of Nominatim website (`nominatim.openstreetmap.org`) after having executed a geocoding search, licensed under GPL, using OpenStreetMap data © OpenStreetMap contributors (ODbL)*

In this example, a place called "Timbuktu" is searched for. The results are listed on the left (two results: the city of Timbuktu, Mali, and its administrative region), and the map of the first result (the city of Timbuktu) is displayed on the right, with a marker placed at the center of the city.

Like Overpass, Nominatim can be used either through its public API and web interface or via private self-hosting (its source code can be downloaded and installed on private servers):

- Public access is recommended for small volumes, testing, and educational purposes.
- Self-hosting is preferred for high volumes, sensitive applications, or when offline access or full control over the data is required.

The Nominatim API will be used to build several prototypes in the remainder of this book.

Nominatim is not the only solution available for accessing geocoding services in the open source sector. Other tools exist and are based on OpenStreetMap data or other sources: Photon, Pelias, and Gisgraphy. These tools will not be used in this book.

Other APIs

Another API, the **Main API**, enables interaction with the OSM database via HTTP/REST requests. Responses are returned in XML format (XML is a flexible text-based format used to store and transport data; it allows users to structure data using custom tags). It can be used to query the OSM database, as well as to add and modify data in it (it is the only API that allows this). In terms of querying, it limits searches to very small areas (squares of less than 5° in latitude and longitude). In practice, this API should only be used to modify data in the OSM database.

This API is documented on the following site: `https://wiki.openstreetmap.org/wiki/API`.

Every week, all the data in the OSM database is copied in various formats to another database called **Planet.osm**, to make it available to those who want to use it. Tools are available to download extracts of this data by selecting geographical zones.

To find out more about Planet.osm, visit `https://wiki.openstreetmap.org/wiki/Planet.osm`.

It should also be noted that the German company **Geofabrik** makes available, in the form of downloadable files, data extracted from OSM for various zones. Some data is available free of charge, while others require payment for advanced services or specific extractions.

To find out more about Geofabrik, see `https://www.geofabrik.de`.

The following table provides a summary of the various APIs that can be used to work with OSM.

Table 1-1. *OSM APIs*

API	Functionality
Overpass API	A general-purpose API widely used for querying data from OSM. It provides a powerful query language that enables complex queries with advanced filters.
Nominatim API	API dedicated to geocoding and reverse geocoding. It is restricted to this specific use case.
Main API	The core OSM API, designed for reading, writing, and editing OSM data. It allows basic object retrieval, but does not support advanced filtering. It is the only API that supports writing and editing OSM data.

Leaflet, Folium, and Other Map-Building Tools

The Overpass API and other OpenStreetMap APIs can be used to query the OpenStreetMap database, but they cannot be used to build maps to present this data. For this, you need to use a complementary tool, and this is where the open source tools Leaflet and Folium come in.

Leaflet

Leaflet.js is a JavaScript library for building interactive maps created in 2011 by Ukrainian developer Volodymyr Agafonkin, then working at CloudMade.

Leaflet lets you build maps using map tiles - small pieces of the map - supplied on demand by a server. The term **tile** is used because a map is built by placing several tiles side by side, until the complete map is obtained (in the same way that a roof is built by juxtaposing tiles).

Leaflet also lets you enrich tile-based maps by adding various elements:

- Add markers to highlight specific points and objects.
- Draw polylines and various geometric figures, as well as delimit and fill areas using polygons.

- Display information in the form of popups (text in a bubble displayed when the user clicks on an object) and tooltips (text in a bubble displayed when the mouse cursor passes over an object).
- Images superimposed on maps.
- Overlay of geographic data in GeoJSON and TopoJSON formats, resulting in markers or polylines being added to the map.
- Construction of choropleth maps (i.e., thematic maps where geographic zones – countries, regions, departments, etc. – are colored according to the value of an indicator, e.g., population density or unemployment rate).

It also stands out for its ability to create highly interactive maps, i.e., offering users the possibility of interacting with the maps.

Leaflet.js is documented on the following site: `https://leafletjs.com/reference.html`, and Chapter 4 is devoted to it later in this book.

For example, the following HTML code, which includes Leaflet statements on lines 11 to 16, will display a map of the world (this code is available in the file `leaflet_maps/page_hello_world.html` on the website `https://github.com/Apress/Modern-Web-Cartography`):

```
1  <html>
2  <head>
3  <link rel="stylesheet" href="https://unpkg.com/leaflet@1.9.4/
   dist/leaflet.css" integrity="sha256-p4NxAoJBhIIN+hmNHrzRCf9tD/
   miZyoHS5obTRR9BMY=" crossorigin=""/>
4  <title>OSM map created with Leaflet</title>
5  </head>
6  <body>
7  <h2>Hello World!</h2>
8  <div id="map" style="width: 1030px; height: 620px;"></div>
9  <script src="https://unpkg.com/leaflet@1.9.4/dist/leaflet.js"
   integrity="sha256-20nQCchB9co0qIjJZRGuk2/Z9VM+kNiyxNV1lvTlZBo="
   crossorigin=""></script>
10 <script type="text/javascript">
11 let map = L.map('map').setView([0, 0], 2);
12
14
```

```
L.tileLayer('https://tile.openstreetmap.org/{z}/{x}/{y}.png', {
    maxZoom: 19,
    attribution: '&copy; <a href="http://www.openstreetmap.org/copyright">OpenStreetMap</a>'
}).addTo(map);
</script>
</body>
</html>
```

After getting this file in your own environment, open it in a browser using the following URL: `file:///C:/modern_web_cartography/leaflet_maps/page_with_simple_map.html` (adjust the path as needed, depending on where you saved the file).

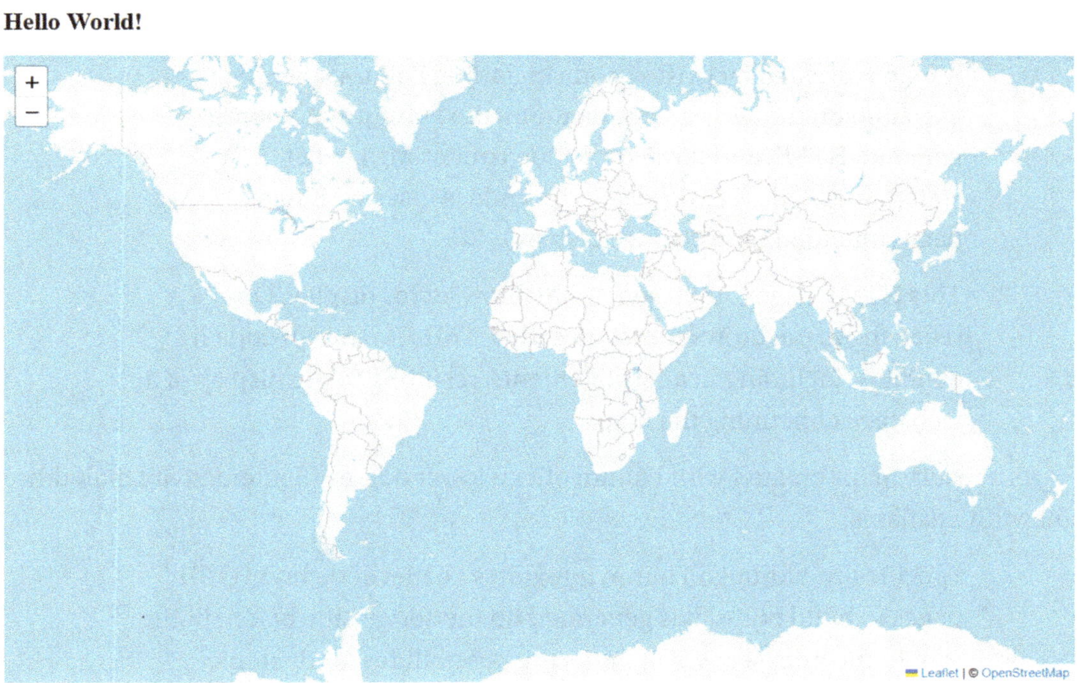

Figure 1-4. *Map resulting from the previous code and created with Leaflet using OpenStreetMap tiles. Data © OpenStreetMap contributors, licensed under the ODbL*

Leaflet is a client-side library executed in the browser through JavaScript code embedded in HTML pages. Its interactive features require a browser environment to function. Although there are JavaScript-enabled web servers such as Node.js, Leaflet is not designed to run in such environments.

Leaflet will be used in many prototypes and on the client side (running in a browser) in the remainder of this book.

Folium

Folium is a Python library created in 2013 by American developer Rob Story, which generates interactive maps based on Leaflet.js. Folium offers functionality similar to that of Leaflet.js, but integrates better into a Python development environment, particularly for implementing data analysis and visualization applications.

Folium is documented on the website `https://python-visualization.github.io/folium/latest/index.html`, and Chapter 5 is devoted to it later in this book.

Folium is linked to the Python language and can only be run on web servers supporting this language. Its operation is based on server-side map generation:

1. A map is created by Python code including Folium statements executed on the server. On execution, the Folium statements generate HTML and JavaScript code containing Leaflet statements. This HTML/JavaScript code is inserted into an HTML page (intended for web page display).

2. This HTML page is then sent to the browser for display. On receiving it, the browser executes the HTML/JavaScript code it contains, including Leaflet statements, resulting in the display of a web page containing the map.

As a result, maps created with Folium offer a lower degree of interactivity than those built with Leaflet.js:

- Specifically, although Folium integrates Leaflet code into HTML pages to build maps, it is generated on the server side before being sent to the browser, which limits the possibilities for dynamic interaction.

- Conversely, Leaflet.js works and builds maps in the browser and offers full latitude, via event management (mouse clicks, key presses, dragging, zooming, etc.), to dynamically modify maps in response to user interactions.

To implement the same level of interactivity with Folium, when the user requests a modification to the map (going beyond what is natively possible with maps created using Folium), it is necessary to inject JavaScript code with Leaflet statements into the map while building it with Folium on the server (before sending it to the browser in an HTML page). Sometimes, when the user interacts with the map, it is also necessary to go back through the Python code on the server and completely regenerate the map. Some prototypes presented in Chapters 7 and 11 will demonstrate this approach.

Folium has the advantage of integrating well with common geospatial Python tools: you can easily visualize data from Pandas/GeoPandas DataFrames and from Shapefiles, after loading them with GeoPandas and passing them to Folium as GeoJSON (Shapefile is a format for representing cartography data created by Esri and used as a de facto standard by Desktop cartography tools). Folium is also a little more concise than Leaflet (it can sometimes do with a single statement what Leaflet requires several to accomplish), which makes it a little easier to use.

Leaflet, on the other hand, is a little more powerful (most of Leaflet functionalities are exposed by Folium, but not all of them) and should be preferred for building maps that offer high levels of interactivity.

In many cases, however, both tools are suitable, and the programmer chooses which he or she prefers. In the remainder of this book, Leaflet and Folium will be used extensively in prototypes and in roughly equal measure.

Other Tools

Leaflet.js and Folium are widely used today, but they are not the only open source tools available for building maps into web applications. Others include:

- **Mapnik**: A powerful, server-side map-building engine written in C++, with bindings available for Python and JavaScript. It is mainly used to generate raster and vector maps with great flexibility.
- **OpenLayers**: A JavaScript library for building and displaying interactive maps in the browser. Occupying the same niche as Leaflet.js, it is more complete than the latter in terms of functionality, but also more complex to master.

- **MapLibre GL JS**: A JavaScript (and TypeScript) library for building and displaying maps in the browser using WebGL (a JavaScript library that enables rendering interactive 2D and 3D graphics directly in web browsers, without needing plugins). It offers high performance and excels at creating maps with vector tiles. Its complexity is intermediate between Leaflet and OpenLayers.

  MapLibre GL JS is forked from Mapbox GL JS after version 1.13.0, when Mapbox changed its license to a non-open-source model. It aims to provide functionalities comparable to Mapbox, without relying on proprietary platforms.

The following table provides a summary of the various map-building tools that can be used with OSM.

Table 1-2. *OSM map-building tools*

Map-building tool	Functionality
Leaflet.js	A widely used JavaScript library that is efficient, lightweight, and easy to use and offers advanced event handling. It is used client-side in a browser.
Folium	A popular Python library designed for server-side use. It wraps Leaflet.js and brings its functionality to Python environments. Folium provides mapping capabilities similar to Leaflet, but its support for event handling is more limited.
Mapnik	A powerful server-side map rendering engine compatible with C++, JavaScript, and Python. It is best suited for advanced map generation tasks, such as high-performance tile rendering and custom cartographic styling.
OpenLayers	A JavaScript library comparable to Leaflet, offering more advanced functionality but with a steeper learning curve. It runs client-side in a browser and is generally used in professional contexts where mapping requirements go beyond what Leaflet can provide.
MapLibre GL JS	An open source JavaScript library for rendering interactive, vector-tile-based maps in the browser. It supports fully client-side styling and rendering. It runs client-side in a browser and is ideal for high-performance and highly customizable maps.

Other OpenStreetMap-Related Tools and Components

The following tools and components are also worth mentioning.

TagInfo

TagInfo is an open source tool that provides a better understanding of the tags used to associate information with objects in the OSM database. As each tag is made up of a key and a value (e.g., "building": "school"), TagInfo lets you find out which keys are used in the OSM database, what values are associated with them, what types of objects use them, etc. TagInfo also provides statistics on these questions. For example: the key "building" is used 638,279,144 times (mostly associated with ways, but sometimes also with nodes and relations), with 7,820 different values. Its three most frequent values are "yes" (79.60%), "house" (9.42%), and "residential" (2.44%).

The aim of TagInfo is to enable OSM contributors and users to better know and understand the content of the OSM database:

- For contributors, to know which keys and values already exist and to avoid creating new ones unnecessarily (which would mean creating virtual duplicates).
- For users (who only query OSM), to know which keys are available and what values are associated with them, in order to determine which selection criteria to use to find the objects they are interested in.

TagInfo features a web user interface (`https://taginfo.openstreetmap.org`) for tag analysis:

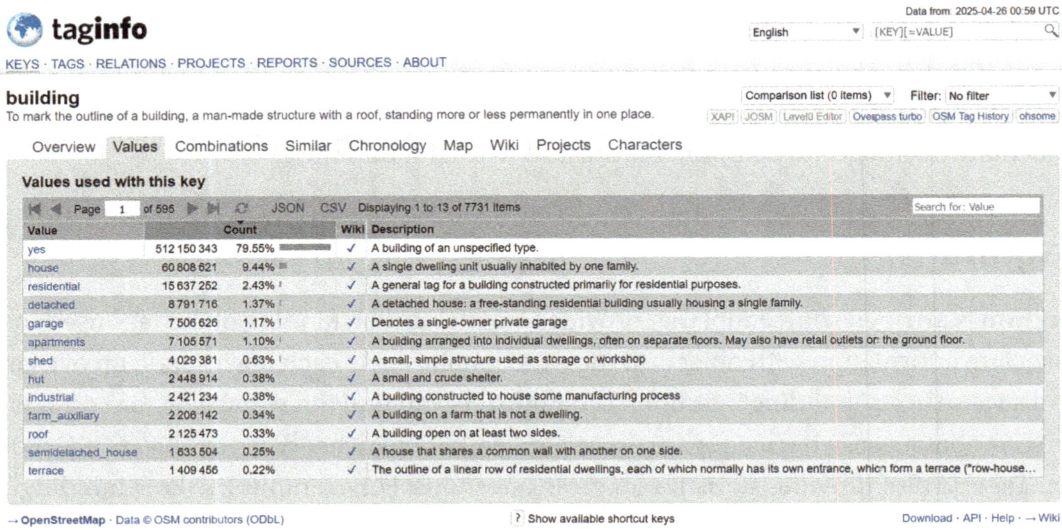

Figure 1-5. *Screenshot of TagInfo (`taginfo.openstreetmap.org`), used under fair use for illustrative purposes, using OpenStreetMap data © OpenStreetMap contributors (ODbL)*

It also features an HTTP API for programmatic querying. For example, the following request:

`https://taginfo.openstreetmap.org/api/4/key/values?key=building&page=1&rp=10&sortname=count_all&sortorder=desc`

returns the values used by the "building" key, limited to the first ten values (because these values are very numerous) and sorted by frequency of use in descending order.

Editors for Creating and Modifying OSM Data

Editors are available for creating and modifying data in the OpenStreetMap database. They handle and exchange data in a format called **OSM XML**, which is an XML-based format designed to represent OSM elements such as `<node>`, `<way>`, `<relation>`, and `<tag>`.

The two main editors are **iD** (a browser-based editor) and **JOSM** (a more advanced desktop editor).

iD

iD is a web-based editor enabling contributors to create and modify data in the OSM database. With this tool, contributors can draw points, lines, and polygons using aerial images. Using tags, they can also add information to the objects they draw.

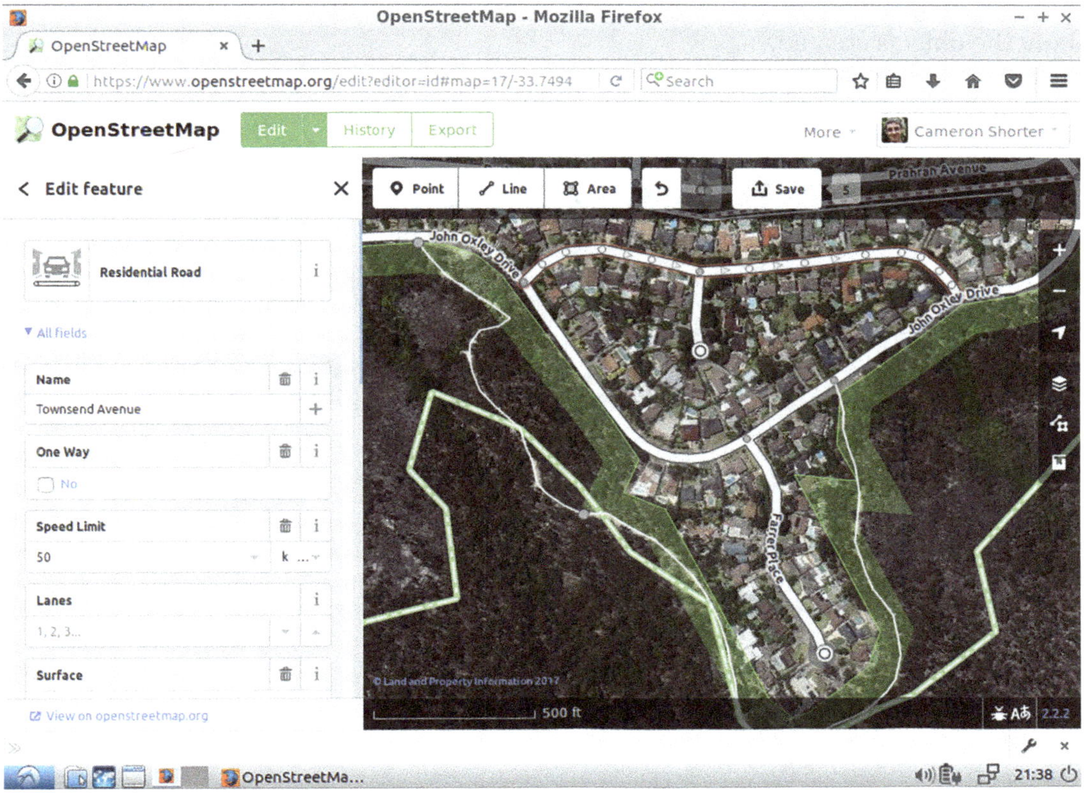

Figure 1-6. Screenshot of iD editor (`openstreetmap.org/edit`), licensed under the ISC license for illustrative purposes, using OpenStreetMap data © OpenStreetMap contributors (ODbL)

It is currently OSM's default editor and the one most used by contributors. It is easy to get to grips with and well suited to new and occasional contributors.

JOSM

JOSM is another editor for experienced OSM contributors. It is implemented as a Java fat client (supported on any environment that can run Java) and offers more advanced features (than iD), particularly when it comes to processing data in batches. It lets you work with GPS tracks and aerial photos. It can also be used to add tags with information about the objects created.

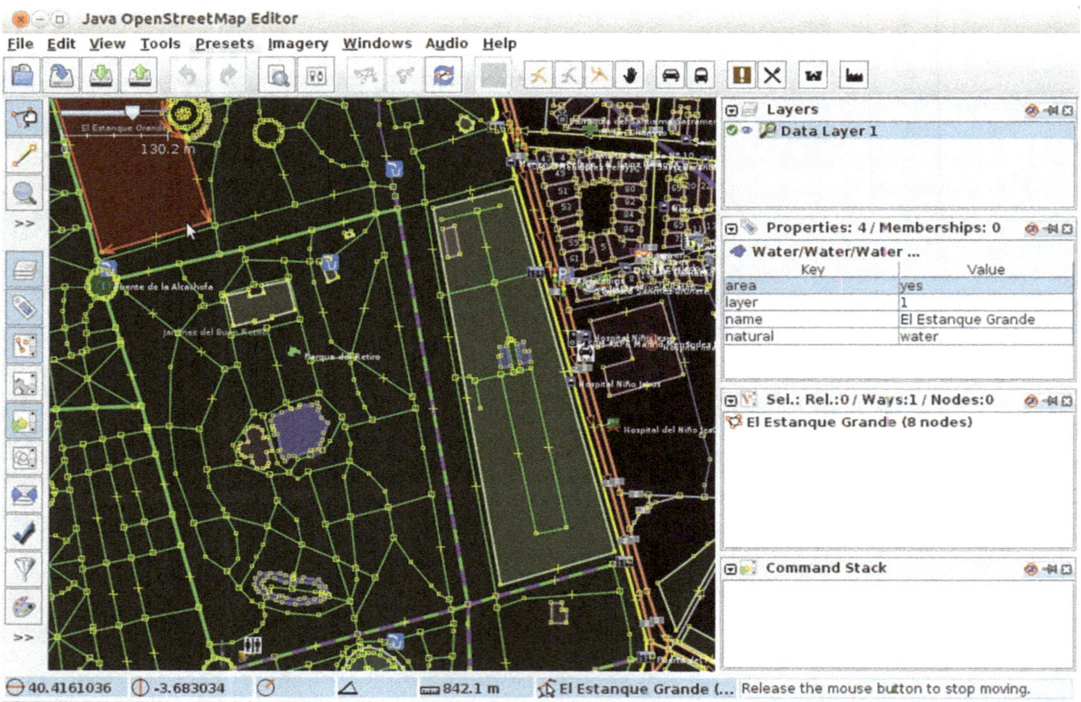

Figure 1-7. *Screenshot of JOSM (`josm.openstreetmap.de`), licensed under GPL, using OpenStreetMap data © OpenStreetMap contributors (ODbL)*

JOSM is also more customizable than iD via plugins. On the other hand, these superior capabilities come at the price of greater complexity of use.

Other Editors

Although iD and JOSM are the two most frequently used editors, the choice of tools available is a little wider. Other options include **RapiD** (a version of iD enhanced with artificial intelligence functions), **StreetComplete** (only on Android cell phones),

Vespucci (also Android), and **Level0** (web-based). **Potlatch** played an important role in the early days of OSM, but is hardly used today.

Tile Servers

Tile servers are used to display maps with OSM and its accompanying tools (Folium and Leaflet.js in particular). As we have seen, the OSM database contains a large amount of geographic data (organized in the form of nodes, ways, and relations), but no images for constructing and displaying maps. Libraries such as Folium and Leaflet.js can generate interactive maps, but they need a source of background images to build them. Tile servers provide these images in the form of small pieces of map, called **tiles**. The term **tile** is used because a map is built by placing several pieces called "tiles" side by side, until a complete map is obtained.

A tile server stores and supplies these tiles on demand, generally in two forms:

- Most often in the form of **raster images**: these can be still images generated from OSM data, satellite photos, etc.
- Sometimes also as **vectors**, for a smoother, more interactive display.

Tile servers offer different map styles:

- General maps (like standard OSM tiles).
- Topographical maps (with relief and contour lines).
- Satellite maps (aerial images).
- Thematic maps (cycling, hiking, weather, etc.).

At `https://leaflet-extras.github.io/leaflet-providers/preview/`, you can view the different map styles available (and choose the right ones for the maps you need to create). The main tile servers, using OSM data to build their tiles, are:

- **OSM Standard Tiles (Mapnik)**: The official OpenStreetMap tiles.
- **Stamen Design**: Artistic tiles based on OSM.
- **Thunderforest**: Thematic tiles (hiking, mountain biking, transport) based on OSM.
- **OpenMapTiles**: Vector tiles based on OSM.

- **Mapbox**: Raster and vector tiles, partly using OSM data (supplemented by their own data and data from other sources).
- **OpenTopoMap**: Provides OSM-based tiles to build maps for outdoor physical activities (especially hiking).
- **OpenCycleMap**: Provides OSM-based tiles for building cycling maps.

The vast majority of these tile providers use the **Web Mercator projection (EPSG:3857)**. This is a map projection that projects the Earth's surface onto a vertical cylinder. It preserves angles and shapes over small areas (making it conformal), but distorts size and area – especially as you move away from the equator.

In Web Mercator, the map scale is not uniform: the further you go from the equator, the greater the scale distortion becomes. For example, landmasses near the poles appear far larger than they are in reality, while equatorial regions are more accurately represented. Greenland, for instance, appears comparable in size to Africa, even though it is about 14 times smaller in real area. This distortion occurs because Web Mercator stretches distances as latitude increases, causing features to be increasingly exaggerated toward the poles.

Web Mercator projection tends to be the standard projection for web mapping. Some rare specialized tile providers offer other projections, but these typically require additional Leaflet plugins or configuration to render properly.

Another coordinate reference system is the World Geodetic System 1984 (EPSG:4326), commonly used with GPS or when working with raw geographic data. It does not use a projection but instead represents locations using true latitude and longitude, measured in degrees. Although highly accurate, it is not well suited for web maps, which require a projected coordinate system to display data on a flat surface.

Suppliers of Immersive or Panoramic Views

Immersive view providers enable mapping applications to obtain road-level photos of different locations. The aim is to enable users consulting a map to "immerse" themselves in it, by displaying actual views of the locations of interest to them. If the views extend in all directions – a full 360° – they are referred to as **panoramic**.

Open source immersive view providers offer alternatives to proprietary solutions such as Google Street View. Below are a few of the better-known providers.

Mapillary, a Swedish company now owned by Meta Platforms, offers more than 2 billion street-level images covering a wide range of countries and regions. These images are primarily contributed by users worldwide, and some are also shared by public institutions or organizations.

While panoramic and spherical images are supported, the majority of contributions are standard (non-panoramic) street-level photos.

Mapillary images can be viewed and used for free, but downloading raw images in bulk for offline use through the public API is not supported. Moreover, the images are licensed under a Mapillary-specific license:

- You may view, share, and use the images in accordance with Mapillary's terms of use and Mapillary API License.
- But you are not allowed to modify or redistribute the raw images unless explicitly permitted.
- And attribution to Mapillary is required when using their imagery.

For further details, see `https://www.mapillary.com/`.

Mapilio is an open platform for sharing street-level images. Users can contribute by uploading their own images, and the platform allows access to these images under an open license.

Images can be viewed, accessed, and downloaded (including in bulk) through the platform or its API. However, current global coverage is limited compared to larger platforms like Mapillary.

For further details, see `https://mapilio.com/`.

Panoramax is an open source project launched in 2022 by the Institut Geographique National (IGN), a French public state administration in charge of producing and maintaining geographical information for France. It offers free downloads of immersive (but not panoramic) views, taken at ground level and accessible via APIs. Coverage of the French territory is uneven: major urban centers, as well as some rural areas, are well covered, while other areas are not covered at all. Internationally, coverage remains limited or nonexistent in many countries.

For further details, see `https://panoramax.ign.fr`.

KartaView (formerly OpenStreetView), now owned by Singapore-based Grab, also offers immersive (but not panoramic) views. The platform is partially open source, notably for images accessed via the Android application. On the other hand, images accessible via web browsers remain proprietary. Coverage of the USA is relatively good (but not complete), while that of other countries is limited.

For further details, see `https://kartaview.org/landing`.

OpenTrailView is a platform for uploading panoramic images of footpaths, cycle paths, and trails, with the goal of building a global library of street-level imagery specifically for hikers and cyclists. It integrates with OpenStreetMap to enable navigation between panoramas along mapped routes.

However, as of now:

- There is no public API provided for programmatic access, search, or bulk image download (at the time of writing this book).
- Images can typically only be browsed via the website interface.

For further details, see `https://www.opentrailview.org`.

Route Planning Services

Route planning services provide APIs that allow applications to calculate routes between two or more geographic locations (optionally with waypoints) directly from application code. Several such services are open source and use OpenStreetMap data. Although they are not official OSM projects, they are considered part of the broader OSM ecosystem.

Examples of open source route planning services include **OSRM**, **Valhalla**, **GraphHopper**, and **BRouter** (specialized in bicycle routing). These engines generally provide a public API for route queries.

Route planning services will be further discussed in Chapter 9, and an example using OSRM will be provided.

Usage Policies

Most of the open source tools discussed above – and offered as public services through APIs (Overpass, Nominatim, tile servers, providers of immersive views, route planning services, etc.) – have associated **usage policies**.

These policies explain how these public services can be used and what is not permitted. In general, these policies state that:

- These public APIs are intended only for demonstration, education, development, testing, and noncommercial web applications.

  They usually **limit request throughput** and throttle requests originating from the same user.

- For production or commercial applications, it is recommended to:

  - Either set up your own server using OpenStreetMap data.
  - Or use a commercial service provider.

These usage policies should always be reviewed before using such resources. More details can be found at `https://operations.osmfoundation.org/policies/`.

Data Licensing and Attribution Best Practices

When building maps, even in the open source environment using tools within the OSM ecosystem, care must be taken regarding licensing and attribution. The rules enforced by OSM are described here: `https://osmfoundation.org/wiki/Licence`.

OpenStreetMap data is licensed under the **Open Database License (ODbL) 1.0**. This means that you are free to use, share, modify, and build upon the data - even commercially - as long as you follow certain conditions:

- **Attribution is required**: You must always credit OpenStreetMap and its contributors. A typical attribution is:

  "*© OpenStreetMap contributors*", with a link to `www.openstreetmap.org`.

- **ShareAlike applies to derived databases**: If you **modify and publicly share** a database based on OSM data, you must also share it under the same ODbL license.

- **Produced works** (such as printed maps or rendered images) do not need to be licensed under ODbL, but **attribution is still required**.

Examples of proper attributions are

- On a website map:

  "Map data © OpenStreetMap contributors", a short and minimal attribution usually shown in the corner of the map with a clickable link.

- In a print publication:

  "Base map © OpenStreetMap contributors, used under ODbL.", a more formal and complete form of attribution.

Additional notes:

- Using OSM data does not necessarily mean querying data directly from OSM via Overpass or Nominatim. As soon as you load map tiles from a tile server using tools like Leaflet or Folium, you are already using OSM data - and proper attribution is required.
- You must attribute even if you are using only a small portion of the map.
- If you combine OSM data with other datasets to create a new database, that new database may also need to be shared under ODbL.
- Using tools like Leaflet or Folium does not affect the OSM license: the license applies to the data, not the mapping library.

  On the other hand, when you build maps with Leaflet or Folium, no specific attribution is required by either library.

- If you use data from other sources alongside OSM data to build maps, you must ensure that you are authorized to use those datasets and comply with their licensing and attribution requirements.

Other Digital Mapping Systems

So as not to create the impression that digital mapping does not exist outside the OpenStreetMap environment, it is worth mentioning other digital mapping systems, both open source and proprietary, which can also be used to obtain geographic data, build maps, and integrate them into web applications.

Open Source Mapping Systems

In addition to the OpenStreetMap project and its ecosystem of tools, other open source mapping initiatives offer similar or complementary services, particularly in terms of geographic data distribution and API provision. These include:

- **USGS National Maps**: This system aims to provide topographic and geological data and aerial photos for the United States. It is a joint project of the United States Geological Survey and the US Federal Government. Data is stored in the USGS database and can be queried via APIs. USGS also provides tile sets for building maps. Unlike OSM, the data is produced mainly by public institutions or professional organizations.

  For more details, see `https://www.usgs.gov/programs/national-geospatial-program/national-map` and `https://www.usgs.gov/sciencebase-statements-and-documentation/api-and-web-services`.

- **OpenAerialMap**: This project is an initiative of the Humanitarian OpenStreetMap Team (HOT) to freely share aerial images useful for humanitarian mapping, development, or emergency response. It essentially provides aerial images of various locations. An API makes it possible to search for images and accompanying data.

  For more details, see `https://openaerialmap.org/` and `https://docs.openaerialmap.org/api/api/`.

There are also a number of systems which, unlike OpenStreetMap, do not offer a public data hosting service with extensive available data, but instead allow you to deploy your own mapping infrastructure and store your own data on a private server. These open source software packages enable geographic data to be stored and shared via standard APIs:

- **QGIS Server**: This is the server component of the QGIS open source mapping software, enabling you to view, edit, print, and analyze geographic data in a variety of formats. With QGIS Server, QGIS projects can be published on the web via open standards (WMS, WFS, WCS), offering data display, consultation, and download services via APIs.

  For more information, see `https://docs.qgis.org/latest/en/docs/server_manual/`.

- **GeoServer**: An open source server for sharing geographic data. Data can be published on the web via open standards (WMS, WFS, WCS, etc.). GeoServer offers an API for querying and manipulating this data, and it is widely used by public institutions, local authorities, and businesses.

  For more information, see `https://geoserver.org`.

- **MapServer**: Another open source server for publishing geographic data on the Internet. Originally created by the University of Minnesota, it can generate dynamic maps from spatial data and supports many open standards (WMS, WFS, etc.).

  For more information, see `https://mapserver.org`.

Proprietary Mapping Systems

There are a number of companies that manage geographic data and offer graphic tools or APIs for querying it. They provide accurate, regularly updated data, but access to this data is generally subject to a fee (either via subscription or pay-per-use billing). However, the boundary between "open source" and "proprietary" is not always clear-cut: many predominantly "proprietary" companies may also provide part of their data and services in open source mode. These proprietary mapping systems include:

- **Google Maps**: Google Maps is a mapping service well known to the general public, thanks in particular to its website (`https://www.google.com/maps`) and its Google Street View feature, offering immersive panoramic views. Google stores its own map data and makes it available via various APIs, enabling geocoding, route finding, elevation search, thematic searches (hotels, parking lots, etc.), access to real-time traffic data, as well as panoramic views from traffic lanes (Google Street View). Use of these APIs is subject to a fee.

  For more information, see `https://developers.google.com/maps`.

- **Microsoft Bing Maps**: Microsoft Bing Maps is an online mapping service integrated in the Bing suite. Its geographic data comes from both Microsoft and various suppliers, including TomTom and OpenStreetMap. Bing Maps offers a web user interface (`https://www.bing.com/maps`) and a range of APIs for accessing its data and various services: interactive map display, geocoding, elevation search, route and travel time calculation, real-time traffic information, thematic location search (gas stations, museums, etc.), and aerial image display. Access to these services is free for occasional use, but chargeable for professional use.

  For more information, see `https://learn.microsoft.com/en-us/bingmaps/rest-services`.

- **Apple Maps**: Apple has been offering a mapping service for its devices (laptops, tablets, phones, watches, etc.) for over 12 years. The company recently announced the availability of this service on the web, and a beta version is already available online at `https://beta.maps.apple.com`. The system is based on Apple's own geographic data, supplemented by data from third-party suppliers, notably TomTom. This data can be accessed via APIs, enabling users to search and display maps, perform geocoding queries, search for routes, obtain traffic information, activate navigation guidance (taking into account real-time traffic and dynamically recalculating routes), perform thematic searches (e.g., shops, schools, etc.), consult satellite views, and explore 3D maps. Use of the APIs is free for developers, but chargeable for production use.

For more information, see `https://developer.apple.com/maps`.

- **Here Technologies**: This mapping system, formerly owned by Nokia, now belongs to the Dutch company Here, controlled by a consortium of German automotive industries. Although primarily designed for car navigation, it is also available for Windows. Through its APIs, Here offers a range of services, such as route finding, geocoding, fleet vehicle tracking and logistics functions, 3D maps, as well as indoor geolocation (for searching for objects or people within a structure). These services are free of charge for those wishing to discover the system, but chargeable for production use.

  For more information, see `https://www.here.com/platform?cid=Developer_Api`.

- **TomTom Maps**: TomTom, a Dutch company specializing in GPS navigation systems, both mobile and on-board, also manages its own geographic data in databases. It offers access to this data via various APIs (under the "TomTom Developer" service): map rendering, location search, geocoding, route search, real-time traffic information, vehicle tracking, and logistics. Use of the APIs is free of charge up to a defined number of requests and chargeable beyond that.

  For more information, see `https://developer.tomtom.com`.

- **Esri ArcGIS**: The American company Esri (Environmental Systems Research Institute) has been managing geographic data and providing software and services for over 50 years and has been a pioneer in digital cartography. It offers a wide range of data, including road maps, aerial and satellite views, topographic information, as well as thematic data (hydrography, agriculture, climate, geology, etc.) and social data (population density, income, health, infrastructure, etc.). Esri provides APIs and web services (notably the ArcGIS REST API) for querying this data. While some services are free for basic queries, most are chargeable.

  For more information, see `https://www.esri.com/fr-fr/arcgis/open-vision/initiatives/developer-api`.

- **Mapbox**: Mapbox is an American company specializing in online mapping, offering interactive maps for integration into websites. The company has its own geographic data, but also integrates external data from a variety of sources. Mapbox provides APIs to access its data, enabling operations such as map search and display, geocoding, route finding, travel time calculation, real-time road traffic information, and 3D map display. Use of these APIs is subject to a fee.

  For more information, see `https://www.mapbox.com`.

- **Geoportail**: Geoportail is a French mapping service developed by the Institut Geographique National (IGN), a French public institution. It features a web user interface (`https://www.geoportail.gouv.fr/`) offering detailed, high-quality maps covering the French territory exclusively. Unlike OpenStreetMap, it is designed more for cartographic visualization than for the storage of raw geographic data.

  Geoportail also provides a number of APIs for map consultation, geolocation, altitude search (based on coordinates), route calculation, and access to cadastral data. These APIs can be used free of charge or for a fee, depending on the type of API and the frequency of requests.

  For more information, see `https://www.geoportail.gouv.fr/`.

- **Yandex**: The Russian company Yandex offers the Yandex web mapping service, which includes a web user interface for displaying maps and searching for routes (`https://yandex.com/maps`). The service also offers APIs for accessing its geographic data, enabling map search and display, map customization and interactivity, geocoding, route finding, and thematic searches. Yandex Maps also offers, via its APIs, street-level panoramic views (similar to Google Street View, but limited to certain areas of Russia and neighboring countries) and real-time traffic data. Basic use of the APIs is free, but charges apply for higher throughputs corresponding to production use.

For more information, see `https://yandex.com/maps-api`.

There are also a number of other players in the field of digital mapping, notably via web and cell phone applications, which unfortunately do not provide APIs to access their services:

- **ViaMichelin**: ViaMichelin is a French company and a division of the Michelin Group dedicated to digital cartography. Michelin, a historic player in the cartography market thanks to its road maps, is continuing this activity in the digital domain with ViaMichelin. ViaMichelin's web user interface (`https://www.viamichelin.com`) enables users to consult road maps and search for itineraries in most countries of the world. ViaMichelin map data is compiled from a variety of sources, including IGN, OpenStreetMap, Here, TomTom, as well as Michelin's own data (tourism, restaurants, hotels, etc.).
- **Mappy**: Mappy is a French mapping and navigation service, currently owned by the RATP group (responsible for public transport in Paris, France). Its web user interface (`https://fr.mappy.com`) enables users to consult maps, search for routes, and obtain real-time traffic information. Map coverage is limited to Europe.
- **Mappy.cz**: Mappy.cz is a Czech company, unrelated to the French company Mappy. It offers online map applications for cell phones and the web (`https://mapy.cz/`). Its maps offer excellent coverage of cycle paths and footpaths, in addition to classic road maps.

Other map and geographic service suppliers, requiring fluency in Mandarin or Korean, have not been mentioned.

CHAPTER 2

Querying OpenStreetMap Data with Overpass

Overpass is a general-purpose API for querying OpenStreetMap. It enables you to extract geographic data from OSM's database simply and efficiently.

A dedicated query language (comparable to SQL, but specially designed for geographic data retrieval) is provided. It allows you to specify the types of elements to be searched, as well as selection criteria: attributes, tags, or targeted geographic areas.

Contrary to popular belief, Overpass does not query the main OpenStreetMap database directly, but instead operates on a dedicated database maintained separately. It relies on data regularly extracted from Planet.osm, the reference file containing all OSM data. This data is incrementally imported into a database optimized for Overpass. This secondary database, structured with data in binary format, speeds up searches and improves performance. Overpass queries are executed using this database.

Overpass is complemented by **Overpass Turbo** (`https://overpass-turbo.eu/`), a web user interface that makes it easy to write, edit, and execute Overpass queries. Results can be displayed on an interactive map or in text form.

In Chapter 6, we will look at how to use the Overpass API in JavaScript or Python code to automate the query of data from OSM, and dynamically integrate it into maps, in order to enhance them with additional map elements.

The list of Overpass functions described below is not exhaustive. For a complete reference, please consult the Overpass documentation: `https://wiki.openstreetmap.org/wiki/Overpass_API`.

At the time of writing this book, the current release of Overpass is version 0.7.61, and it is the version described below and used for all the examples.

P. Marie, *Modern Web Cartography*, https://doi.org/10.1007/979-8-8688-2335-0_2

Building and Refining Queries with Overpass Turbo

Overpass queries designed to extract geographic data can be tested and adjusted interactively using the **Overpass Turbo** interface, available online at `https://overpass-turbo.eu`.

This tool is an excellent introduction to the Overpass Query Language. It offers:

- An input zone for writing, editing, and submitting queries.
- An interactive map displaying query results directly.
- And a text output panel listing the elements found.

Here is an example showing how to use it to execute a request.

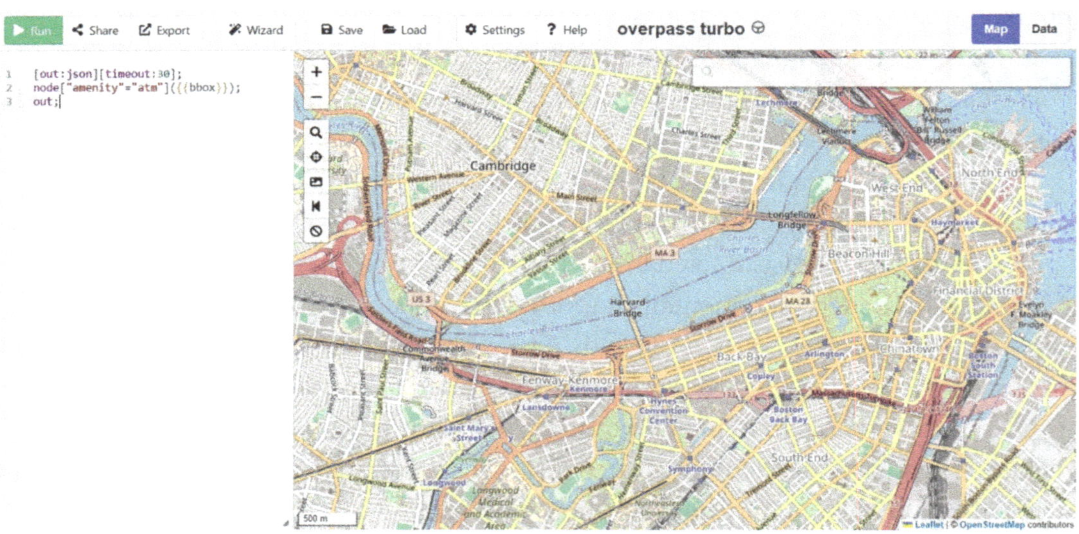

Figure 2-1. *Screenshot of Overpass Turbo (overpass-turbo.eu) before submitting a request. Interface licensed under MIT. Map data © OpenStreetMap contributors (ODbL)*

The query entered in the input box on the left-hand side of the Overpass Turbo screen will search for all ATMs located in the geographical area visible on the map on the right (the Boston, Massachusetts, area). Moreover, the results are requested in JSON format, and the request must be completed within 30 seconds (otherwise, the processing will be aborted). The final `out;` statement will trigger the retrieved data to be returned in the response.

Once the query has been entered or copied into the appropriate zone, simply click the **Run** button (green arrow icon in the top left-hand corner of the interface). A couple of seconds later, the map is automatically updated: ATMs found are indicated by small blue circles. If you click on one of these circles, a popup opens, providing additional details about this ATM from the OSM database, such as the name of the bank, opening hours, or other information, where available.

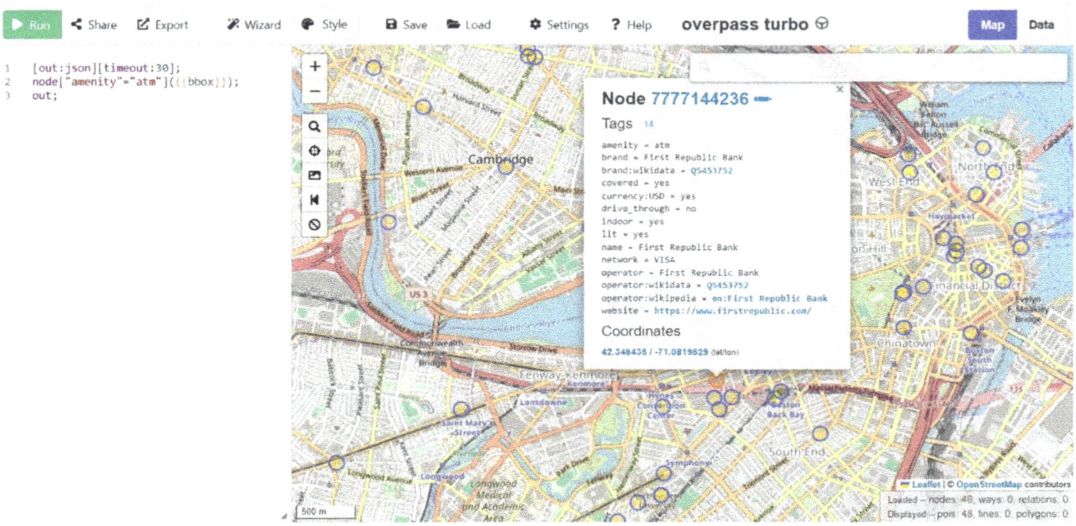

Figure 2-2. *Screenshot of Overpass Turbo (overpass-turbo.eu) after submitting a request and clicking on one of the nodes found. Interface licensed under MIT. Map data © OpenStreetMap contributors (ODbL)*

By clicking the **Data** button in the top right-hand corner of the Overpass Turbo interface, you can view the raw data returned by the query. Below, this data is displayed in JSON format (a structured format widely used to represent complex and hierarchical objects and easily exploited by computer programs), as specified in the request.

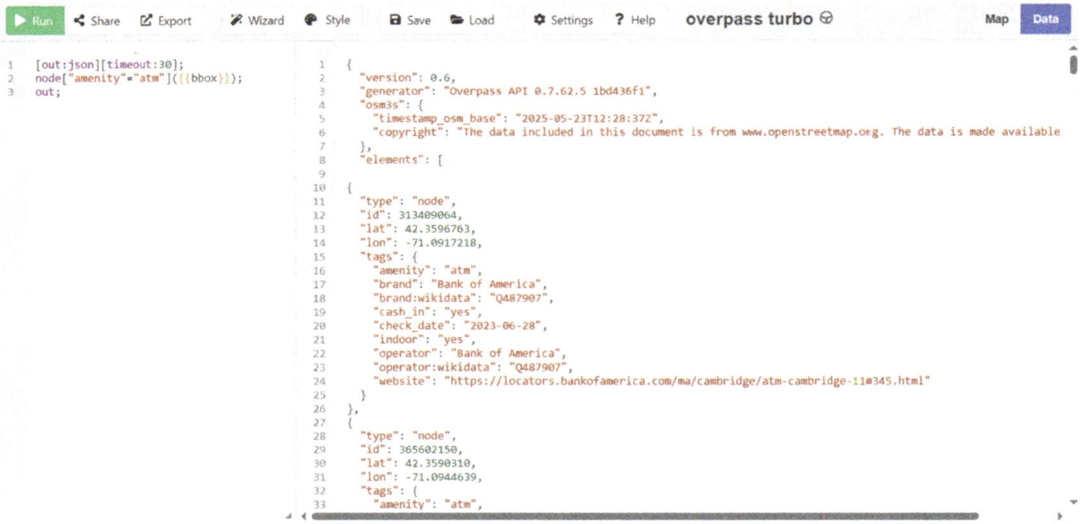

***Figure 2-3.** Screenshot of Overpass Turbo (`overpass-turbo.eu`) after submitting a request and selecting the Data rendering. Interface licensed under MIT. Data © OpenStreetMap contributors (ODbL)*

Please note:

- The filter (`{{bbox}}`) used in the previous query indicates that the search should be limited to the area currently displayed on the Overpass Turbo map (right-hand side of the screen). This filter is **specific to the Overpass Turbo interface**: it dynamically injects the coordinates of the map's bounding box into the query. However, this shortcut does not work if Overpass is queried directly via its API, from JavaScript, Python, or any other client. In this case, it is necessary to explicitly indicate the geographical area. This is possible using various options, such as
    - Providing bounding box coordinates (minimum latitude, minimum longitude, maximum latitude, maximum longitude)
    - Or specifying an administrative zone identified by its name or OSM identifier

The next section of this chapter will detail these various geographical delimitation options.

- When a geographical delimitation filter other than `({{bbox}})` is used, the elements returned by the query may not be immediately visible on the map, because Overpass Turbo does not automatically recenter the view. To focus the map on the retrieved elements, click the Loop icon ("zoom to data").

Overpass Query Language

Overpass Query Language is general-purpose and designed to search for all types of elements in the OpenStreetMap database and extract their data. This query language is documented on several websites (including `https://wiki.openstreetmap.org/wiki/Overpass_API/Overpass_QL`, `https://wiki.openstreetmap.org/wiki/Overpass_API/Language_Guide`, and `https://dev.overpass-api.de/overpass-doc/en/`).

The following paragraphs attempt to provide an explanatory overview of Overpass Query Language. All the queries included as examples can be easily tested with Overpass Turbo. For more details on the Overpass Query Language, please refer to the sites mentioned above.

The elements stored in the OpenStreetMap database and searchable with Overpass are of the following three types:

- **Nodes** are geographical points, with latitude and longitude coordinates. They can also include tags with various attributes characterizing their nature and providing information about them. They are used to register all kinds of locations: buildings, crosswalks, ATMs, road intersections, restaurants, public mailboxes, cities (via their central point), localities, mountain peaks, etc.

- **Ways** are polylines (lines made up of contiguous segments). They are defined by the points constituting their vertices and their extremities. Ways can be open or closed. They can also include tags with various attributes characterizing their nature and providing information about them. They are used to record roads or other communication routes and rivers (open ways), but also the contours of particular areas: cities, provinces, regions, countries, etc. (closed ways).

- **Relations** are more complex elements, made up of a set of points, ways, or even other relations. They can also include tags with various attributes characterizing their nature and providing information about them. They are used to record itineraries or long-distance routes (such as freeways), large buildings, and territorial entities (cities, regions, countries, etc.).

Query Grammar

This section presents the grammar of the Overpass Query Language.

When testing the following query examples with Overpass Turbo:

- Be sure to add a final `out geom;` statement after the query statements to ensure that the results are returned and displayed on the map on the right.

  If you want the response data to be returned in JSON format, start your request with `[out:json]` (otherwise, you will get the response in XML format).

  Therefore, the recommended structure for your test requests is the following:

  ```
  [out:json]
  (Your query statements to be tested)
  out geom;
  ```

- After executing a query, do not forget to click the Loop icon ("zoom to data") to focus the map on the retrieved elements.

Query Statements and Filters

All Overpass **query statements** search for nodes, ways, or relations and return them in objects named **sets**. For example, `node;` would search for all the nodes in the world (but this query should be avoided, as there are currently over eight billion nodes in the OSM database), and `nwr;` would search for all the nodes, ways, and relations, i.e., all the elements in the OSM database (so this query should also be avoided).

Overpass queries can return a lot of data, and even fail to produce a response, if too much data is found and the search takes too much time. This is why it is essential to select specific elements and limit the amount of data to be returned in the response. To achieve this, **filters** must be used. Several filters can be combined in a query statement to select only the elements matching all the filters. These filters can specify:

- The geographical area in which the elements are searched. This geographical area can be specified in particular:
  - In the form of a bounding box, that is a rectangle, defined by its limits in all directions. For example:

    ```
    nwr(51.477,-0.001,51.478,0.001);
    ```

    searches for all elements in the bounding box defined by:

    - 51.477 as southern limit (minimal latitude),
    - -0.001 as western limit (minimum longitude),
    - 51.478 as northern limit (maximum latitude),
    - 0.001 as eastern limit (maximum longitude).
  - In the form of a circle of given radius around a point which coordinates you specify. For example:

    way(around:1000,51.477,0.0);

    searches for all ways around the point `[51.477, 0.0]` within a one-kilometer radius.
  - In the form of a polygon defined by the coordinates of its vertices. For example:

    ```
    node(poly: "51.47 -0.01 51.477 0.01 51.484 -0.01");
    ```

    finds all the nodes in the triangle defined by these three points.
  - In the form of a territorial entity (district, town, region, country, etc.) found in the OSM database using relations, ways, and nodes. For example:

    ```
    area[name="Truth or Consequences"]->.searchArea;
    way(area.searchArea);
    ```

searches for all ways in the Truth or Consequences, New Mexico, area.

Some explanations about these statements:

- ***area*** is not one of the basic element types stored in the OSM database (such as **node, way,** and **relation**). Instead, it is a derived concept (areas are constructed dynamically from closed ways or relations tagged with boundary information) that can be used to define and select geographic areas.
- In the first statement, the matching areas are stored in a named set (a set assigned a specific name, which will be explained later).
- In the second statement, this named set is used to restrict the geographic area in which **ways** will be searched.

However, the results of such searches are not always what the programmer expects:

- The area selected using this type of filter does not always correspond exactly to the official administrative boundaries of the city or designated entity, but may cover a larger area. For finer filtering, taking better account of the municipality area boundaries, the filter can be completed as follows:

  ```
  area[name="Truth or Consequences"][admin_level="8"]
  [boundary="administrative"]->.searchArea;
  ```

  `admin_level="8"` specifies that we are looking for a city (not a county or a region), and `boundary="administrative"` indicates that we want to restrict the search to the administrative boundaries of the city.

- This type of filter may select several cities (as well as counties, provinces, or countries) with the same name. To keep only the one you are interested in, you can add an additional selection criterion. For example:

  ```
  area[name="Toledo"][wikipedia="es:Toledo"]->.searchArea;
  ```

  `wikipedia="es:Toledo"` selects only the Toledo city located in Spain.

- With the Overpass Turbo user interface, you can also limit the search area with a `{{bbox}}` filter (meaning "bounding box") provided by Overpass Turbo and corresponding to the map portion displayed on the right-hand side of the screen. For example:

  ```
  relation({{bbox}});
  ```

  searches only for relations on the map currently displayed in Overpass Turbo.

  But beware, the `({{bbox}})` filter is **specific to the Overpass Turbo user interface**. It cannot be used when Overpass is invoked programmatically via the API.

- Tag values on which elements are to be selected. For example:
  - `node[amenity="atm"]({{bbox}});`

    searches for ATMs in the geographical area defined by Overpass Turbo's bounding box (i.e., in the map displayed on the right-hand side of the page). More specifically, this request searches for nodes that have a tag with the key `amenity` and the value "atm", located within the bounding box.

  - `node[place="town"]({{bbox}});`

    searches for all medium-sized towns (identified by the `place` tag key and "town" value in the OSM database) in the geographic area defined by the Overpass Turbo's bounding box.

- A conditional filter (`if:`) can be used to evaluate a condition and retain only the elements for which this condition returns `true`. For example:

  ```
  node[place="town"](if:(is_number(t["population"]) &&
  number(t["population"])>=5000))({{bbox}});
  ```

  searches for all medium-sized towns (identified by the place tag key and "town" value in the OSM database) where the population (found by the population tag key) is greater than or equal to 5000 in the geographic area defined by the Overpass Turbo's bounding box.

The expression:

```
(if:(is_number(t["population"]) && number(t["population"])>=5000))
```

means having a tag with key population whose value is numeric and greater than or equal to 5000.

`is_number()` and `t[]` are evaluators. Evaluators will be explained in the next section.

The example above uses the && (and) boolean operator. Other available boolean operators are || (or) and ! (not).

This example also uses the >= comparison operator. Other comparison operators are: ==, !=, >, <, and <=.

This query looks for nodes and therefore returns the center points of the cities found. To obtain other information, such as city outlines, you need to search for the cities as relations. Also, only medium-sized cities are found, as large cities have a place tag with the value "city" and small towns with the value "village".

- It is also possible to search elements by their ID in OSM, when it is known. For example:
    - `node(777);`

      searches for the node with ID "777", which happens to be located in Norway.
    - `way(id:3998029,3998240,4065354,4065364,4065421);`

      searches for the ways with IDs "3998029", "3998240", "4065354", "4065364", and "4065421", which happen to be located in Germany.

Evaluators

Evaluators are used within `if:` conditional filters to compute or evaluate expressions based on the tags or attributes of OSM elements. They allow more advanced filtering logic than simple tag matching.

In the preceding section, we used the following conditional filter:

```
(if:(is_number(t["population"]) && number(t["population"])>=5000))
```

- `is_number(...)` is an evaluator that returns `true` if the argument passed is numeric.
- `t["key"]` is an evaluator that retrieves the value of the tag with key "key".

Other evaluators are the following:

- `length(...)`: Returns the length of a string (e.g., a tag value).
- `number(...)`: Converts a string (e.g., a tag value) to a numeric value.
- `type()`: Returns the type of the element currently being evaluated ("node", "way", or "relation").
- `count_tags()`: Returns the number of tags of the element currently being evaluated.
- `id()`: Returns the ID of the element currently being evaluated.
- `user()`: Returns the username of the user who last modified the element.
- `timestamp()`: Returns the timestamp of the last modification to the element.
- `version()`: Returns the version number of the element.

Regular Expressions

Overpass does not offer a wildcard per se, but the equivalent can be obtained via **regular expressions** on tag values.

For those unfamiliar with regular expressions, let's start with some basics about regular expression syntax. When checking whether a pattern matches a line or string:

- ^ anchors the match to the **beginning** of the line or string. For example, "`^the`" will match "`the end`", but not "`at the beginning`" (because "the" is not located at the beginning of the string).

- `$` anchors the match to the **end** of the line or string. For example, "`end$`" will match "`the end`", but not "`the end of the month`" (because "end" is not located at the end of the string).
- `.` matches **any single character** (except a new line by default). For example, "`e.d`" will match "`sending`", but not "`edit`" (because there is no character between "e" and "d") nor "`expanded`" (because there is more than one character between "e" and "d").
- `.*` matches **zero or more** of any character. For example, "`e.*d`" will match "`edit`", "`sending`", and "`expanded`".
- `.+` matches **one or more** of any character. For example, "`e.+d`" will not match with "`edit`" (because there is no character between "e" and "d"), but will match "`sending`" and "`expanded`".

Now, let's look at how to use regular expressions in Overpass QL filters through a few examples:

- The `[name~"^East"]` filter limits the search to elements whose `name` tag has a value that starts with "East".
- The `[name~"Beach$"]` filter limits the search to elements whose `name` tag has a value that ends with "Beach".
- The `[name~"^East Atl.ntic Beach$"]` filter limits the search to elements whose `name` tag has a value that starts with "East Atl", ends with "ntic Beach", and contains exactly one character between these two parts.
- The `[name~"^East.*Beach$"]` filter limits the search to elements whose `name` tag has a value that starts with "East" and ends with "Beach", with any number (including 0) of characters in between.

  If at least one character is needed between the first and the second parts, the filter must be modified as follows: `[name~"^East.+Beach$"]`.
- The `[name~"Atlantic"]` filter limits the search to elements whose `name` tag has a value that includes the string "Atlantic".

Regular expressions can also be used on keys. For example, the filter:

```
[~"am"~"^East Atlantic Beach$"]
```

limits the search to elements that have a tag whose key contains the string "am" and whose value is exactly equal to "East Atlantic Beach".

In addition:

- You can search for all elements containing a tag with a given key, whatever its value. For example, the `[power]` filter selects elements related to the production, transport, or distribution of electricity.
- It is also possible to search for all elements containing a tag with a given value, regardless of the associated key. For example, the filter `[~".*"~"school"]` selects elements with a tag whose value includes the string "school", regardless of the key associated with that value.

Overpass supports regular expressions compatible with POSIX Extended syntax. For more details on the grammar of these regular expressions, please refer to sites such as `https://www.boost.org/doc/libs/latest/libs/regex/doc/html/boost_regex/syntax/basic_extended.html`.

Named Sets

The results of queries such as those of the preceding sections are written to a default set (whose name is _). However, it is also possible to write query results to a named set (a set with a specific name chosen by the programmer). For example:

```
way(around:1000,51.477,0.0)->.greenwich_ways;
```

searches for the ways located around the point `[51.477, 0.0]` within a one-kilometer radius and stores them in a set named `greenwich_ways`. Then, the next statement:

```
way.greenwich_ways[highway="footway"];
```

uses the `greenwich_ways` set as input and selects only the ways that are footpaths (writing the result to the default set _).

Unions, Differences, and Intersections

Several queries on nodes, ways, and relations with particular filters can be grouped together in **unions**. The returned answer is then the (deduplicated) sum of the answers to each of these queries. For example, the following statement:

```
(
    node[place="city"]({{bbox}});
    node[place="town"]({{bbox}});
    node[place="village"]({{bbox}});
);
```

searches for all cities (nodes with a `place` tag with value "city", "town", or "village" in the OSM database) in the geographic area defined by the Overpass Turbo bounding box.

Similarly, it is possible to use a **difference** to subtract the set returned by a query from the set returned by another. For example, the following statement:

```
(
    node[place="village"](12.5, 77, 13.5, 78);
    - node[place="village"](12.8, 77.3, 13.2, 77.7);
);
```

searches for villages located within a first bounding box in the Bangalore region, India, then for villages located within a second bounding box (which is smaller and contained within the first), and finally removes the villages found by the second search from the set returned by the first search.

Finally, it is possible to use an **intersection** to select the elements common to two sets returned by queries. For example, the following statements:

```
node[place="village"](33, -5.5, 34, -4.5)->.a;
node[place="village"](33.5, -5, 34.5, -4)->.b;
node.a.b;
```

search for villages located in the Fez region, Morocco, within a first bounding box, then for villages also located in the same area, within a second bounding box which partially overlaps the first, and finally select the villages present in both sets.

Note that the same result can be obtained by combining several filters in a single query statement. For example, the preceding query could also be written as follows:

```
node[place="village"](33, -5.5, 34, -4.5)(33.5, -5, 34.5, -4);
```

Overpass also provides statements such as `if`, `foreach`, and `for` to implement more complex logic in queries. This is beyond the scope of this book, but detailed documentation can be found at `https://wiki.openstreetmap.org/wiki/Overpass_API/Overpass_QL`.

Settings

Global declarations called **settings** can be placed at the beginning of the request, before the query statements:

- The `out` declaration is used to specify the format in which the response is expected. Several formats are supported: XML, JSON, CSV, etc. For example, `[out:json];` is used to request that the response be returned in JSON format. The default is XML.
- The `timeout` declaration is used to specify how long the request is allowed to run before being interrupted. For example: `[timeout:30];` limits this time to 30 seconds. If the request processing is not completed within this time, it is aborted and an error returned. The default is 25 seconds.
- The `maxsize` declaration is used to specify the maximum memory size allowed for the query on the server. If the query needs more memory than this value, the request may be aborted by the server. The default is 512 MB.

Standalone Statements

The `out` statement is usually placed at the end of the request, after the query block. It triggers the output of response data and specifies its level of detail (determining which attributes of the elements found are included or not in the response). For example, `out geom;` requests that data concerning the element's complete geometry (in particular, the coordinates of each node in a way or relation) be included in the response. Other possible values are:

- `out;` (or the equivalent `out body;`) is the default response return statement. In this case, way and relation geometry is not fully included in the response (as it is with `out geom;`).
- `out count;` returns only the number of elements found.
- `out skel;` returns only the identifier and type of the elements found and the node coordinates.
- `out ids;` returns only the identifier and type of the elements found (the most minimal response possible).
- `out tags;` returns only the identifier and the tags of the elements found (not the coordinates).
- `out meta;` returns all available information for the elements found (such as version, changeset ID, timestamp, user data, etc.).

Here is a summary, with the `out` options ordered from the least to the most verbose:

Table 2-1. *Overpass out options*

Statement	Output
`out count`	Only the number of matching elements – no actual data
`out ids`	Only the element IDs (node/way/relation IDs)
`out skel`	Only basic structure (IDs and references) – no tags or geometry
`out / out body`	Main data (tags) of the elements, but not related geometry
`out geom`	Full geometry (node coordinates for ways and relations)
`out meta`	All available metadata (timestamps, version numbers, user info, etc.)

The `>;` statement is the **recurse down** statement and can be used to complete a set returned by the query block. For example, in the following statement sequence:

```
[out:json]
(query block)
out body;
>;
out geom;
```

- `out body;` returns elements selected by the query block.
- `>;` is a **recurse up** statement. It receives as input the elements returned by the previous statement and completes them with the elements they reference. More specifically:
    - If a way references nodes, these nodes are also included in the result set.
    - If a relation references nodes and ways, these are also included in the result set (and the nodes referenced by these ways are also included). However, if the relation references other relations, they are not included in the result set.
- `out geom;` returns all elements resulting from the previous statement, with their complete geometry.

The `>>;` statement (**recurse down (deep)** statement) is similar to the `>;` statement, with a key difference: it recursively explores relations (i.e., relations referenced by other relations are followed all the way down), including all referenced elements (nodes, ways, and relations) in the result set.

The Overpass Query Language also provides:

- A **recurse up** statement (`<;`): Searches for the elements that reference those in the input set and adds them to the result set, in addition to the original input elements. For example:

    ```
    [out:json]
    (
       node(51.249,7.148,51.251,7.152);
       <;
    );
    out geom;
    ```

    returns all the nodes within the specified bounding box, along with all the ways and relations that reference these nodes.
- A **recurse up (deep)** statement (`<<;`): Similar to the `<;` statement, but the search for referencing elements continues recursively all the way up, including all such elements in the result set.

Performance Best Practices

Depending on how they are written, Overpass queries can take a long time and may not complete within the time allowed by the `timeout` setting at the beginning of the query (or by the server's timeout setting) - in which case they will be aborted. The factors that affect performance include:

- **The number of objects returned**, either in the final result set or intermediate sets. For example, a query returning 5,000 objects will consume significantly more resources and take much longer that a request returning just 10 objects.
- **The level of detail of the output.** For instance, using `out ids;` is much faster than `out geom;`, since it avoids transferring full geometries.
- **The number of separate queries in the query block.** When possible, it is best to combine multiple queries into one using more refined filters.

Best practices for writing efficient Overpass queries:

- **Restrict the geographical area** as much as possible using bounding boxes or area definitions.
- **Use strong and specific tag filters** for each query to reduce the number of matched elements to only those you truly need. You can test whether your queries are too broad by running them individually with `out count;` in Overpass Turbo.
- **Minimize the number of individual queries.** Whenever possible, combine them using conditional filters, regular expressions, or evaluators.
- **Use the `out` statement efficiently**, specifying only the level of detail you actually need (`out ids;`, `out skel;`, `out body;`, or `out geom;`).
- **Be cautious with recursive operators** like > and <, as they can significantly increase processing time and result size.

On self-hosted Overpass servers, it is possible to configure a **throttling mechanism** at the server level. This limits how many requests users can make within a certain time window (e.g., requests per IP per minute). This setting is defined by the system administrator and is not configurable by those writing Overpass QL queries.

However, developers building applications that use Overpass should still take throttling into account, for example, by adding appropriate delays between queries to avoid exceeding server limits.

Bringing It All Together

To conclude this brief description of the Overpass Query Language, here is an example of a complete query combining several of the settings and statements described in the preceding sections:

```
1   [out:json][timeout:30];
2   (
3       node[place="city"](if:(is_number(t["population"]) &&
        number(t["population"])>=5000))(-34, -72, -32.5, -70);
4       node[place="town"](if:(is_number(t["population"]) &&
        number(t["population"])>=5000))(-34, -72, -32.5, -70);
5       node[place="village"](if:(is_number(t["population"]) &&
        number(t["population"])>=5000))(-34, -72, -32.5, -70);
6   );
7   out;
```

This query searches for all towns with a population greater than or equal to 5,000 in the area of Santiago and Valparaiso in Chile. More precisely:

- **Line 1**: The response is requested in JSON format, and processing time should not exceed 30 seconds.
- **Lines 2 to 6**: The query searches for nodes with a tag with key `place` containing the value "city", "town", or "village" and with another tag with key `population` containing a numeric value greater than or equal to 5000 inhabitants. The search is restricted to a bounding box defined by its lower-left and upper-right coordinates.
- **Line 7**: The response must contain the data for the nodes found.

This same query can also be written by using a named set and factoring the population filter:

```
[out:json][timeout:30];
(
    node[place="city"](-34, -72, -32.5, -70);
    node[place="town"](-34, -72, -32.5, -70);
    node[place="village"](-34, -72, -32.5, -70);
)->.a;
node.a(if:(is_number(t["population"]) && number(t["populati
on"])>=5000));
out;
```

This version of the query is less efficient, as it operates in two stages: first, it searches for all communities within the surrounding rectangle (returning more elements than the previous version), and then it applies the population-based filter.

The same query can also be formulated using a regular expression:

```
1  [out:json][timeout:30];
2  node[place~"^(city|town|village)$"](if:(is_number(t["population"]) &&
   number(t["population"])>=5000))(-34, -72, -32.5, -70);
3  out;
```

- **Line 2**: The filter `[place~"^(city|town|village)$"]` selects only nodes with a "place" tag whose value is exactly "city", "town", or "village".

This version of the query is more efficient than the two previous ones, as it performs a single search pass instead of three or more.

The answer returned by this query comprises 1048 lines and 57 nodes. Here is the beginning:


```
{
  "version": 0.6,
  "generator": "Overpass API 0.7.62.5 1bd436f1",
  "osm3s": {
    "timestamp_osm_base": "2025-04-24T19:35:15Z",
    "copyright": "The data included in this document is from www.
    openstreetmap.org. The data is made available under ODbL."
```

```
},
"elements": [
    {
        "type": "node",
        "id": 50016356,
        "lat": -33.4377756,
        "lon": -70.6504502,
        "tags": {
            "admin_level": "2",
            "capital": "yes",
            "capital_ISO3166-1": "yes",
            "ele": "521",
            "name": "Santiago",

            ...

            "official_name": "Santiago de Chile",
            "old_name": "Santiago de Nueva Extremadura",
            "place": "city",
            "population": "6139087",
            "population:date": "2017",
            "postal_code": "8320000",
            "short_name": "Stgo",
            "source:name:oc": "Lo Congrès",
            "source:population": "INE",
            "wikidata": "Q2887",
            "wikipedia": "es:Santiago de Chile"
        }
    },
    {
        "type": "node",
        "id": 50026147,
        "lat": -33.0244535,
        "lon": -71.5517636,
        "tags": {
            "alt_name:ar": "فينا دل مار",
```

```
                "capital": "8",
                "name": "Viña del Mar",
                "name:ar": "فينيا ديل مار",
                "name:ru": "Винья-дель-Мар",
                "place": "city",
                "population": "332875",
                "population:date": "2017",
                "source:population": "INE",
                "wikidata": "Q184345",
                "wikipedia": "es:Viña del Mar"
            }
        },
        {
            "type": "node",
            "id": 214180385,
            "lat": -33.7537387,
            "lon": -70.9039325,
            "tags": {
                "capital": "8",
                "name": "Isla de Maipo",
                "place": "town",
                "population": "14176",
                "population:date": "2017",
                "source:population": "INE"
            }
        },
        ...
```

In order to build correct and relevant Overpass queries, Overpass Turbo is indispensable. The use of TagInfo (presented in the previous chapter) is also essential. With TagInfo, we can determine on which keys and with which values to carry out the search in order to obtain the information we are interested in.

Finally, it should be remembered that Overpass queries are limited to querying OSM data. They do not support adding or updating data in the OSM database.

However, please note that Overpass allows you to track changes in OSM data using the **Augmented Diffs** functionality. This feature is not part of the Overpass Query Language, but instead relies on the XML-based query format (described next). For more information about Augmented Diffs, see `https://wiki.openstreetmap.org/wiki/Overpass_API/Augmented_Diffs`.

As previously mentioned, Chapter 6 will explore how to use Overpass queries within Python and JavaScript programs in web applications and how to display their results on maps.

Overpass Query XML

Another option, less commonly used than the Overpass QL, for querying OpenStreetMap with Overpass is Overpass Query XML. In this format, queries are written using XML syntax. For example, to search for cafés in the center of Vienna, Austria:

```
<osm-script output="json">
  <query type="node">
    <has-kv k="amenity" v="cafe"/>
    <bbox-query s="48.205" w="16.365" n="48.215" e="16.375"/>
  </query>
  <print/>
</osm-script>
```

It is documented at `http://wiki.openstreetmap.org/wiki/Overpass_API/Language_Guide`.

Overpass Query XML requests can also be tested using Overpass Turbo.

Figure 2-4. *Overpass Turbo – after executing the preceding XML request (cafés in Vienna)*

Overpass XML was the original query language provided by Overpass. Overpass QL was introduced later and is generally preferred because it is more compact and human-readable. However, some applications and tools may still require queries in XML format.

Overpass Query API

In order to use the Overpass Query Language explained in the previous section from a program (typically, JavaScript or Python code, as will be done in the following chapters), an API is necessary. Overpass provides an HTTP query API for this purpose.

This API is documented at `https://wiki.openstreetmap.org/wiki/Overpass_API`. HTTP requests sent to Overpass consist of:

- The following URL: `https://overpass-api.de/api/interpreter`.

  This is the endpoint of the public Overpass API, intended for demonstration, education, and low-volume testing.

Another public Overpass API endpoint that can be used as an alternative is `https://overpass.kumi.systems/api/interpreter`. The same limitations apply to this endpoint as well.

For production use and high-throughput applications, it is recommended to self-host an Overpass server – either by setting up the infrastructure within your organization or by using a third-party Overpass service. In such cases, the API URL will differ.

- The `POST` method.
- A `header` specifying the content type.
- A `body` containing an Overpass query written in the Overpass Query Language explained earlier.

Here is an HTML example using JavaScript to invoke an Overpass HTTP request, searching for ATMs in the area of Bonn, Germany:

```
<html>
<head>
<title>Overpass example</title>
</head>
<body>
<h2>Overpass HTTP request example</h2>
<p>Fetching ATMs from OpenStreetMap (bounding box: 50.6, 7.0, 50.8, 7.3)</p>
<h3>Raw JSON Response:</h3>
<p id="response">Loading...</p>
<script>
    fetch("https://overpass-api.de/api/interpreter", {
        method: "POST",
        headers: { "Content-Type": "application/x-www-form-urlencoded" },
        body: "data=" + encodeURIComponent(`[out:json];
        node[amenity=atm](50.6, 7.0, 50.8, 7.3); out;`)
    })
    .then(response => response.json())
    .then(data => {
```

```
18            const responseElement = document.getElementById('response');
19            responseElement.textContent = JSON.stringify(data, null, 2);
20        })
21        .catch(error => {
22            document.getElementById('response').textContent = 'Error: '
              + error;
23        });
24    </script></body>
25    </html>
```

- **Line 11**: `fetch` is a JavaScript function used to send HTTP requests.
- **Lines 13 and 14**: The `encodeURIComponent()` function is used to encode the query for sending data with the `Content-Type: application/x-www-form-urlencoded` format, which is the format expected by the Overpass server.

When opened in a browser, this file produces the following output (truncated here):

Overpass HTTP request example

Fetching ATMs from OpenStreetMap (bounding box: 50.6, 7.0, 50.8, 7.3)

Raw JSON Response:

{ "version": 0.6, "generator": "Overpass API 0.7.62.8 e802775f", "osm3s": { "timestamp_osm_base": "2025-10-21T12:16:02Z", "copyright": "The data included in this document is from www.openstreetmap.org. The data is made available under ODbL." }, "elements": [{ "type": "node", "id": 248616250, "lat": 50.7191133, "lon": 7.1571286, "tags": { "amenity": "atm", "brand": "Postbank", "brand:wikidata": "Q708835", "brand:wikipedia": "de:Postbank", "cash_in": "no", "opening_hours": "24/7", "operator": "Postbank", "operator:wikidata": "Q708835", "operator:wikipedia": "de:Postbank", "wheelchair": "yes" } }, { "type": "node", "id": 253064288, "lat": 50.741322, "lon": 7.1230285, "tags": { "amenity": "atm", "brand": "Postbank", "brand:wikidata": "Q708835", "brand:wikipedia": "de:Postbank", "cash_in": "no", "check_date": "2025-10-14", "network": "CashPool", "operator": "Postbank", "operator:wikidata": "Q708835", "operator:wikipedia": "de:Postbank", "toilets:wheelchair": "no", "wheelchair": "yes" } }, { "type": "node", "id": 253074555, "lat": 50.7346529, "lon": 7.0989215, "tags": { "amenity": "atm", "brand": "Postbank", "brand:wikidata": "Q708835", "brand:wikipedia": "de:Postbank", "cash_in": "no", "check_date": "2024-02-05", "drive_through": "no", "indoor": "yes", "network": "CashGroup", "operator": "Postbank", "operator:wikidata": "Q708835", "operator:wikipedia": "de:Postbank", "wheelchair": "yes" } }, { "type": "node", "id": 253838875, "lat": 50.7398894, "lon": 7.121844, "tags": { "amenity": "atm", "brand": "Volksbank Köln Bonn eG", "brand:wikidata": "Q41680844", "brand:wikipedia": "de:Volksbank Köln Bonn", "cash_in": "yes", "check_date": "2024-11-02", "indoor": "yes", "operator": "Volksbank Köln Bonn eG", "operator:wikidata": "Q41680844", "operator:wikipedia": "de:Volksbank Köln Bonn", "wheelchair": "no" } }, { "type": "node", "id": 253916602, "lat": 50.7479124, "lon": 7.1174425, "tags": { "amenity": "atm", "cash_in": "no", "check_date": "2023-11-19", "indoor": "yes", "note": "SB Banking Center mit Automaten von KSK und Volksbank", "opening_hours": "Mo-Su,PH 05:00-24:00", "operator": "Sparkasse KölnBonn und Volksbank", "wheelchair": "yes" } }, { "type": "node", "id": 254309936, "lat": 50.7111148, "lon": 7.0665734, "tags": { "amenity": "atm", "note": "War mal eine Filiale, jetzt nur noch Geldautomat an Hauswand", "opening_hours": "24/7", "operator": "Sparkasse KölnBonn", "wheelchair": "yes" } }, { "type": "node", "id": 256802219, "lat": 50.7330213, "lon": 7.0662578, "tags": { "amenity": "atm", "check_date": "2025-04-13", "description": "SB-Stelle Auf dem Hügel", "note": "Geldautomat in der Fassade eingelassen", "operator": "Sparkasse KölnBonn" } }, { "type": "node", "id": 266400373, "lat": 50.7885918, "lon": 7.1832855, "tags": { "addr:city": "Sankt Augustin", "addr:country": "DE", "addr:housenumber": "11", "addr:postcode": "53757", "addr:street": "Einsteinstraße", "amenity": "atm", "bic": "COKSDE33", "name": "Kreissparkasse Köln SB-Stelle", "operator": "Sparkasse", "wheelchair": "yes" } }, { "type": "node", "id": 269550329, "lat": 50.7196256, "lon": 7.0423293, "tags": { "amenity": "atm", "brand": "Sparkasse", "disused:shop": "yes", "operator": "Sparkasse" } }, { "type": "node", "id": 271891642, "lat": 50.7981326, "lon": 7.2093212, "tags": { "amenity": "atm", "cash_in": "yes", "check_date": "2024-12-15", "indoor": "yes", "name": "Kreissparkasse Köln", "website": "https://www.ksk-koeln.de", "wheelchair": "limited" } }, { "type": "node", "id": 310635672, "lat": 50.6655278, "lon": 7.1731377, "tags": { "addr:housenumber": "66", "addr:street": "Drachenburgstraße", "amenity": "atm", "opening_hours": "Mo-Su 05:00-24:00", "operator": "Sparkasse KölnBonn", "wheelchair": "yes" } }, { "type": "node", "id": 392601280, "lat": 50.7441096, "lon": 7.1524969, "tags": { "amenity": "atm", "brand": "Volksbank Köln Bonn eG", "brand:wikidata": "Q41680844", "brand:wikipedia": "de:Volksbank Köln Bonn", "check_date": "2025-06-09", "operator": "Volksbank Köln Bonn eG", "operator:wikidata": "Q41680844", "operator:wikipedia": "de:Volksbank Köln Bonn", "wheelchair": "yes" } }, { "type": "node", "id": 416078519, "lat": 50.7551865, "lon": 7.008999, "tags": { "addr:city": "Bornheim", "addr:country": "DE", "addr:housenumber": "36", "addr:postcode": "53332", "addr:street": "Siegesstraße", "addr:suburb": "Roisdorf", "amenity": "atm", "check_date": "2024-01-09", "location": "indoor", "name": "Kreissparkasse Köln", "opening_hours": "Mo-Su,PH 05:30-23:00", "operator": "Kreissparkasse Köln", "wheelchair": "yes" } }, { "type": "node", "id":

***Figure 2-5.** Output*

Complete examples of Overpass requests used with JavaScript or Python, including responses rendered on maps, are available in Chapter 6.

Such HTTP requests cannot be tested directly using a browser's URL input field because they use the POST method. However, they can be tested using tools like cURL, Postman, or soapUI. Here is an example showing how to test them with cURL.

This command:

```
curl -X POST https://overpass-api.de/api/interpreter -H "Content-Type:
application/x-www-form-urlencoded" --data-urlencode "data=[out:json];
node[amenity=atm](50.6, 7.0, 50.8, 7.3); out;"
```

returns the following result (partially shown below):


```
{
  "version": 0.6,
  "generator": "Overpass API 0.7.62.8 e802775f",
  "osm3s": {
    "timestamp_osm_base": "2025-10-21T10:26:30Z",
    "copyright": "The data included in this document is from www.
    openstreetmap.org. The data is made available under ODbL."
  },
  "elements": [

{
  "type": "node",
  "id": 248616250,
  "lat": 50.7191133,
  "lon": 7.1571286,
  "tags": {
    "amenity": "atm",
    "brand": "Postbank",
    "brand:wikidata": "Q708835",
    "brand:wikipedia": "de:Postbank",
    "cash_in": "no",
    "opening_hours": "24/7",
    "operator": "Postbank",
    "operator:wikidata": "Q708835",
    "operator:wikipedia": "de:Postbank",
    "wheelchair": "yes"
```

```
  }
},
{
  "type": "node",
  "id": 253064288,
  "lat": 50.7413220,
  "lon": 7.1230285,
  "tags": {
    "amenity": "atm",
    "brand": "Postbank",
    "brand:wikidata": "Q708835",
    "brand:wikipedia": "de:Postbank",
    "cash_in": "no",
    "check_date": "2025-10-14",
    "network": "CashPool",
    "operator": "Postbank",
    "operator:wikidata": "Q708835",
    "operator:wikipedia": "de:Postbank",
    "toilets:wheelchair": "no",
    "wheelchair": "yes"
  }
},

...

{
  "type": "node",
  "id": 13108462416,
  "lat": 50.7022836,
  "lon": 7.1769341,
  "tags": {
    "amenity": "atm",
    "currency:EUR": "yes",
    "network": "Volksbank Köln Bonn eG",
```

```
    "survey:date": "2025-08-30",
    "wheelchair": "yes"
  }
}

  ]
}
```

CHAPTER 3

Geocoding and Reverse Geocoding with Nominatim

Nominatim is an open source geocoding engine based on OpenStreetMap data. It is a full-fledged project within the OSM ecosystem, enabling two types of operation:

- **Geocoding**, or **direct geocoding**, which consists in finding the geographic coordinates of a location using a name, address, or zip code. For example:
    - Find all cities named "New York" and their respective coordinates.
    - Identify the town associated with the zip code "10004" in the USA, and obtain its location.
    - Search for the place named "Statue of Liberty".

  For each result, Nominatim returns a set of information from the OSM database, including the geographical coordinates of the location(s).

- And **reverse geocoding**, which consists in finding an address using geographic coordinates. For example:
    - Get the full address (country, city, street, number) of the point with latitude "50" and longitude "24".
    - Determine in which country the point with latitude "45" and longitude "57" is located.

P. Marie, *Modern Web Cartography*, https://doi.org/10.1007/979-8-8688-2335-0_3

Nominatim then provides a more or less detailed address, depending on the query parameters: you can limit the response to a country or region, or go as far as a specific building number.

Like Overpass, Nominatim does not directly query the OpenStreetMap database. At regular intervals, OSM data is extracted and imported into an optimized database specially designed for geocoding (with indexing adapted to address or coordinate search queries). It is this database that is used to process the queries received by Nominatim.

Nominatim offers:

- A user-friendly Web interface (`https://nominatim.openstreetmap.org/ui/`), for testing geocoding and reverse geocoding queries.
- An API accessible via scripts or applications, allowing automated searches and facilitating the integration of results with map-building tools such as Leaflet or Folium.

The list of Nominatim functions described below is not exhaustive. For a complete reference, please consult the Nominatim documentation: `https://nominatim.org/release-docs/latest/`.

At the time of writing this book, the current release of Nominatim is version 5.1.0, and it is the version described below and used for all the examples.

Nominatim-UI

The Nominatim Web user interface can be accessed via the link `https://nominatim.openstreetmap.org/ui/` and displays the page below.

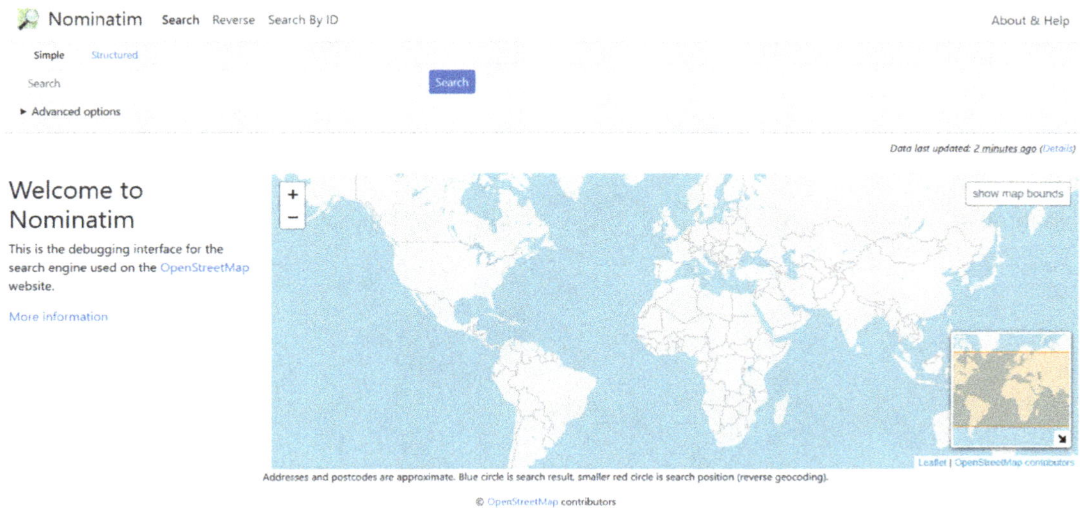

Figure 3-1. *Screenshot from Nominatim (`nominatim.openstreetmap.org`), licensed under GPL, using OpenStreetMap data © OpenStreetMap contributors (ODbL)*

Geocoding

On the page above, you can search for a location by providing a variable number of criteria. The search can be:

- Simple, by entering a free text designating the place you are looking for (e.g., “Beverly Hills”, “Mount Everest”, or “Öresund Bridge”).
- Or structured, with detailed information on the various components of the address: country, region, city, zip code, street, etc.

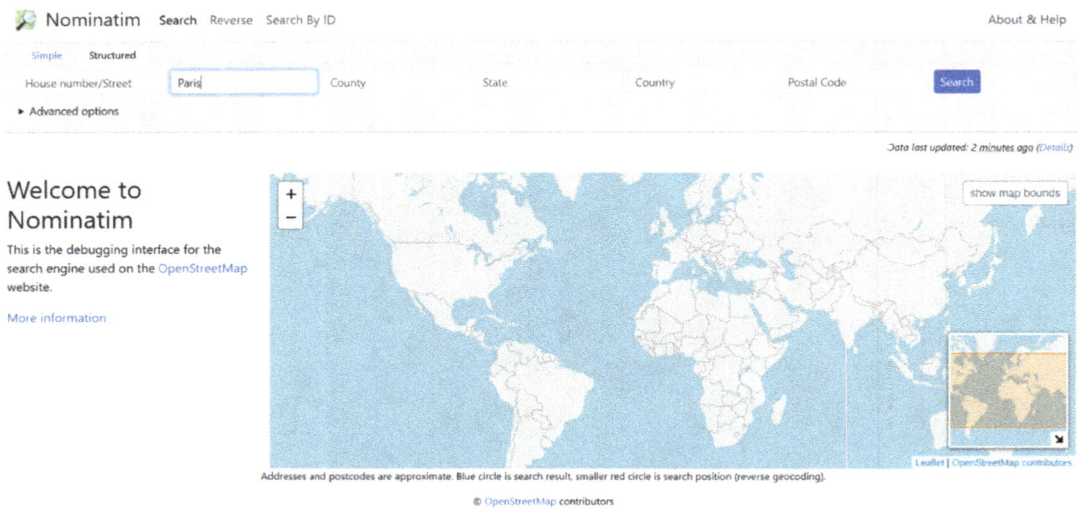

Figure 3-2. *Screenshot from Nominatim (`nominatim.openstreetmap.org`), licensed under GPL, using OpenStreetMap data © OpenStreetMap contributors (ODbL)*

For example, by entering "Paris" in the "City" field of the structured search page and clicking the **Search** button (the blue button on the right), all cities with this name are found and displayed as a list on the left-hand side of the results page.

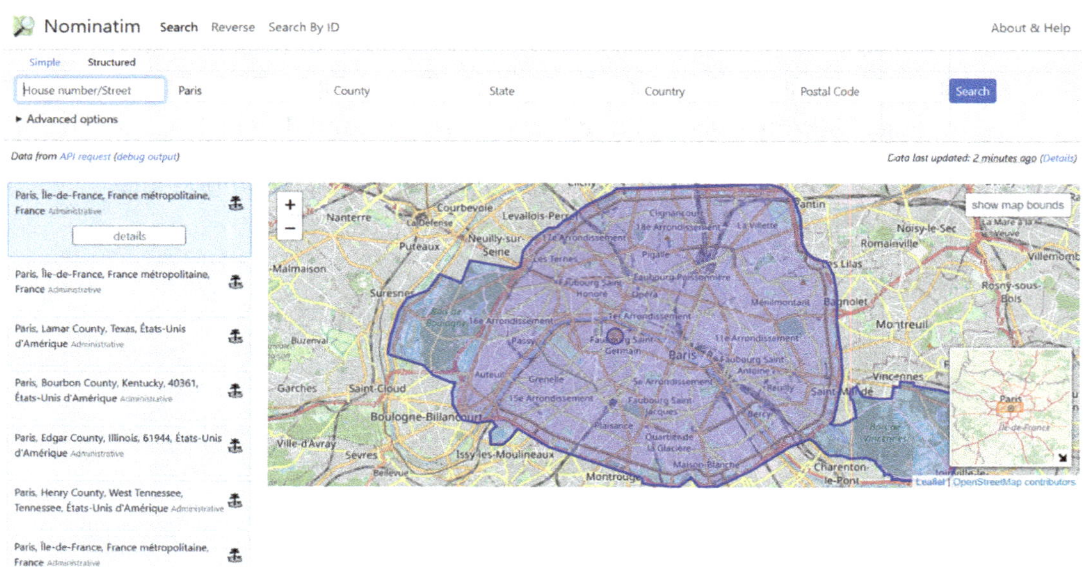

Figure 3-3. *Screenshot from Nominatim (`nominatim.openstreetmap.org`), licensed under GPL, using OpenStreetMap data © OpenStreetMap contributors (ODbL)*

By clicking on a city in the list, its map is displayed on the right-hand side of the page, and its territory is outlined and filled in with a semi-transparent color, allowing you to visualize both the city limits and the underlying map elements.

The **Details** button displays the OpenStreetMap data returned by the query for this city (only the first few lines are shown below):

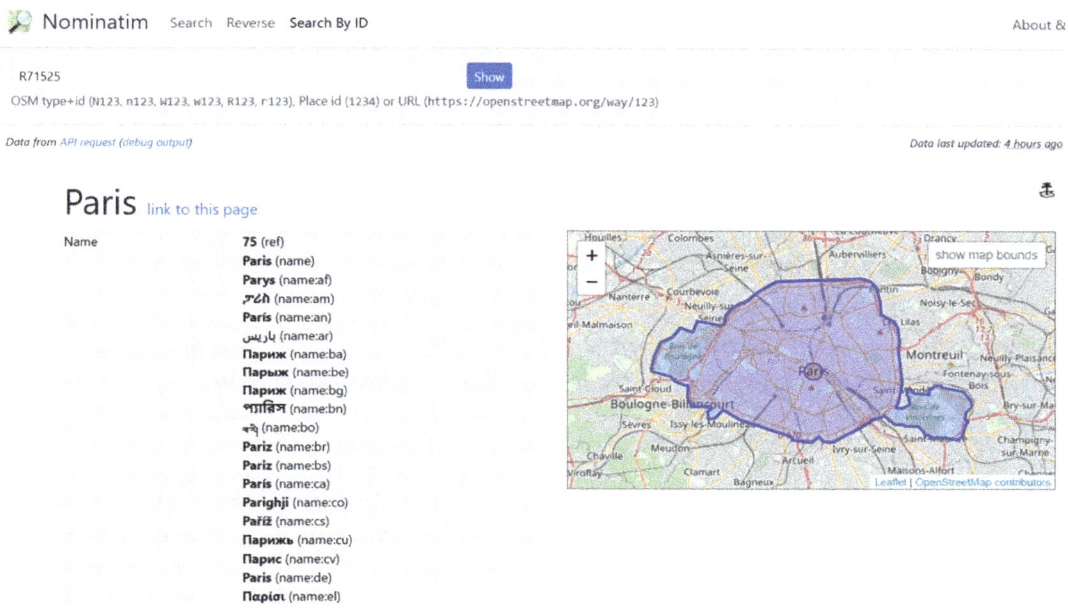

Figure 3-4. *Screenshot from Nominatim (`nominatim.openstreetmap.org`), licensed under GPL, using OpenStreetMap data © OpenStreetMap contributors (ODbL)*

Reverse Geocoding

Select the **Reverse** option (in the action bar at the top of the page) to access reverse geocoding, which lets you search for the address of a location using its geographic coordinates.

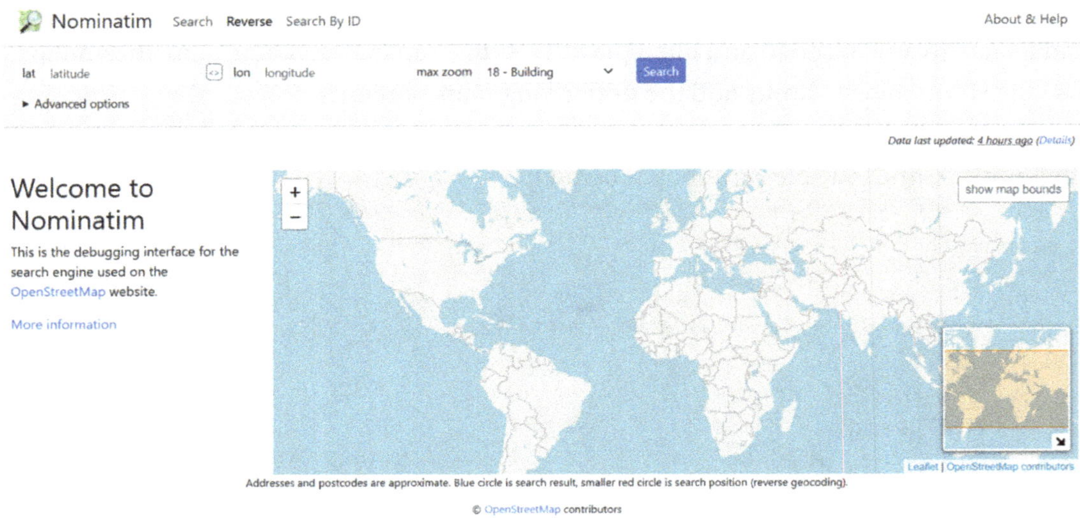

Figure 3-5. *Screenshot from Nominatim (`nominatim.openstreetmap.org`), licensed under GPL, using OpenStreetMap data © OpenStreetMap contributors (ODbL)*

On this page, you can enter the coordinates of a point to be searched for and specify the level of detail expected concerning its address ("max zoom" drop-down list). For example:

- The search for the point at latitude 50 and longitude 24, with high precision (18), identifies a place called Wiesenberg, in the oblast (province) of Lviv in Ukraine.

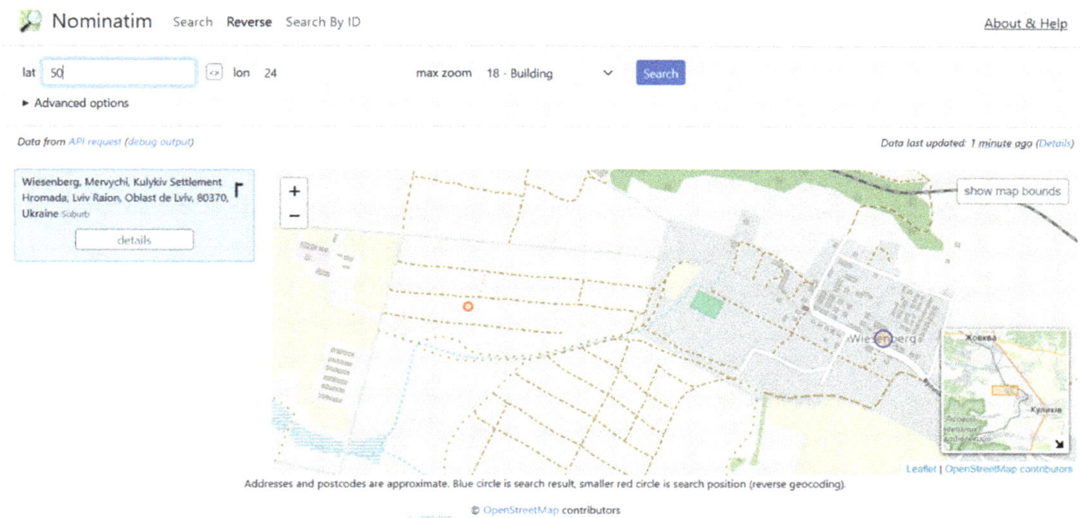

***Figure 3-6.** Screenshot from Nominatim (`nominatim.openstreetmap.org`), licensed under GPL, using OpenStreetMap data © OpenStreetMap contributors (ODbL)*

The point searched for by its coordinates is indicated by a small red circle. The blue circle indicates the nearest point for which an address has been found in OSM (and it is this address that has been returned as the answer).

The details button displays the OSM information of the nearest point identified with a blue circle. In the screen below, only the first few lines are visible.

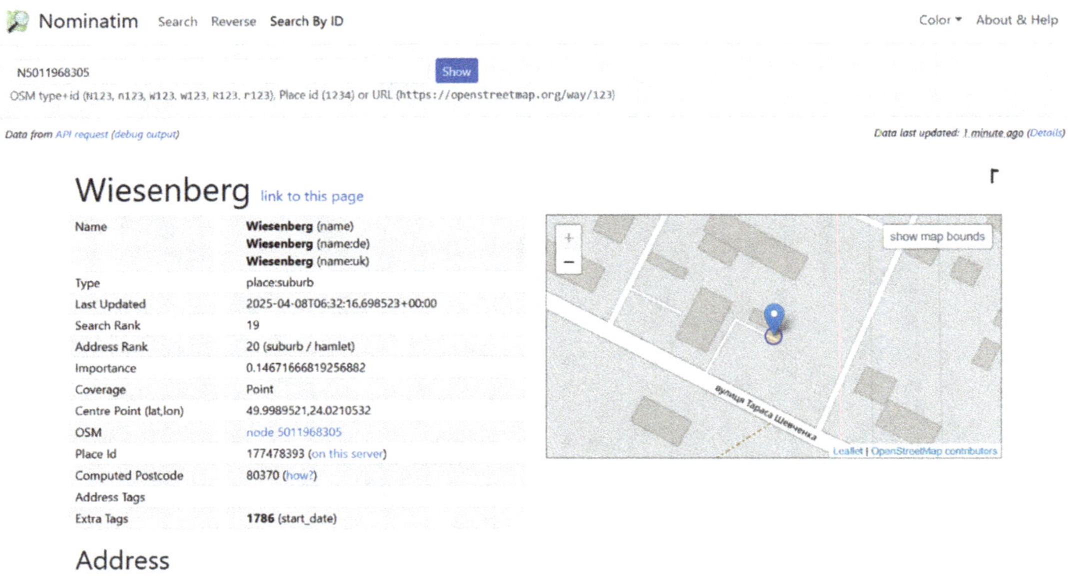

Figure 3-7. *Screenshot from Nominatim (`nominatim.openstreetmap.org`), licensed under GPL, using OpenStreetMap data © OpenStreetMap contributors (ODbL)*

- The search for the point with latitude 45 and longitude 57, with low precision (3), identifies this location as being in Uzbekistan (not far from the border with Kazakhstan).

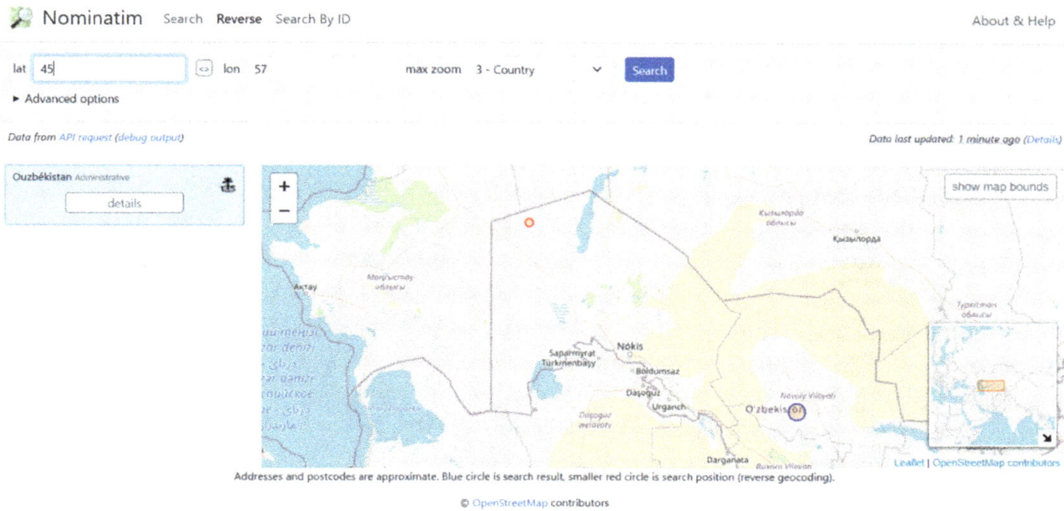

Figure 3-8. *Screenshot from Nominatim (`nominatim.openstreetmap.org`), licensed under GPL, using OpenStreetMap data © OpenStreetMap contributors (ODbL)*

The point searched for by its coordinates is indicated by a small red circle, and the blue circle marks the search result, that is, the center of Uzbekistan (because "3 - Country" was selected for the max zoom search parameter).

The details button displays the associated OSM information. In the screen below, only the first few lines are visible, and the data relates to the country, as the search precision level (3) corresponds to the country.

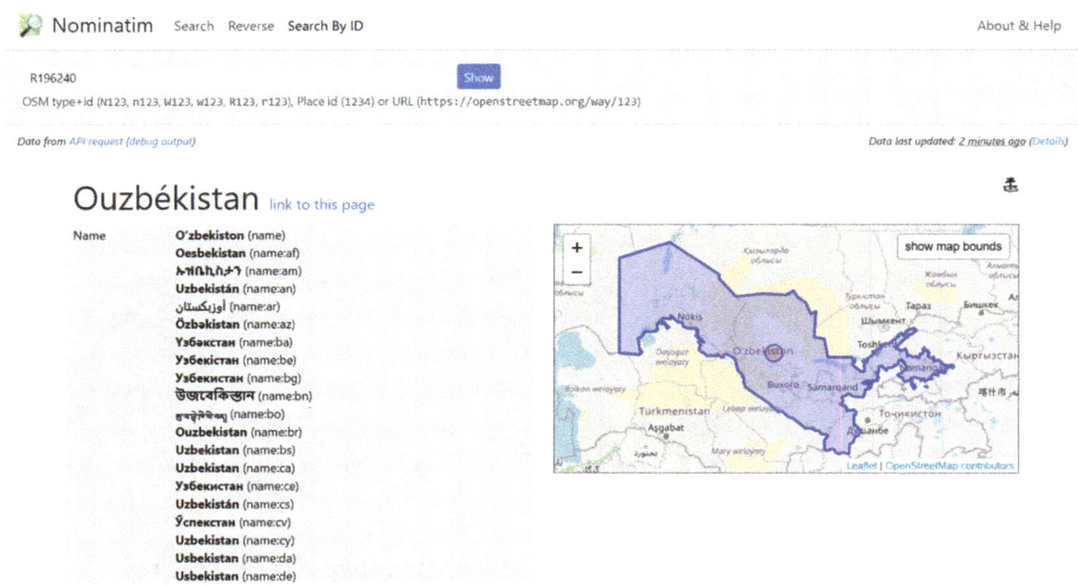

***Figure 3-9.** Screenshot from Nominatim (`nominatim.openstreetmap.org`), licensed under GPL, using OpenStreetMap data © OpenStreetMap contributors (ODbL)*

Nominatim's API

Nominatim's API enables geocoding (finding the coordinates of a location from an address) and reverse geocoding (finding an address from geographic coordinates) searches, all via HTTP requests.

This API is documented at `https://nominatim.org/release-docs/latest/api/Overview/`.

The public endpoint of the Nominatim API is `https://nominatim.openstreetmap.org/`. It is intended for demonstration, education, and low-volume testing. For production use and high-throughput applications, it is recommended to self-host a Nominatim server – either by setting up the infrastructure within your organization or by using a third-party Nominatim service. In such cases, the API URL will differ.

This URL must be completed by an action (e.g., `/search`), by search criteria (e.g., `q=Statue+of+Liberty` or `city=New York`), and by an output format (such as `format=jsonv2`). For example:

```
https://nominatim.openstreetmap.org/search?q=Statue+of+Liberty&for
mat=jsonv2
```

This query searches for a place called “Statue of Liberty” and returns the corresponding results in JSON v2 format. The following response is obtained:


```
[
  {
    "place_id": 329344069,
    "licence": "Data © OpenStreetMap contributors, ODbL 1.0.
    http://osm.org/copyright",
    "osm_type": "way",
    "osm_id": 32965412,
    "lat": "40.6892532",
    "lon": "-74.0445482",
    "category": "tourism",
    "type": "attraction",
    "place_rank": 30,
    "importance": 0.614408947655079,
    "addresstype": "tourism",
    "name": "Statue of Liberty",
    "display_name": "Statue of Liberty, Flagpole Plaza, Manhattan, New York
    County, New York, 10004, United States",
    "boundingbox": [
      "40.6888049",
      "40.6896741",
      "-74.0451069",
      "-74.0439637"
    ]
  },
  {
    "place_id": 159721779,
    "licence": "Data © OpenStreetMap contributors, ODbL 1.0.
    http://osm.org/copyright",
    "osm_type": "node",
    "osm_id": 7214128361,
    "lat": "63.0961840",
    "lon": "21.6155510",
    "category": "historic",
```

```
    "type": "monument",
    "place_rank": 30,
    "importance": 0.3269037816797396,
    "addresstype": "historic",
    "name": "Statue of Liberty",
    "display_name": "Statue of Liberty, Hovioikeudenpuistikko, Keskusta 3,
    Vaasa, Vaasa sub-region, Ostrobothnia, Mainland Finland, 65101,
    Finland",
    "boundingbox": [
      "63.0961340",
      "63.0962340",
      "21.6155010",
      "21.6156010"
    ]
  },
  {
    "place_id": 50199313,
    "licence": "Data © OpenStreetMap contributors, ODbL 1.0. http://osm.
    org/copyright",
    "osm_type": "node",
    "osm_id": 2579439967,
    "lat": "39.1058609",
    "lon": "26.5631164",
    "category": "historic",
    "type": "memorial",
    "place_rank": 30,
    "importance": 0.2535713079130456,
    "addresstype": "historic",
    "name": "Statue of Liberty",
    "display_name": "Statue of Liberty, Αργύρη Εφταλιώτη, Κιόσκι,
    Mytilene, Mytilene Municipal Unit, Municipality of Mytilene, Lesbos
    Regional Unit, Northern Aegean, Aegean, 811 00, Greece",
    "boundingbox": [
      "39.1058109",
      "39.1059109",
```

```
      "26.5630664",
      "26.5631664"
    ]
  }
]
```

Three places were found. The first one is the Statue of Liberty in New York. The other two are other statues with the same name in Finland and in Greece.

The Nominatim API offers several main actions:

- `/search`, to search for locations by name or address and obtain their geographic coordinates (direct geocoding).
- `/reverse`, to obtain the address (more or less detailed) of a location from its geographic coordinates (reverse geocoding).
- Other less commonly used actions are also available: `/lookup`, `/status`, etc.

Nominatim requests can be easily tested with a browser (by entering the request in the URL input field of the browser). They can also be tested with tools such as cURL, Postman, or soapUI.

Note also that, after performing a search with Nominatim's web user interface (described in the previous section of this chapter), you can view the HTTP request generated (as well as the response data) by clicking on the **API request** link on the left-hand side of the page.

Geocoding Requests

These are queries that use the `/search` action to search for places by name or address. The main location search parameters are detailed below.

Table 3-1. */search queries*

Parameter	Description
q	Free text search (e.g., "Pima Air & Space Museum").
street	Detailed address (e.g., "6000 E. Valencia Rd").
city	City (e.g., "Tucson").
county	County, province, or department depending on the country (e.g., "Pima County").
state	State, region, or province depending on the country (e.g., "Arizona").
country	Country (e.g., "USA").
postalcode	Postal code (e.g., "85756").
amenity	Nature of location or type of equipment (e.g., "museum", "parking", "restaurant").
format	Expected response format (xml, json, jsonv2, geojson, etc.). This parameter is mandatory.

Queries that use the `q` parameter are called *unstructured,* whereas queries that use parameters such as `street`, `city`, and `county` are called *structured.* Structured queries are more precise, while unstructured queries are more flexible.

The following parameters are also available:

- `polygon_geojson`: Includes in the response a polygon in GeoJSON format corresponding to the administrative boundaries of the location searched for (for a city, region, or country, the boundaries of its territory). This option is useful if you intend to represent the contours of the geographical entity found on a map.
- `dedupe`: Eliminates duplicates in the response. Indeed, some queries may return elements representing the same thing (e.g., different sections of the same road).
- `addressdetails`: The level of detail of the address in the response:
    - 0 = Complete address in a single line.
    - 1 = Detailed address split into several fields.

Default: 0

- `extratags`: Specifies whether or not additional properties of the elements found should be included in the response:
  - 0 = Additional properties are not included,
  - 1 = Additional properties are included.

  Default: 0
- `limit`: Specifies the maximum number of matches to be returned in the answer.

  Default: 10

Please note that Nominatim queries do not always return all places matching the search criteria.

When multiple matches are found, Nominatim prioritizes them and may return only those with higher importance. This occurs especially when using unstructured queries with many matches. If you cannot retrieve the place that you are looking for, try switching to a structured query and specifying additional parameters such as `city`, `state`, and `country`.

Here is a query example:

```
https://nominatim.openstreetmap.org/search?street=Times+Square&country=USA&
postalcode=10036&format=jsonv2
```

This query searches for a place with address "Times Square" and zipcode "10036" in the USA and requests that the result be returned in JSONv2 format. When submitted with a browser, the following response is obtained, including the latitude and longitude of the location found:


```
[
  {
    "place_id": 332723154,
    "licence": "Data © OpenStreetMap contributors, ODbL 1.0. http://osm.
    org/copyright",
    "osm_type": "relation",
    "osm_id": 14942838,
    "lat": "40.7570122",
    "lon": "-73.9859709",
    "category": "highway",
```

```
    "type": "pedestrian",
    "place_rank": 26,
    "importance": 0.5936829858927538,
    "addresstype": "road",
    "name": "Times Square",
    "display_name": "Times Square, Theater District, Manhattan Community
    Board 5, Manhattan, New York County, New York, 10036, United States",
    "boundingbox": [
      "40.7558313",
      "40.7591362",
      "-73.9870666",
      "-73.9845108"
    ]
  }
]
```


It is also possible to test the request using cURL:

```
curl -s -G "https://nominatim.openstreetmap.org/search" ^
--data-urlencode "street=Times Square" ^
--data-urlencode "country=USA" ^
--data-urlencode "postalcode=10036" ^
--data-urlencode "format=jsonv2" ^
--user-agent "curl 8.13.0 (Windows)"
```

A User-Agent header must be included to comply with Nominatim's identification usage policy (see the "Usage Policy" section below).

This command uses a Windows syntax. With Linux or macOS, the line continuation character is not ^ but \.

This command returns the following result:


```
[{"place_id":333041682,"licence":"Data © OpenStreetMap contributors, ODbL
1.0. http://osm.org/copyright","osm_type":"relation","osm_id":14942838,
"lat":"40.7570095","lon":"-73.9859724","category":"highway","type":
"pedestrian","place_rank":26,"importance":0.5936829858927538,"addresstype":
"road","name":"Times Square","display_name":"Times Square, Theater
```

```
District, Manhattan Community Board 5, Manhattan, New York County, City of
New York, New York, 10036, United States of America","boundingbox":
["40.7558313","40.7591362","-73.9870666","-73.9845108"]}]
```


Here is an HTML example using JavaScript to invoke a Nominatim geocoding HTTP request, searching for the geographic coordinates of the same place (Times Square):

```
1  <html>
2  <head>
3  <title>Nominatim example</title>
4  </head>
5  <body>
6  <h2>Nominatim geocoding HTTP request example</h2>
7  <p>Geocoding Times Square with Nominatim</p>
8  <h3>Raw JSON Response:</h3>
9  <p id="response">Loading...</p>
10 <script>
11     fetch("https://nominatim.openstreetmap.org/search?street=Times+Squ
       are&country=USA&postalcode=10036&format=jsonv2")
12     .then(response => response.json())
13     .then(data => {
14         const responseElement = document.getElementById('response');
15         responseElement.textContent = JSON.stringify(data, null, 2);
16     })
17     .catch(error => {
18         document.getElementById('response').textContent = 'Error: '
           + error;
19     });
20 </script></body>
21 </html>
```

- **Line 11**: fetch is a JavaScript function used to send HTTP requests.

When opened in a browser, this file produces the following output:

Nominatim geocoding HTTP request example

Geocoding Times Square with Nominatim

Raw JSON Response:

[{ "place_id": 333041682, "licence": "Data © OpenStreetMap contributors, ODbL 1.0. http://osm.org/copyright", "osm_type": "relation", "osm_id": 14942838, "lat": "40.7570095", "lon": "-73.9859724", "category": "highway", "type": "pedestrian", "place_rank": 26, "importance": 0.5936829858927538, "addresstype": "road", "name": "Times Square", "display_name": "Times Square, Theater District, Manhattan Community Board 5, Manhattan, New York County, New York, 10036, United States", "boundingbox": ["40.7558313", "40.7591362", "-73.9870666", "-73.9845108"] }]

Reverse Geocoding Requests

These are queries that use the /reverse action to search for the address (more or less detailed depending on the parameters) of a location based on its latitude and longitude. The main search parameters are detailed below.

Table 3-2. */reverse queries*

Parameter	Description
`lat`	Latitude of searched point (e.g., "-46.59").
`lon`	Longitude of searched point (e.g., "-70.92").
`zoom`	Level of detail: 3 = country, 5 = state, 8 = county, 10 = city, 18 = complete address, Default: 18.
`format`	Expected response format (json, jsonv2, xml, geojson, etc.). Default: xml (for historical reasons). Use format=jsonv2 or format=json to request JSON. This chapter uses jsonv2 in examples.
`addressdetails`	Level of address detail in response: 0 = Complete address in a single line set in a field named "display_name", 1 = A detailed address is added in an "address" block. It is split into several fields: "house_number", "road", "village", "municipality", "state_district", etc. Default: 1.

The example below searches for the address of the location with coordinates -46.59 and -70.92:

```
https://nominatim.openstreetmap.org/reverse?lat=-46.59&lon=-70.92&zoom=18&f
ormat=jsonv2
```

When submitted with a browser, the following response is returned:


```
{
  "place_id": 253606,
  "licence": "Data © OpenStreetMap contributors, ODbL 1.0. http://osm.org/
  copyright",
  "osm_type": "way",
  "osm_id": 322914100,
  "lat": "-46.5899667",
  "lon": "-70.9200013",
  "category": "place",
  "type": "house",
  "place_rank": 30,
  "importance": 9.99999999995449e-06,
  "addresstype": "place",
  "name": "",
  "display_name": "591, Saavedra, Perito Moreno, Lago Buenos Aires, Santa
  Cruz, 9040, Argentina",
  "address": {
    "house_number": "591",
    "road": "Saavedra",
    "village": "Perito Moreno",
    "municipality": "Perito Moreno",
    "state_district": "Lago Buenos Aires",
    "state": "Santa Cruz",
    "ISO3166-2-lvl4": "AR-Z",
    "postcode": "9040",
    "country": "Argentina",
    "country_code": "ar"
  },
```

```
  "boundingbox": [
    "-46.5900167",
    "-46.5899167",
    "-70.9200513",
    "-70.9199513"
  ]
}
```

This place is located in the town of Perito Moreno in Argentina.

It is also possible to test the request using cURL:

```
curl -s -G "https://nominatim.openstreetmap.org/reverse" ^
--data-urlencode "lat=-46.59" ^
--data-urlencode "lon=-70.92" ^
--data-urlencode "zoom=18" ^
--data-urlencode "format=jsonv2" ^
--user-agent "curl 8.13.0 (Windows)"
```

A User-Agent header must be included to comply with Nominatim's identification usage policy (see the "Usage Policy" section below).

This command uses a Windows syntax. With Linux or macOS, the line continuation character is not ^ but \.

This command returns the following result:


```
{"place_id":190661,"licence":"Data © OpenStreetMap contributors, ODbL 1.0.
http://osm.org/copyright","osm_type":"way","osm_id":322914100,"lat":
"-46.5899667","lon":"-70.9200013","category":"place","type":"house",
"place_rank":30,"importance":9.99999999995449e-06,"addresstype":"place",
"name":"","display_name":"591, Saavedra, Perito Moreno, Lago Buenos Aires,
Santa Cruz, 9040, Argentina","address":{"house_number":"591","road":
"Saavedra","village":"Perito Moreno","municipality":"Perito Moreno",
"state_district":"Lago Buenos Aires","state":"Santa Cruz","ISO3166-2-lvl4":
"AR-Z","postcode":"9040","country":"Argentina","country_code":"ar"},
"boundingbox":["-46.5900167","-46.5899167","-70.9200513","-70.9199513"]}
```

Here is an HTML example using JavaScript to invoke a Nominatim reverse geocoding HTTP request to search for the address of the same location (coordinates -46.59 and -70.92):

```
1   <html>
2   <head>
3   <title>Nominatim example</title>
4   </head>
5   <body>
6   <h2>Nominatim reverse geocoding HTTP request example</h2>
7   <p>Reverse geocoding the location [-46.59, -70.92] with Nominatim</p>
8   <h3>Raw JSON Response:</h3>
9   <p id="response">Loading...</p>
10  <script>
11      fetch("https://nominatim.openstreetmap.org/reverse?lat=-46.59
        &lon=-70.92&zoom=18&format=jsonv2")
12      .then(response => response.json())
13      .then(data => {
14          const responseElement = document.getElementById('response');
15          responseElement.textContent = JSON.stringify(data, null, 2);
16      })
17      .catch(error => {
18          document.getElementById('response').textContent = 'Error: '
            + error;
19      });
20  </script></body>
21  </html>
```

- **Line 11**: `fetch` is a JavaScript function used to send HTTP requests.

When opened in a browser, this file produces the following output:

Nominatim reverse geocoding HTTP request example

Reverse geocoding the location [-46.59, -70.92] with Nominatim

Raw JSON Response:

{ "place_id": 253606, "licence": "Data © OpenStreetMap contributors, ODbL 1.0. http://osm.org/copyright", "osm_type": "way", "osm_id": 322914100, "lat": "-46.5899667", "lon": "-70.9200013", "category": "place", "type": "house", "place_rank": 30, "importance": 0.00000999999999995449, "addresstype": "place", "name": "", "display_name": "591, Saavedra, Perito Moreno, Lago Buenos Aires, Santa Cruz, 9040, Argentina", "address": { "house_number": "591", "road": "Saavedra", "village": "Perito Moreno", "municipality": "Perito Moreno", "state_district": "Lago Buenos Aires", "state": "Santa Cruz", "ISO3166-2-lvl4": "AR-Z", "postcode": "9040", "country": "Argentina", "country_code": "ar" }, "boundingbox": ["-46.5900167", "-46.5899167", "-70.9200513", "-70.9199513"] }

Usage Policy

The usage policy of Nominatim's public API is detailed at `https://operations.osmfoundation.org/policies/nominatim/`. This policy notably enforces the following rules:

- **Rate Limiting**: Heavy use is not permitted (no more than one request per second). Bulk or automated requests are explicitly forbidden. For such use cases, users should consider setting up a private instance or using a paid geocoding service.
- **Identification**: Each request must include a valid **HTTP Referer** or **User-Agent** header clearly identifying the application. Requests without proper identification will be rejected.

In the previous sections, Nominatim requests were submitted to the public endpoint without explicitly including the User-Agent header. However, the browser automatically added several standard HTTP headers, including a User-Agent, and therefore, the requests were accepted. On the other hand, when the requests were submitted using cURL, no browser was involved, and the User-Agent header had to be included explicitly.

As previously mentioned, Chapter 7 will explore how to use Nominatim requests within Python and JavaScript programs in web applications and how to display their results on maps.

Other Geocoding Tools

As already explained, other geocoding solutions exist in the open source market:

- **Photon**: Based on OpenStreetMap (data import source) and ElasticSearch (database and search engine), it offers faster direct and reverse geocoding searches than Nominatim, but not as detailed results.

  It offers a Web interface (`https://photon.komoot.io`) and a public search API (e.g., `https://photon.komoot.io/api?q=Berlin`).

- **Pelias**: A geocoding engine which data is imported from OpenStreetMap and other sources (OpenAddresses, Who's on First, etc.). It uses ElasticSearch for data storage and as a search engine to answer the direct and reverse geocoding queries it receives.

  It offers no public API and requires its code to be installed on a server. Some companies (such as Geocode Earth, Stadia Maps, and Jawg Places) offer geocoding services based on Pelias, for a fee.

  For more information, see `https://pelias.io/`.

- **Gisgraphy**: A geocoding engine written in Java which imports data from GeoNames, OpenStreetMap, and other sources. It provides a REST API for geocoding (and reverse geocoding) requests.

  For more information, see `https://www.gisgraphy.com/`.

- **GeoNames**: A geocoding website (`https://www.geonames.org`) using its own data (not based on OSM, but on government data sources from various countries). It allows direct and reverse geocoding searches, but on city and place names only (not addresses).

  It offers a public API (`http://api.geonames.org/`), but requires prior registration of a user account.

  GeoNames is not exactly open source, although it is an open-data service. More precisely:

  - The data and basic API services are free for light use, and the returned data are open and reusable under the Creative Commons Attribution 4.0 License.

  - However, the underlying server code is not open sourced, which means it is not possible to self-host a GeoNames server.

Finally, it is worth mentioning that proprietary geocoding (and reverse geocoding) services are also offered by several companies, providing high precision and wide coverage of territories for searches. These services are charged starting from a certain number of queries. These include Google Maps Geocoding API, Microsoft Bing Maps API, Mapbox Geocoding API, Here Geocoding API, Tomtom Geocoding API, and others.

CHAPTER 4

Creating Interactive Maps with Leaflet and JavaScript

Leaflet.js is a JavaScript library for building interactive maps and displaying them in a browser. To do this, Leaflet communicates with tile servers, requesting the individual tiles needed to render the map. Maps can be enhanced by adding various elements such as:

- **Markers** to highlight specific points or objects.
- **Polylines** to highlight ways or rivers and **geometric shapes** to delineate areas and fill them with colors.
- **Popups** and **tooltips** to display additional information as text.
- **Images** superimposed on maps.

Geographic data in GeoJSON or TopoJSON format can be added to the maps, generating elements like markers, lines, and polygons.

Choropleth maps can be built to visualize statistical data across geographic regions.

Moreover, **event handlers** can be used to make Leaflet maps highly interactive by responding to user actions.

Leaflet's use is client-side, where its code is executed directly in the browser as JavaScript embedded in HTML pages. Its interactive capabilities require a browser to be fully leveraged. Although there are JavaScript-enabled web servers such as Node.js, Leaflet is not designed to run in such environments.

P. Marie, *Modern Web Cartography*, https://doi.org/10.1007/979-8-8688-2335-0_4

The list of Leaflet functions described below is not exhaustive. In particular, the many options available for Leaflet objects are not detailed or explained. For a complete reference, please consult the Leaflet documentation: `https://leafletjs.com/index.html`.

At the time of writing this book, the current release of Leaflet is version 1.9.4, and it is the version used for all the examples.

The code for the examples in this chapter can be found on the website `https://github.com/Apress/Modern-Web-Cartography` and in the `leaflet_maps` directory.

Basic Principles of Leaflet

Using Leaflet begins with creating a base map. This involves specifying the map's center coordinates, a zoom level, and a tile set to render the map's visual style. This initial setup forms the **base layer**.

Additional elements can then be overlaid on the map to highlight places, communities, routes, regions, or other geographic entities. These are overlay layers and may include:

- Markers, optionally enhanced with popups or tooltips.
- Polylines, polygons, and other geometric shapes.
- Images.
- Alternative tile sets.
- Etc.

Elements added to the map can be created through programming statements or generated from imported data in formats such as **GeoJSON**, **TopoJSON**, or **GPX**.

These map elements can be grouped using **layer groups** or **feature groups** for easier control and management (in particular, to allow the user to show or hide them).

Controls enable users to interact with the map – for example, by zooming in or out or toggling layer visibility (enabling the user to show or hide groups of overlaid elements). They can also be used to display supplementary information such as a scale bar or attribution notice (sort of copyright embedded in the map). Some controls are included by default (but customizable), while others must be explicitly added by the developer.

Leaflet also supports event handling on the map and on most overlay elements. Events include:

- Clicking on the map or a map element (marker, polyline, polygon, etc.).
- Moving the mouse over or out of a map element.
- Pressing the mouse or keyboard buttons.
- Etc.

Developers can define **custom actions** in response to these events, making Leaflet a powerful tool for creating highly interactive maps.

Environment Used to Build and Run the Following Examples

In this chapter, we will learn how to build maps with Leaflet.js and JavaScript by creating a number of elementary maps. At this stage, these examples will be developed and tested in the simplest possible way: Leaflet and JavaScript statements will be placed in HTML files, then these files will be opened in a browser (using the `file://` protocol) for testing. An Internet connection is required to access the Leaflet.js code and tile servers.

If you are not comfortable using the `file://` protocol or encounter security restrictions in your browser, you can instead serve your test files through a local web server. For example, using Python, run the following command from the directory where the files are located to start a simple local server:

```
python -m http.server 8000
```

Then, open the following URL in your browser:

```
http://localhost:8000/
```

This allows you to access your test files without running into local file access limitations.

Although not essential, it is advisable to use VS Code to implement these prototypes, in order to benefit from its editor (which detects and reports syntax errors) and debugger (which facilitates error diagnosis in JavaScript code).

Note that this very simple configuration, using only HTML files, is limited to unit testing: in a more realistic web application development context, a more complete configuration (such as the one described in Appendixes A and B) is required.

Most of the HTML files for the prototypes that we will create share the same structure. Only one example, using an asynchronous function, will differ in its structure for error handling (it will not include a `try/catch` block). Below is a typical structure for these HTML files (the parts that may vary depending on the examples are shown in **bold**):

```
<html>
<head>
<link rel="stylesheet" href="https://unpkg.com/leaflet@1.9.4/dist/leaflet.css" integrity="sha256-p4NxAoJBhIIN+hmNHrzRCf9tD/miZyoHS5obTRR9BMY=" crossorigin=""/>
<title>OSM map created with Leaflet</title>
</head>
<body>
<h2>(Specific title of the map)</h2>
<div id="map" style="width: 1030px; height: 620px;"></div>
<script src="https://unpkg.com/leaflet@1.9.4/dist/leaflet.js" integrity="sha256-20nQCchB9co0qIjJZRGuk2/Z9VM+kNiyxNV1lvTlZBo=" crossorigin=""></script>
<script type="text/javascript">
window.onerror = function(message, url, linenumber) {
    if (typeof map !== 'undefined' && map) {map.remove();}

    const div = document.getElementById("map")
    if (div) {div.remove();}

    const elemDiv = document.createElement('div');
    elemDiv.innerHTML = `<p>${message}, line ${linenumber}</p>`;
    document.body.appendChild(elemDiv);
}
</script>
<script type="text/javascript">
let map;
try {

(Specific statements used to build a map)
```

```
31    } catch(e) {
32        if (typeof map !== 'undefined' && map) {map.remove();}
33
34        const div = document.getElementById("map")
35        if (div) {div.remove();}
36
37        const elemDiv = document.createElement('div');
38        elemDiv.innerHTML = '<p>' + e + '</p>';
39        document.body.appendChild(elemDiv);
40    }
41    </script>
42    </body>
43    </html>
```

- **Line 3**: A link to the style sheet used by Leaflet.js. This statement is essential to Leaflet.
- **Line 7**: The map header is set. It is specific to each example.
- **Line 8**: A division with ID `map`, designed to host the map built by Leaflet.js. The `width` and `height` properties specify the amount of space required to contain the map.

  In HTML, a division is a container used to group and organize content on a web page. Here, it is used to specify the place where Leaflet will insert the map.
- **Line 9**: A link to the site where the Leaflet.js code can be found. This statement is essential to Leaflet.
- **Lines 11 to 20**: A JavaScript function dedicated to error handling is defined. It is responsible for processing all JavaScript errors that have not been intercepted by the `catch` part (on lines 31 to 40) of the `try/catch` block (notably syntax errors). The processing consists of deleting the map and the division containing it and adding a new division, in which an error message is inserted.
- **Line 23**: A `map` variable, which will contain our map, is declared.

- **Line 24**: This line starts a `try/catch` block. The `try` part contains the statements for building the map. These statements will be executed as long as no error is encountered.
- **From line 25 on**: Leaflet statements must be added to create a map, which will be placed in the division with ID `map` (on line 8).
- **Lines 31 to 40** (these line numbers are arbitrary, as there can be any number of map creation statements): The `catch` part (of the `try/catch` block), which statements will be executed only if an error occurs during execution of the `try` part. Processing consists of deleting the map and the division containing it, and adding a new division, in which an error message is inserted.

  Please note that some errors are not caught by the `catch` block, which is why additional error handling is defined in lines 11 to 20.

The architecture used by the examples of this chapter is the following:

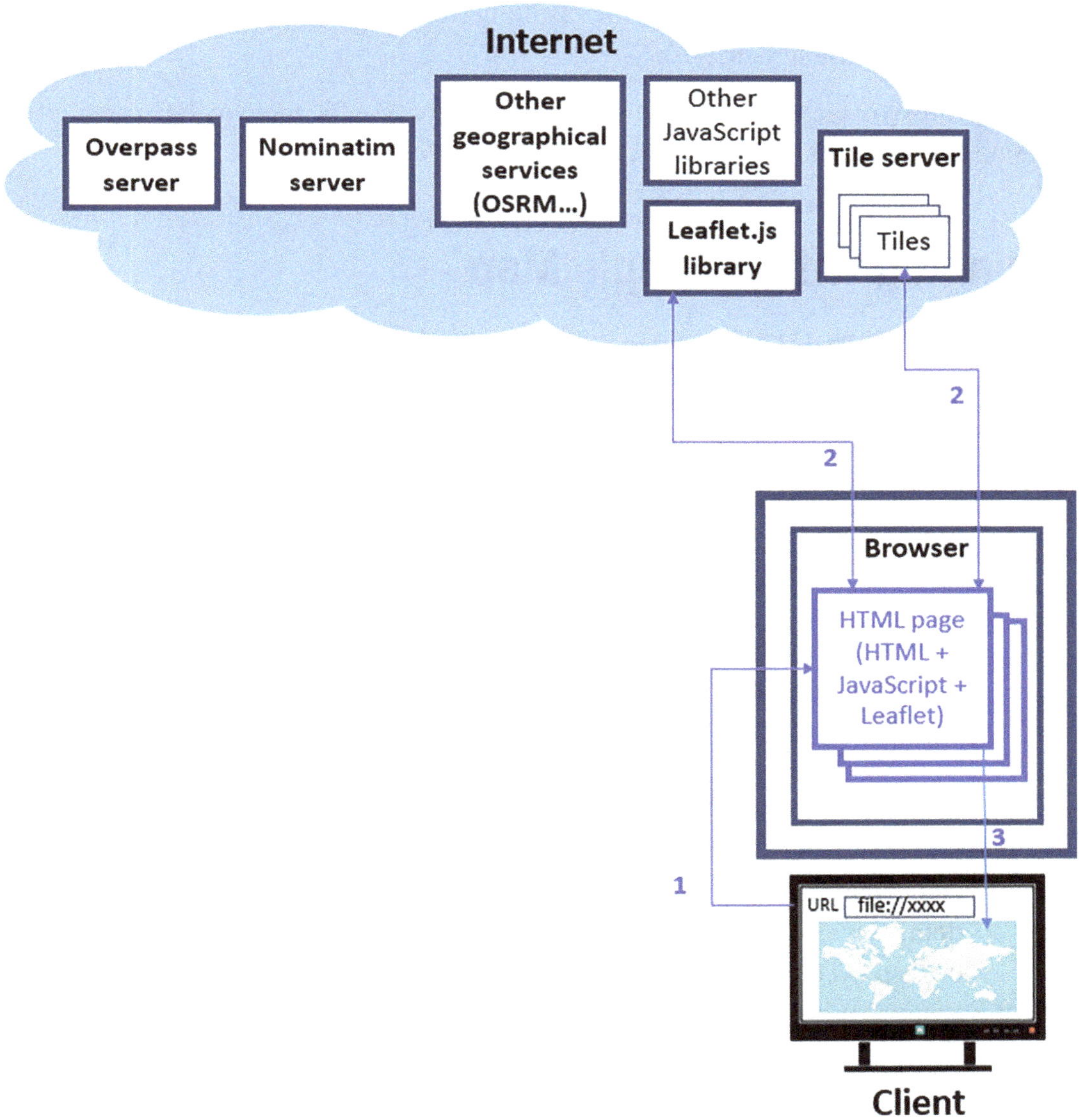

Figure 4-1. *Architecture used to test maps created with Leaflet*

1. The user invokes a URL from the browser. This URL points to an HTML file located in the local file system.
2. The browser loads this HTML file, executes the embedded JavaScript and Leaflet code, and downloads the tiles necessary to render the map.

3. The browser then displays the resulting web page, including a map, on the screen.

Every time you test a new map, do not forget to save the file that you have created or modified to produce it before opening it with a browser.

Beginning with a Simple Map

Let's start by creating a simple map of the area around a point defined by a latitude and a longitude.

It is very easy to create such a map with Leaflet using an `L.map` object. For example:

```
1   map = L.map('map').setView([51.505, -0.09], 13);
2
3   L.tileLayer('https://tile.openstreetmap.org/{z}/{x}/{y}.png', {
4       maxZoom: 19,
5       attribution: '&copy; <a href="http://www.openstreetmap.org/copyright">OpenStreetMap</a>'
6   }).addTo(map);
```

- **Line 1**: A map is created using an `L.map` object supplied by the Leaflet library. It will be placed in a division (of the HTML page) with ID `map`. This map is centered on latitude 51.505 and longitude -0.09 and defined with an initial zoom level of 13. At this stage, our map cannot be displayed yet, as it has not received any tiles.

- **Lines 3 to 6**: A tile layer object is created. To do this, we supply the reference of a tile set to be used, the maximum zoom level allowed on the map (19), and the attribution notice (a kind of copyright) to be inlaid in the bottom right-hand corner of the map. Leaflet will ask the tile server (corresponding to the chosen tile set) for the tiles it needs to build the map around the point on which it has been centered.

  The available tile sets are presented on the website `https://leaflet-extras.github.io/leaflet-providers/preview/`. This site lets you visualize the rendering, check the area covered by each tile set, and choose the ones that suit you best. It also indicates the reference required to use a tile set and the attribution notice (equivalent to a copyright) to be included using the `attribution` option.

- Finally, the tile layer is added to the map object.

Please note that the tiles used above (`https://tile.openstreetmap.org/{z}/{x}/{y}.png` – OSM Standard Tiles – Mapnik), and all other tiles used in the following examples, are public tiles and intended only for demonstration, education, development, testing, and noncommercial web applications. For production or commercial applications, it is recommended to:

- Either set up your own tile server using OpenStreetMap data.
- Or use a commercial tile provider.

For more details, see `https://operations.osmfoundation.org/policies/tiles/`.

We will now create a map of London, United Kingdom. First, create the directory where the HTML file for this prototype will be stored (along with those for the other prototypes in this chapter). For example: `C:/modern_web_cartography/leaflet_maps` (any other path and name may be used). Then, create an HTML file with the following contents, and save it as `page_with_simple_map.html` in the directory just created (`leaflet_maps`).

```
<html>
<head>
<link rel="stylesheet" href="https://unpkg.com/leaflet@1.9.4/dist/leaflet.css" integrity="sha256-p4NxAoJBhIIN+hmNHrzRCf9tD/miZyoHS5obTRR9BMY=" crossorigin=""/>
<title>OSM map created with Leaflet</title>
</head>
<body>
<h2>Simple map of central London</h2>
<div id="map" style="width: 1030px; height: 620px;"></div>
<script src="https://unpkg.com/leaflet@1.9.4/dist/leaflet.js" integrity="sha256-20nQCchB9co0qIjJZRGuk2/Z9VM+kNiyxNV1lvTlZBo=" crossorigin=""></script>
<script type="text/javascript">
window.onerror = function(message, url, linenumber) {
```

```
12      if (typeof map !== 'undefined' && map) {map.remove();}
13
14      const div = document.getElementById("map")
15      if (div) {div.remove();}
16
17      const elemDiv = document.createElement('div');
18      elemDiv.innerHTML = `<p>${message}, line ${linenumber}</p>`;
19      document.body.appendChild(elemDiv);
20  }
21  </script>
22  <script type="text/javascript">
23  let map;
24  try {
25      map = L.map('map').setView([51.505, -0.09], 13);
26
27      L.tileLayer('https://tile.openstreetmap.org/{z}/{x}/{y}.png', {
28          maxZoom: 19,
29          attribution: '&copy; <a href="http://www.openstreetmap.org/
            copyright">OpenStreetMap</a>'
30      }).addTo(map);
31
32  } catch(e) {
33      if (typeof map !== 'undefined' && map) {map.remove();}
34
35      const div = document.getElementById("map")
36      if (div) {div.remove();}
37
38      const elemDiv = document.createElement('div');
39      elemDiv.innerHTML = `<p>${e}</p>`;
40      document.body.appendChild(elemDiv);
41  }
42  </script>
43  </body>
44  </html>
```

- **Line 7**: A specific header is set for the page containing the map.

- **Line 8**: This is the division where the map will be placed by Leaflet.
- **Line 25**: A map object is created with Leaflet. The parameters have already been explained just above. The `map` ID refers to the HTML `<div>` defined on line 8, and the coordinates correspond to central London.
- **Lines 27 to 30**: A tile layer is added to the map object. The parameters have already been explained just above. The tile set chosen is OpenStreetMap Standard - Mapnik.

Open this HTML file with a browser, using URL `file:///C:/modern_web_cartography/leaflet_maps/page_with_simple_map.html` (adjust the path to match the location where you stored the file in your environment). The browser will show the following web page (remember that a connection to the Internet is needed; otherwise, the browser will not be able to access the Leaflet code or download the necessary tiles).

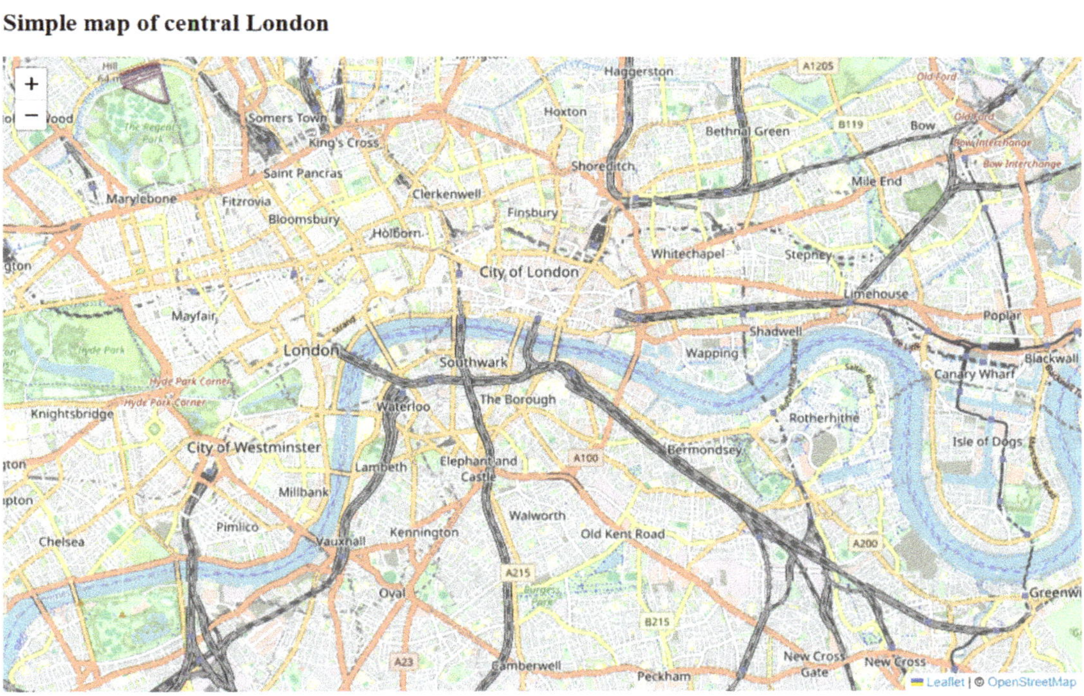

Figure 4-2. Map created with Leaflet using OpenStreetMap tiles. Data © OpenStreetMap contributors, licensed under the ODbL

This map is not static:

- You can zoom in and out at will by double-clicking, using the "+" and "-" buttons (top left of the map) or the mouse wheel.
- You can also move the map by dragging it with the mouse.

For example, by zooming out and moving it to the southeast, you can see the following:

Simple map of central London

Figure 4-3. Map created with Leaflet using OpenStreetMap tiles. Data © OpenStreetMap contributors, licensed under the ODbL

Adding Markers, Popups, and Tooltips

The simple map just created can be enhanced by overlaying additional elements.

Markers

Leaflet lets you add **markers** to maps to highlight specific locations. Various **icons** can be associated with these markers, and text can be displayed in **popups** (the text appears in a bubble when the marker is clicked) or in **tooltips** (the text appears in a bubble when the mouse cursor passes over the marker). They can also be assigned an opacity level, so that the map underneath is still visible through transparency.

Markers can be added to a Leaflet map using `L.marker` objects. For example:

```
1   L.marker([51.50853, -0.07613]).addTo(map)
2   .bindPopup('Tower of London');
```

- **Line 1**: A marker is created using an `L.marker` object. Its geographic position is set by providing coordinates. Then it is added to the map.
- **Line 2**: A popup is associated with the marker. It will appear when the user clicks on the marker.

To create this new map, we will build upon the code developed in the previous section: new statements will be inserted in this code to add some markers indicating tourist places. Before adding these new statements, copy the code of the previous file `leaflet_maps/page_with_simple_map.html` in a new file `leaflet_maps/page_with_markers.html` (that will be used in this section). Then, insert these new statements (shown in **bold** below) into the new file:

```
...
        }).addTo(map);

        L.control.scale().addTo(map);

        L.marker([51.50853, -0.07613]).addTo(map)
        .bindPopup('Tower of London');

        L.marker([51.49628, -0.14520]).addTo(map)
        .bindPopup('Buckingham Palace');

        L.marker([51.50073, -0.12463]).addTo(map)
```

```
41          .bindPopup('Big Ben');
42
43          L.marker([51.50914, -0.16123]).addTo(map)
44          .bindPopup('Hyde Park');
45
46      } catch(e) {
    ...
```

- **Line 32**: A scale control is added. It will appear in the bottom left-hand corner of the map (this statement is unrelated to the markers added next).
- **Line 34**: A marker is created at a geographic position, and it is added to the map.
- **Line 35**: A popup is associated with this marker.
- Then three more markers are created with popups using the same method.

Opening this file in a browser (using URL `C:\modern_web_cartography\leaflet_maps\page_with_markers.html`) produces the following page:

Simple map of central London

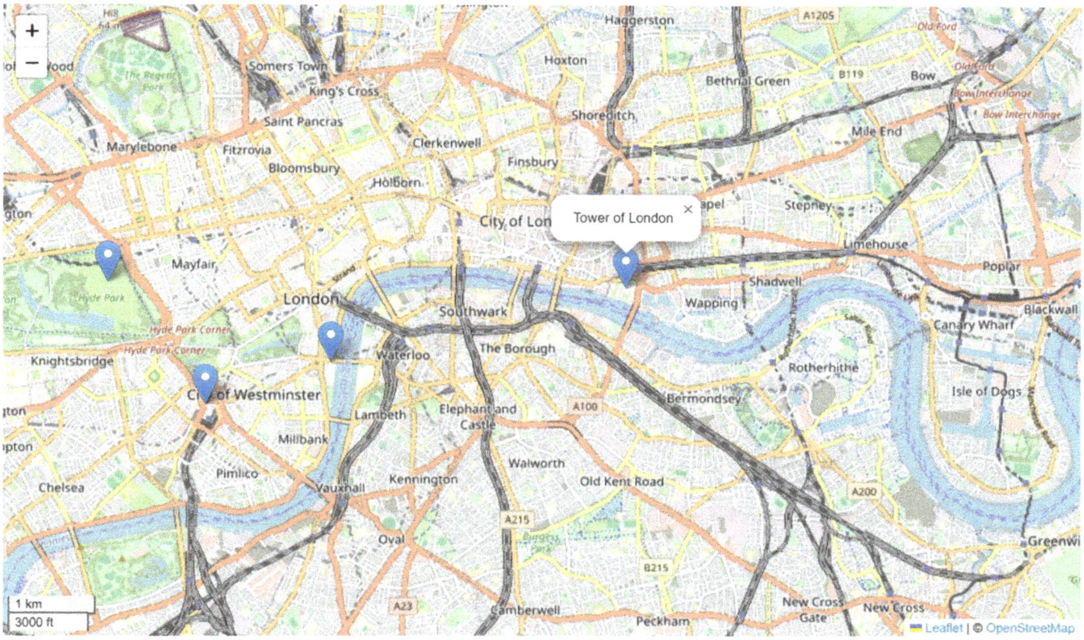

Figure 4-4. *Map created with Leaflet using OpenStreetMap tiles. Data © OpenStreetMap contributors, licensed under the ODbL*

A scale control has been added at the bottom left and you have to click on the markers to display their popups, and you can only see one at a time.

We could also have associated tooltips with markers using statements such as:

```
.bindTooltip('<Tooltip text>')
```

Now we will customize the markers to better distinguish them: red markers with a monument icon inside will be used for three of the sites, and a green marker with a tree icon inside for the fourth. The following statements must be inserted or modified in the file `leaflet_maps/page_with_markers.html` (shown in **bold**):

```
<html>
<head>
<link rel="stylesheet" href="https://unpkg.com/leaflet@1.9.4/
dist/leaflet.css" integrity="sha256-p4NxAoJBhIIN+hmNHrzRCf9tD/
miZyoHS5obTRR9BMY=" crossorigin=""/>
```

```
<link rel="stylesheet" href="https://cdnjs.cloudflare.com/ajax/libs/
font-awesome/4.7.0/css/font-awesome.min.css">
<link rel="stylesheet" href="https://cdn.jsdelivr.net/npm/leaflet-
extra-markers@1.2.1/dist/css/leaflet.extra-markers.min.css">
<title>Simple OSM map created with Leaflet</title>
</head>
<body>
<h2>Simple map of central London</h2>
<div id="map" style="width: 1030px; height: 620px;"></div>
<script src="https://unpkg.com/leaflet@1.9.4/dist/leaflet.js" in
tegrity="sha256-20nQCchB9co0qIjJZRGuk2/Z9VM+kNiyxNV1lvTlZBo="
crossorigin=""></script>
<script src="https://cdn.jsdelivr.net/npm/leaflet-extra-markers@1.2.1/
dist/js/leaflet.extra-markers.min.js"></script>
<script type="text/javascript">
window.onerror = function(message, url, linenumber) {

...

}
</script>
<script type="text/javascript">
let map;
try {
    map = L.map('map').setView([51.505, -0.09], 13);

    L.tileLayer('https://tile.openstreetmap.org/{z}/{x}/{y}.png', {
        maxZoom: 19,
        attribution: '&copy; <a href="http://www.openstreetmap.org/
        copyright">OpenStreetMap</a>'
    }).addTo(map);

    L.control.scale().addTo(map);

    const monumentMarker = L.ExtraMarkers.icon({
        icon: 'fa-university',
```

```
39              markerColor: 'red',
40              shape: 'circle',
41              prefix: 'fa'
42          });
43
44          L.marker([51.50853, -0.07613], {icon: monumentMarker}).addTo(map)
45          .bindPopup('Tower of London');
46
47          L.marker([51.49628, -0.14520], {icon: monumentMarker}).addTo(map)
48          .bindPopup('Buckingham Palace');
49
50          L.marker([51.50073, -0.12463], {icon: monumentMarker}).addTo(map)
51          .bindPopup('Big Ben');
52
53          const gardenMarker = L.ExtraMarkers.icon({
54              icon: 'fa-tree',
55              markerColor: 'green',
56              shape: 'circle',
57              prefix: 'fa'
58          });
59
60          L.marker([51.50914, -0.16123], {icon: gardenMarker}).addTo(map)
61          .bindPopup('Hyde Park');
62
63      } catch(e) {
        ...
71      }
72      </script>
73      </body>
74      </html>
```

- **Lines 4 and 5**: Two links to style sheets are defined. They are required to customize markers.

- **Line 12**: A link to a JavaScript library, needed to customize markers, is defined.
- **Lines 37 to 42**: An `ExtraMarkers.icon` object is created. It defines the icon to be inserted in the marker, the marker color, its shape, and a prefix (to complete the icon name).

  A list of available icons can be found at `https://fontawesome.com/v4/cheatsheet/`, and the supported colors are `red`, `blue`, `green`, `orange`, `yellow`, `purple`, `darkred`, `white`, `black`, and `cadetblue`.

  This object will be used to define the style of markers indicating monuments.
- **Lines 44, 47, and 50**: The preceding `ExtraMarkers.icon` object is associated with the three monument markers, to customize them.
- **Lines 53 to 58**: Another `ExtraMarkers.icon` object is defined, this time for the marker indicating a park.
- **Line 60**: This object is associated with the Hyde Park marker.

Opening the file again in a browser produces the following page, with custom markers:

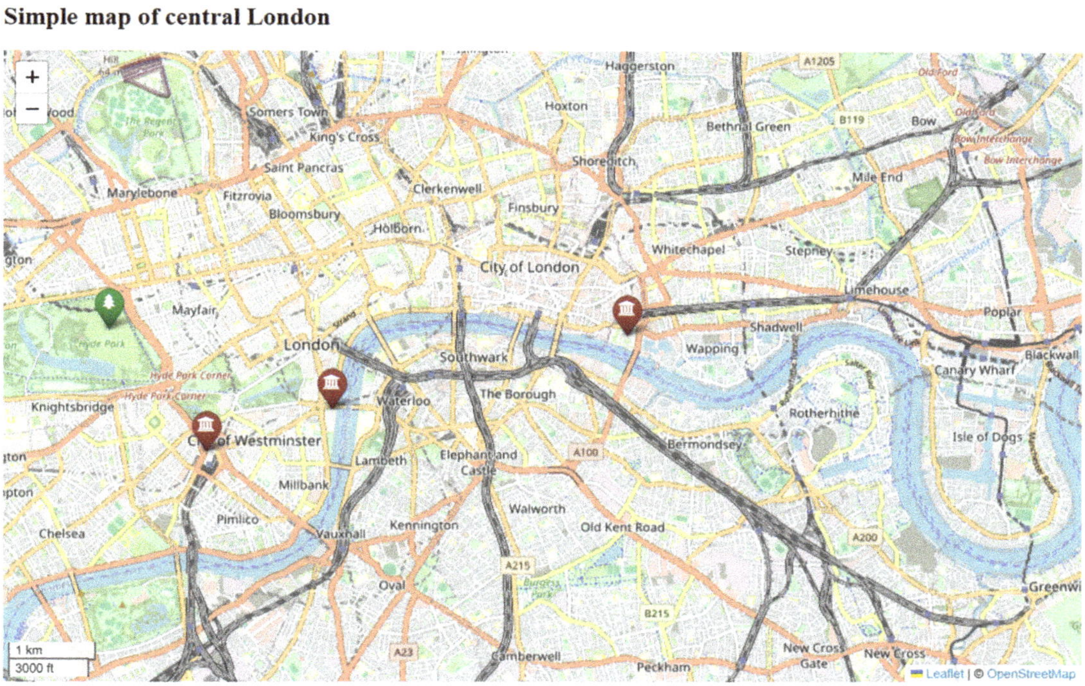

***Figure 4-5.** Map created with Leaflet using OpenStreetMap tiles. Data © OpenStreetMap contributors, licensed under the ODbL*

We have just customized the markers by embedding an icon inside the marker symbol. It is also possible to replace the default marker symbol with an icon of your choice (e.g., use a mug icon to indicate bars or a flower icon for florists). For example:

```
1   const barIcon = L.icon({
2       iconUrl: 'mug.jpg',
3       iconSize: [32, 32],
4   });
5
6   L.marker([44.284, 5.034], { icon: barIcon })
7   .addTo(map)
8   .bindPopup('Really cool bar!')
9   .openPopup();
```

- **Lines 1 to 4**: An icon is defined with an image file and a size in pixels. You must previously have found an image representing a mug and stored it in the same directory as the HTML file (`leaflet_maps` in our case) with the name “mug.jpg”.

- **Line 6**: A marker is created with geographic coordinates and the preceding icon.
- **Lines 7 to 9**: It is added to the map, assigned a popup, then the popup is opened.

Another possibility is to use circle markers (markers in the shape of a circle). Here is an example:

```
1   L.circleMarker([44.174, 5.278], {
2       radius: 10,
3       color : 'red',
4       opacity: 1,
5       fillOpacity: 0
6   })
7   .addTo(map)
8   .bindPopup('Beautiful viewpoint!')
```

- **Lines 1 to 6**: A circle marker is defined with geographic coordinates, a radius of ten pixels, a red stroke color, full stroke opacity, and no fill opacity (meaning the background map will remain visible inside the circle).
- **Lines 7 and 8**: It is then added to the map and assigned a popup.

Popups

Above, popups have been used in conjunction with markers. But popups and tooltips can also be independent objects. For example, we will add a popup to the previous map by inserting the following code (following the marker definitions):

```
    ...
        .bindPopup('Hyde Park');

        L.popup()
```

```
64        .setLatLng([51.50775, -0.04616])
65        .setContent("River Thames")
66        .openOn(map);
67
68    } catch(e) {
   ...
```

- **Line 63**: An `L.popup` object is created.
- **Line 64**: A geographical position is assigned to the popup.
- **Line 65**: The popup text is defined.
- **Line 66**: The popup is opened on the map.

Opening the file again in a browser produces the following page:

Simple map of central London

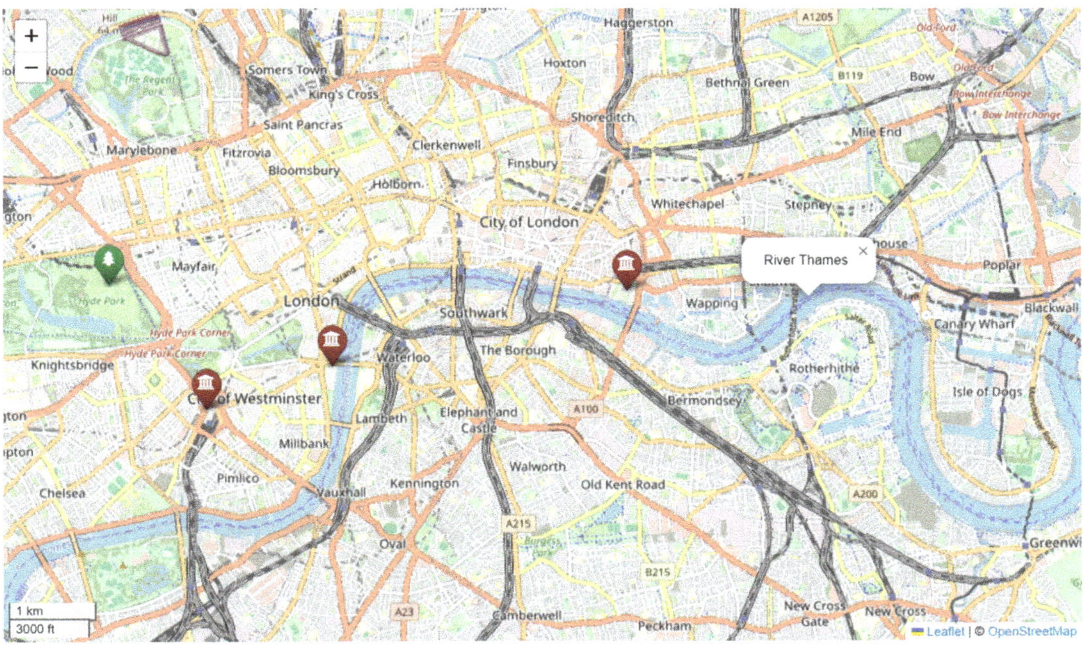

Figure 4-6. *Map created with Leaflet using OpenStreetMap tiles. Data © OpenStreetMap contributors, licensed under the ODbL*

Grouping Map Elements and Controlling Their Display

Elements added to maps (markers, popups, tooltips, etc.) can be associated within **groups**. Layer controls can then be used to show or hide elements in certain groups. More precisely, elements added to the map (markers, popups, tooltips, etc.) are layers superimposed on the map, and a **layer control** lets you choose to show or hide them.

Similarly, **tile layers** that can be added to the map are layers overlaid on it, and a layer control can also be used to show or hide them.

Here is an example showing how to group objects and control their display:

```
1   const mark1 = L.marker([51.50853, -0.07613]).addTo(map)
2   .bindPopup('Tower of London');
3
4   const mark2 = L.marker([51.49628, -0.14520]).addTo(map)
5   .bindPopup('Buckingham Palace');
6
7   const mark3 = L.marker([51.50073, -0.12463]).addTo(map)
8   .bindPopup('Big Ben');
9
10  const monument_group = L.layerGroup([mark1, mark2, mark3]).addTo(map);
11
12  const overlayGroups = {
13      "Monuments": monument_group
14  };
15
17  L.control.layers(overlayGroups).addTo(map);
```

- **Lines 1 to 8**: Three markers are created with popups and added to the map.
- **Line 10**: An `L.layerGroup` object comprising the three markers is created and added to the map.
- **Lines 12 to 14**: An object comprising the previous layer group, associated with a key, is created.

- **Line 17**: An `L.control.layers` object is created with the previous object, then it is added to the map. It will enable the map users to show or hide the three markers.

Here is an example extending the one of the preceding section. We will put the markers and the popup in different groups and add a layer control. Before any change, copy the code of the previous file `leaflet_maps/page_with_markers.html` in a new file `leaflet_maps/page_with_groups_and_controls.html` (that will be used in this section). Then, insert new statements and modify existing statements in this new file as shown in **bold** below:

```
...

try {
    map = L.map('map').setView([51.505, -0.09], 13);

    L.tileLayer('https://tile.openstreetmap.org/{z}/{x}/{y}.png', {
        maxZoom: 19,
        attribution: '&copy; <a href="http://www.openstreetmap.org/copyright">OpenStreetMap</a>'
    }).addTo(map);

    L.control.scale().addTo(map);

    const monumentMarker = L.ExtraMarkers.icon({
        icon: 'fa-university',
        markerColor: 'red',
        shape: 'circle',
        prefix: 'fa'
    });

    const mark1 = L.marker([51.50853, -0.07613], {icon: monumentMarker}).addTo(map)
    .bindPopup('Tower of London');

    const mark2 = L.marker([51.49628, -0.14520], {icon: monumentMarker}).addTo(map)
```

```
    .bindPopup('Buckingham Palace');

    const mark3 = L.marker([51.50073, -0.12463], {icon: monumentMarker}).addTo(map)
    .bindPopup('Big Ben');

    const monument_group = L.layerGroup([mark1, mark2, mark3]).addTo(map);

    const gardenMarker = L.ExtraMarkers.icon({
        icon: 'fa-tree',
        markerColor: 'green',
        shape: 'circle',
        prefix: 'fa'
    });

    const mark4 = L.marker([51.50914, -0.16123], {icon: gardenMarker})
    .bindPopup('Hyde Park');

    const park_group = L.layerGroup([mark4]);

    const popup = L.popup()
    .setLatLng([51.50775, -0.04616])
    .setContent("River Thames");

    const river_group = L.layerGroup([popup]);

    const overlayGroups = {
        "Monuments": monument_group,
        "Parks": park_group,
        "Rivers": river_group
    };

```

```
79          L.control.layers(overlayGroups).addTo(map);
80
81      } catch(e) {
    ...
```

- **Line 53**: A layer group comprising the three previous markers (monuments) is created and added to the map.
- **Line 62**: Remove the `.addTo(map)` statement at the end of the line, so that the marker for Hyde Park is not visible when the map is opened.
- **Line 65**: A layer group containing the previous marker (park) is created. It is not added to the map so it is not selected (in the layer control) when the map is opened.
- **Lines 67 to 70**: Delete the instruction on line 70 (opening the popup on the map): the popup for River Thames will no longer be visible when the map is opened.
- **Line 71**: A layer group containing the popup (river) is created. It is not added to the map so it is not selected (in the layer control) when the map is opened.
- **Lines 73 to 77**: An object comprising the three previous layer groups, each one associated with a key, is created.
- **Line 79**: A layer control is created with the object containing the three groups, then it is added to the map.

Opening this file in a browser produces the following result:

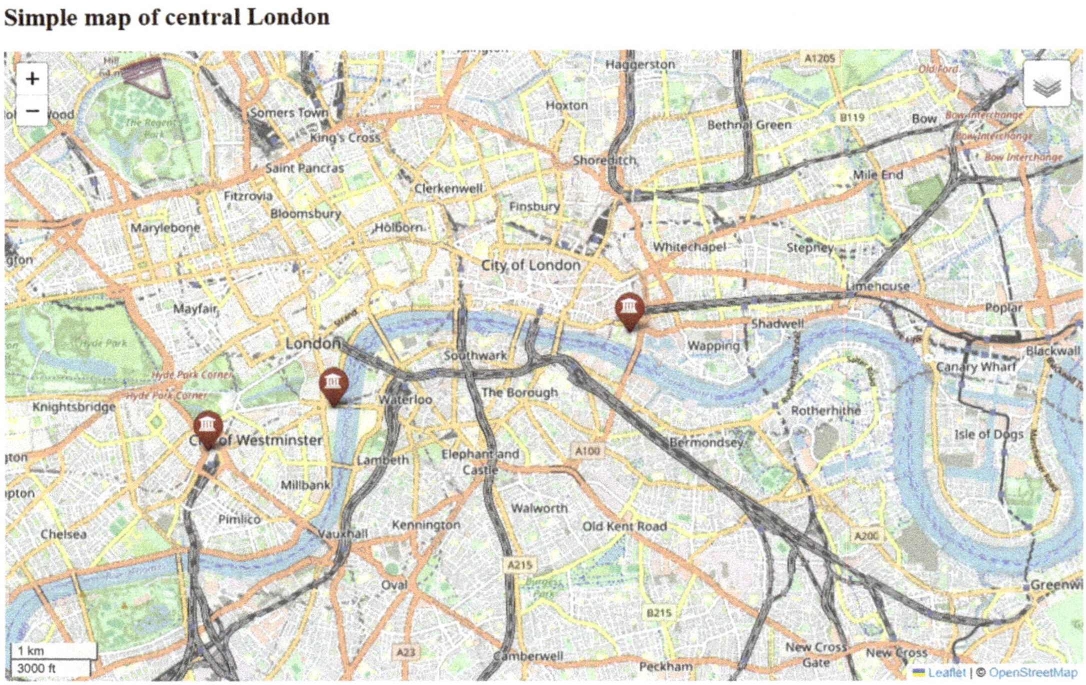

Figure 4-7. *Map created with Leaflet using OpenStreetMap tiles. Data © OpenStreetMap contributors, licensed under the ODbL*

A layer control icon has been added to the top right-hand corner of the map. By hovering over it and selecting one of the menu entries, only the map elements corresponding to that entry will be shown.

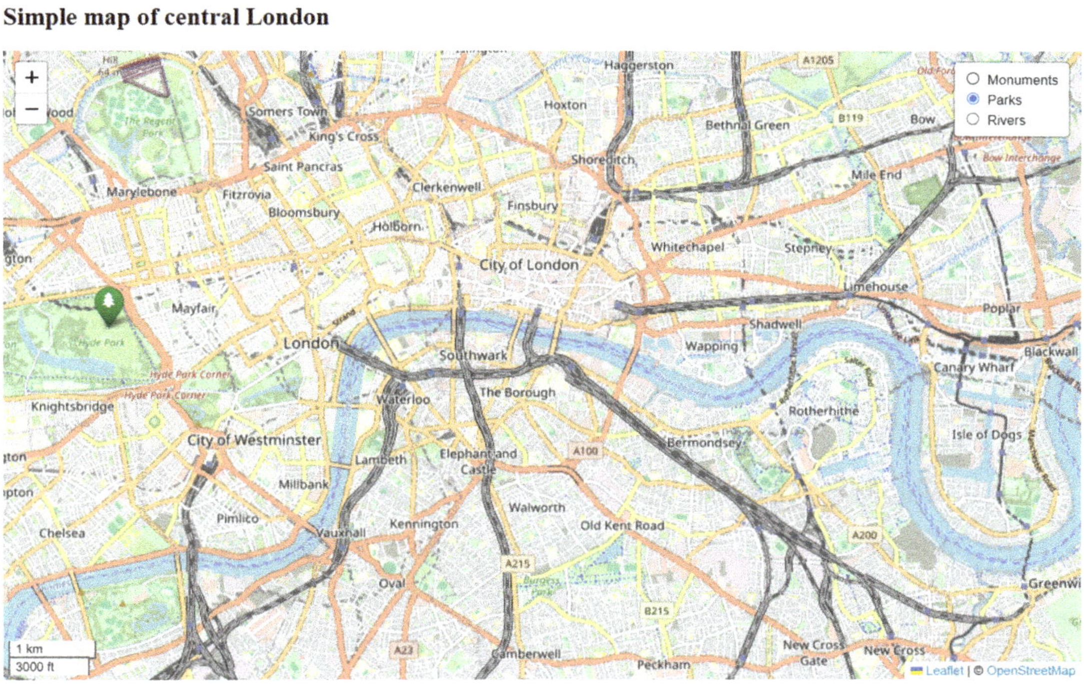

Figure 4-8. *Map created with Leaflet using OpenStreetMap tiles. Data © OpenStreetMap contributors, licensed under the ODbL*

You can choose to display map elements of the first group (the monument markers), the second group (the park marker), or the third group (the river popup), but you can only see one group of elements at a time.

Now let's modify the `L.control.layers()` statement, along with a few others (changes shown in **bold**), to allow multiple groups of elements to be visible at the same time:

```
...

const mark4 = L.marker([51.50914, -0.16123],
{icon: gardenMarker}).addTo(map)
.bindPopup('Hyde Park');

const park_group = L.layerGroup([mark4]).addTo(map);

const popup = L.popup()
```

```
68        .setLatLng([51.50775, -0.04616])
69        .setContent("River Thames")
70        .openOn(map);
71
72        const river_group = L.layerGroup([popup]).addTo(map);

          ...

80        L.control.layers({}, overlayGroups).addTo(map)

          ...
```

- **Lines 62**: The marker for Hyde Park is added to the map, so it is visible when the map is opened.
- **Line 65**: The layer group for Hyde Park is added to the map so it is selected (in the layer control) when the map is opened.
- **Lines 70**: The popup for River Thames is opened on the map, so it is visible when the map is opened.
- **Line 72**: The layer group for River Thames is added to the map so it is selected (in the layer control) when the map is opened.
- **Line 80**: The layer control has been modified to allow you to select all three groups together (or two, or one, or none).

When opening the file in a browser again, it is possible to select more than one layer group.

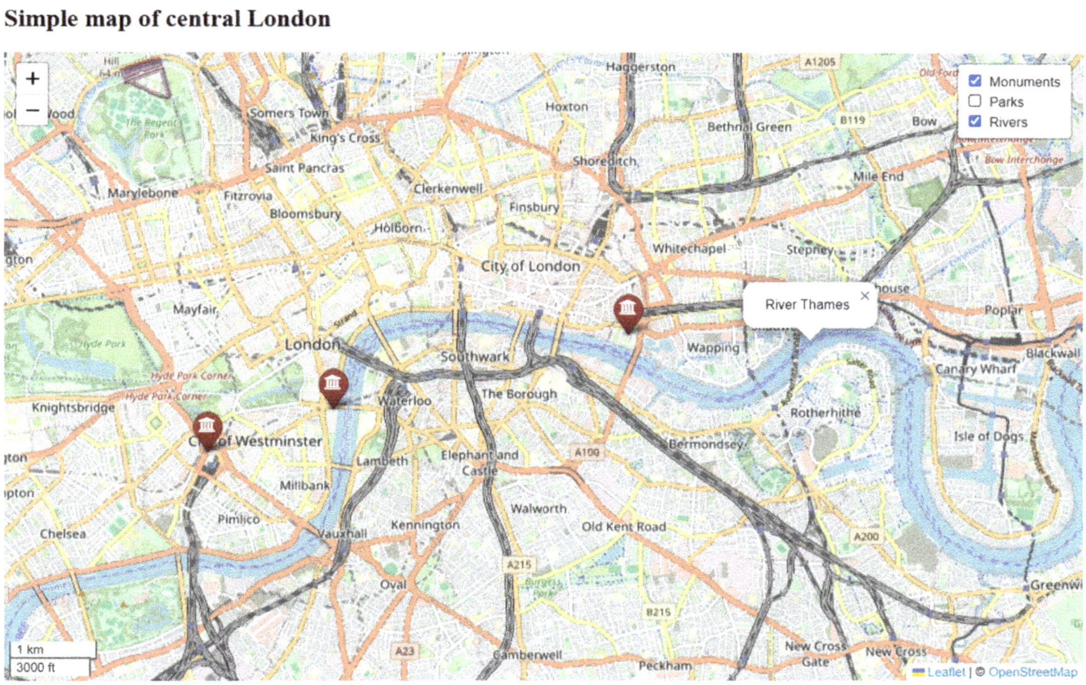

Figure 4-9. *Map created with Leaflet using OpenStreetMap tiles. Data © OpenStreetMap contributors, licensed under the ODbL*

The rule is as follows:

- When a group is passed as the first argument to the `L.control.layers()` function, radio buttons are used and compel an exclusive choice.
- When a group is passed as the next argument to the `L.control.layers()` function, or added with the `addOverlay()` method, check boxes are used and allow multiple selections.

Although the preceding code works fine, it includes a questionable design choice: the `L.popup` class is not a subclass of `L.Layer` and therefore should not be used with a layer control. A more standards-compliant version of the code that produces the same map is (changes shown in **bold**):

```
...

      const park_group = L.layerGroup([mark4]).addTo(map);

```

```
67        const invisibleCircle = L.circleMarker([51.50775, -0.04616], {
68            radius: 0,
69            opacity: 0,
70            fillOpacity: 0
71        }).bindPopup("River Thames").addTo(map);
72
73        invisibleCircle.openPopup();
74
75        map.on('overlayadd', function(e) {
76            if (e.name === "Rivers") {
77                invisibleCircle.openPopup();
78            }
79        });
80
81        const river_group = L.layerGroup([invisibleCircle]).addTo(map);
82
83        const overlayGroups = {
84            "Monuments": monument_group,
85            "Parks": park_group,
86            "Rivers": river_group
87        };
    ...
```

- **Lines 67 to 71**: An invisible (because fully transparent) circle marker is added to the map and will be used as a container for the popup.

  A popup with the text “River Thames” is then bound to it.

- **Line 73**: The popup is explicitly opened (otherwise it would not appear automatically).

- **Lines 75 to 79**: The popup will be reopened each time the invisible circle marker is added back to the map (this ensures it becomes visible again after being hidden using the layer control).

The map obtained with this new code is identical to the previous one.

We will now add another tile layer to the map and define a layer control to switch between tile layers. To do this, still using the same file (`leaflet_maps/page_with_groups_and_controls.html`), modify or insert statements as shown below (changes in **bold**):

```
...

    map = L.map('map').setView([51.505, -0.09], 13);

    const tiles1 = L.tileLayer('https://tile.openstreetmap.org/{z}/{x}/{y}.png', {
        maxZoom: 19,
        attribution: '&copy; <a href="http://www.openstreetmap.org/copyright">OpenStreetMap</a>'
    }).addTo(map);

    const tiles2 = L.tileLayer('https://{s}.tile.opentopomap.org/{z}/{x}/{y}.png', {
       maxZoom: 17,
        attribution: 'Map data: &copy; <a href="https://www.openstreetmap.org/copyright">OpenStreetMap</a> contributors, <a href="http://viewfinderpanoramas.org">SRTM</a> | Map style: &copy; <a href="https://opentopomap.org">OpenTopoMap</a> (<a href="https://creativecommons.org/licenses/by-sa/3.0/">CC-BY-SA</a>)'
    });

    const baseLayers = {
        "OSM standard": tiles1,
        "OpenTopoMap": tiles2
    };

    L.control.scale().addTo(map);
...
```

```
93        const overlayGroups = {
94              "Monuments": monument_group,
95              "Parks": park_group,
96              "Rivers": river_group
97        };
98
99        L.control.layers(baseLayers, overlayGroups).addTo(map);
100
101   } catch(e) {
...
```

- **Lines 35 to 38**: Another tile layer ("OpenTopoMap", for hiking, cycling, and climbing) is defined. It is not added to the map yet (it will be added when requested using the layer control).
- **Lines 40 to 43**: An object containing the two tile layers, associated with keys, is created.
- **Line 99**: The object containing the tile layers is added to the layer control as a parameter. As the object containing the marker groups is not the first argument, check boxes are used to select the marker groups (instead of radio buttons).

By opening this file in a browser and hovering over the layer control to open it, you can select a tile layer to display (i.e., choose the style of the map).

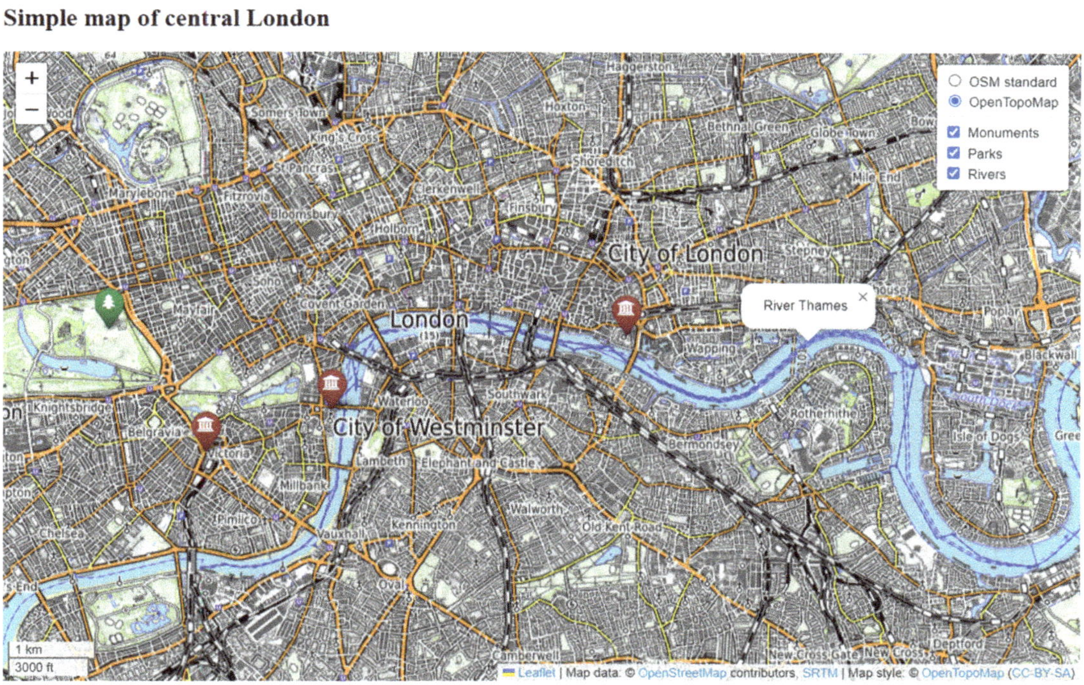

Figure 4-10. *Map created with Leaflet using OpenTopoMap tiles. Tiles © OpenTopoMap, licensed under CC BY-SA 3.0. Map data © OpenStreetMap contributors, licensed under the Open Database License (ODbL)*

We could also have used two separate `L.control.layers()` statements, one for tile layers and another for marker groups, as shown below:

```
...
    }

    L.control.layers(baseLayers).addTo(map);
    L.control.layers({}, overlayGroups).addTo(map);

} catch(e) {
...
```

This would have created two independent layer controls.

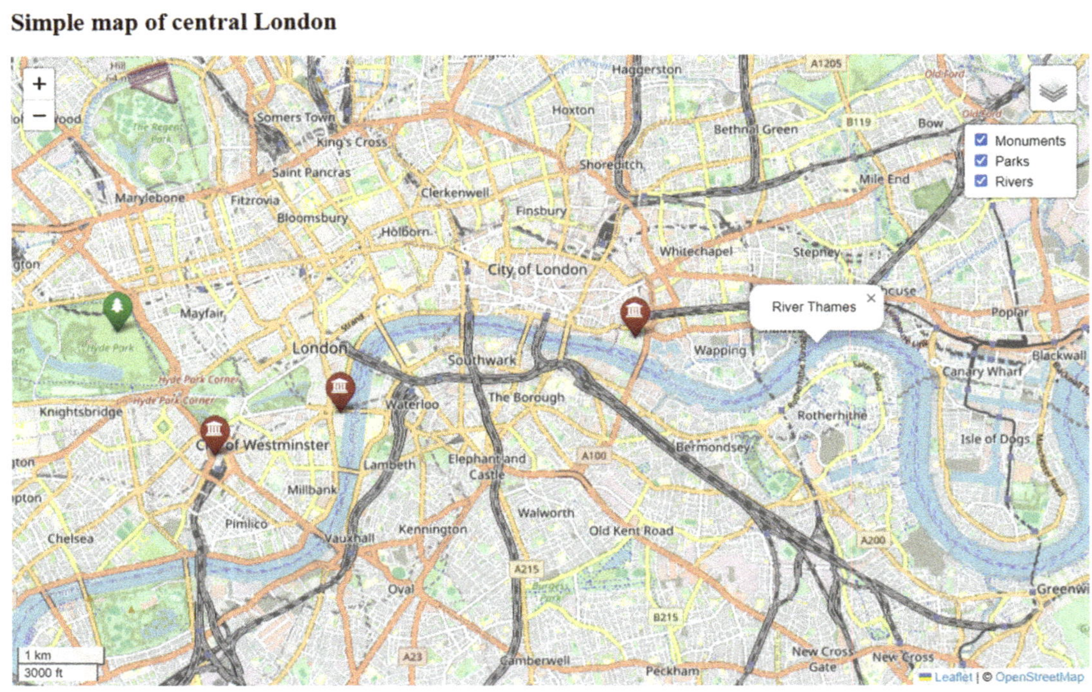

Figure 4-11. *Map created with Leaflet using OpenStreetMap tiles. Data © OpenStreetMap contributors, licensed under the ODb*

Overlaying Images

Leaflet also lets you **overlay images** on maps. An `L.imageOverlay` object can be used for that purpose. For example:

```
1   L.imageOverlay(
'https://upload.wikimedia.org/wikipedia/commons/1/16/Londres_blason.jpg',
2       L.latLngBounds([[51.522, -0.168], [51.537, -0.137]])
3   ).addTo(map);
```

- **Line 1**: An image overlay is created using an `L.imageOverlay` object and a URL to download the image from.
- **Line 2**: Via an `L.latLngBounds` object, a rectangle in which to place and adjust the image on the map is specified. This rectangle is defined by the coordinates of its bottom left and top right corners.

- **Line 3**: The image overlay is added to the map.

We will now take the previous example (map of London) and overlay the coat-of-arms of the city of London on the map.

To do this, the statements below (in **bold**) must be inserted into the previous code. Before adding new statements, copy the code of the previous file `leaflet_maps/page_with_groups_and_controls.html` in a new file `leaflet_maps/page_with_image.html` (that will be used in this section).

```
...
99      L.control.layers(baseLayers, overlayGroups).addTo(map);
100
101     L.imageOverlay('https://upload.wikimedia.org/wikipedia/
        commons/1/16/Londres_blason.jpg',
102         L.latLngBounds([[51.522, -0.168], [51.537, -0.137]]),
103         {
104             opacity: 0.7,
105             errorOverlayUrl: 'https://cdn-icons-png.flaticon.
                com/512/110/110686.png',
106             alt: 'Coat of arms of the city of London',
107             interactive: true
108         }
109     ).addTo(map);
110
111 } catch(e) {
...
```

- **Lines 101 to 109**: An object of type `L.imageOverlay` is created, then added to the map.
    - A URL defines where to find the image of the coat of arms. Here, we fetch the image from the Internet, but we could also use a local file.

      In case this file is no longer available on the Internet, it has also been downloaded and can be found at `https://github.com/Apress/Modern-Web-Cartography` (file: `leaflet_maps/London_coat_of_arms.jpg`).

- Then, via an `L.latLngBounds` object, a rectangle in which to place and adjust the image is specified (this rectangle is defined by the coordinates of its bottom left and top right corners).
- The `opacity` option defines the opacity level of the image (enabling the underlying map to be seen through).
- The `errorOverlayUrl` option specifies another image to be used, if the requested image cannot be loaded.
- The `alt` option defines the HTML "alt" attribute, i.e., a text to be used instead of the image in the event that the browser is unable to display images.
- The `interactive` option determines whether or not the image will generate events (in response to user interactions).

Opening the file in a browser produces the following result:

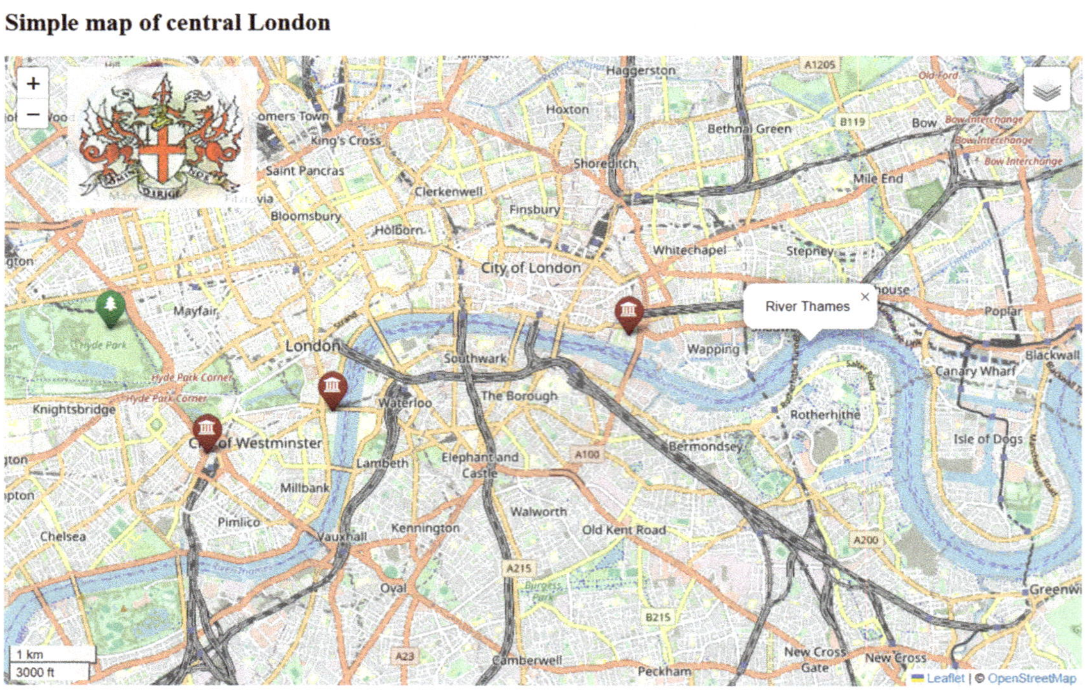

Figure 4-12. *Map created with Leaflet using OpenStreetMap tiles. Data © OpenStreetMap contributors, licensed under the ODb. Image: Heraldry of London, by Ichigo53 — in the public domain*

Drawing Polylines and Geometric Figures

A **polyline** is a line made up of several contiguous segments. These lines are defined by the coordinates of the points constituting their vertices and extremities. Leaflet lets you draw polylines on maps. A color, line thickness, smoothing level, opacity level, and text to be displayed in a popup or tooltip can be associated with a polyline.

Geometric figures such as circles, rectangles, or polygons can also be added to the maps. Line color, thickness and opacity, area fill color and opacity, and text to be displayed in a popup or tooltip can be specified for these figures.

Polylines can be added to a Leaflet map using `L.polyline` objects. For example:

```
1       const latlngs = [
2           [51.0933, 1.1343],
3           [51.1115, 1.2405],
4           [51.0151, 1.4903],
5           [50.9284, 1.7578],
6           [50.9135, 1.8059]
7       ];
8
9       L.polyline(latlngs)
10      .addTo(map);
```

- **Lines 1 to 7**: An array containing coordinate pairs (latitudes and longitudes) is defined. These will be the points constituting the vertices and extremities of the polyline.
- **Lines 9 and 10**: A polyline is created using an `L.polyline` object and the previous array of points. Then it is added to the map.

Here is an example of an HTML file creating a polyline and adding it to a map. Store this code in a new file `leaflet_maps/page_with_polyline.html` (in the `leaflet_maps` directory).

```
<html>
<head>
<link rel="stylesheet" href="https://unpkg.com/leaflet@1.9.4/dist/leaflet.css" integrity="sha256-p4NxAoJBhIIN+hmNHrzRCf9tD/miZyoHS5obTRR9BMY=" crossorigin=""/>
```

```
<title>Simple OSM map created with Leaflet</title>
</head>
<body>
<h2>Simple map of the Channel Tunnel</h2>
<div id="map" style="width: 1030px; height: 620px;"></div>
<script src="https://unpkg.com/leaflet@1.9.4/dist/leaflet.js" integrity="sha256-20nQCchB9co0qIjJZRGuk2/Z9VM+kNiyxNV1lvTlZBo=" crossorigin=""></script>
<script type="text/javascript">
window.onerror = function(message, url, linenumber) {
    if (typeof map !== 'undefined' && map) {map.remove();}

    const div = document.getElementById("map")
    if (div) {div.remove();}

    const elemDiv = document.createElement('div');
    elemDiv.innerHTML = `<p>${message}, ligne ${linenumber}</p>`;
    document.body.appendChild(elemDiv);
}
</script>
<script type="text/javascript">
let map;
try {
    map = L.map('map').setView([51.065, 1.408], 14);

    L.tileLayer('https://tile.openstreetmap.org/{z}/{x}/{y}.png', {
        maxZoom: 19,
        attribution: '&copy; <a href="http://www.openstreetmap.org/copyright">OpenStreetMap</a>'
    }).addTo(map);

    const latlngs = [
        [51.0933, 1.1343],
        [51.0970, 1.1714],
        [51.1115, 1.2405],
```

```
36              [51.1100, 1.2803],
37              [51.1065, 1.3046],
38              [51.0667, 1.4086],
39              [51.0151, 1.4903],
40              [50.9845, 1.5916],
41              [50.9637, 1.6809],
42              [50.9499, 1.7317],
43              [50.9284, 1.7578],
44              [50.9135, 1.8059]
45          ];
46
47          const polyline = L.polyline(latlngs, {
48              color: 'red'
49          })
50          .bindTooltip('Channel Tunnel')
51          .addTo(map);
52
53          map.fitBounds(polyline.getBounds());
54
55      } catch(e) {
56          if (typeof map !== 'undefined' && map) {map.remove();}
57
58          document.getElementById("map").remove();
59
60          const elemDiv = document.createElement('div');
61          elemDiv.innerHTML = `<p>${e}</p>`;
62          document.body.appendChild(elemDiv);
63      }
64      </script>
65      </body>
66      </html>
```

- **Line 25**: A map is created covering the north of France and the southeast of England.
- **Lines 27 to 30**: A tile set is associated with it.

- **Lines 32 to 45**: An array containing coordinate pairs (latitudes and longitudes of the points constituting the vertices and extremities of the polyline to be drawn) is defined.
- **Lines 47 to 49**: A polyline is created with the coordinate array defined above and assigned the color red.
- **Line 50**: A tooltip is associated with the polyline.
- **Line 51**: The polyline is added to the map.
- **Line 53**: The zoom level of the map is adjusted to the polyline.

Opening this file in a browser and placing the mouse cursor over the polyline produces the following result:

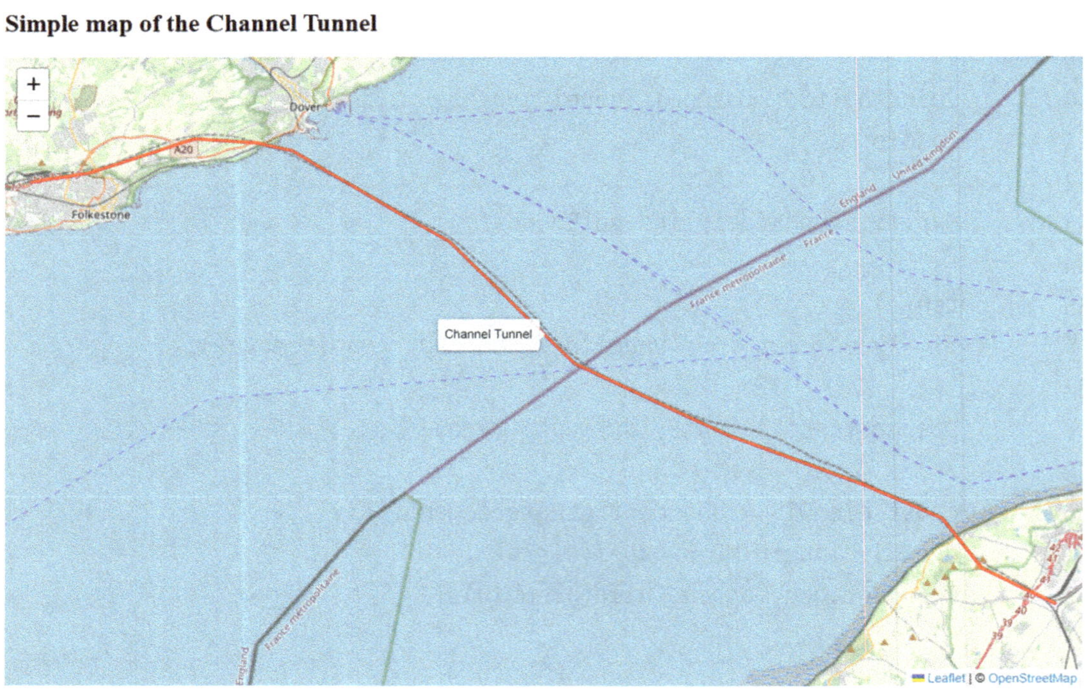

***Figure 4-13.** Map created with Leaflet using OpenStreetMap tiles. Data © OpenStreetMap contributors, licensed under the ODbL*

Our red polyline roughly follows the route of the Channel Tunnel between France and England.

Let's complete this map by adding a circle and a triangle at each end of the polyline. To do this, we will insert the following code (in **bold**) into the previous HTML file (`leaflet_maps/page_with_polyline.html`):

```
...

51        .addTo(map);
52
53        L.circle([51.0933, 1.1343],
54            {color: 'darkblue',
55            fillColor: 'blue',
56            fillOpacity: 0.5,
57            radius: 5000})
58        .bindPopup('Folkestone')
59        .addTo(map);
60
61        L.polygon([
62            [50.8794, 1.7070],
63            [50.8794, 1.9148],
64            [50.9897, 1.8059]],
65            {color: 'darkgreen',
66            fillColor: 'green'})
67        .bindPopup('Peuplingues')
68        .addTo(map);
69
70        map.fitBounds(polyline.getBounds());
...
```

- **Lines 53 to 59**: A circle is created with a radius of 5 km around the English end of the tunnel, colored dark blue, filled with blue, and with an opacity level of 50%. Then a popup ("Folkestone") is attached to the circle, and the circle is added to the map.
- **Lines 61 to 68**: A triangle surrounding the French end of the tunnel is created (to do this, the coordinates of its three vertices are supplied). It is defined with color dark green and fill color green. Then a popup

("Peuplingues") is attached to the triangle, and the triangle is added to the map.

Opening this file in a browser produces the following page.

Simple map of the Channel Tunnel

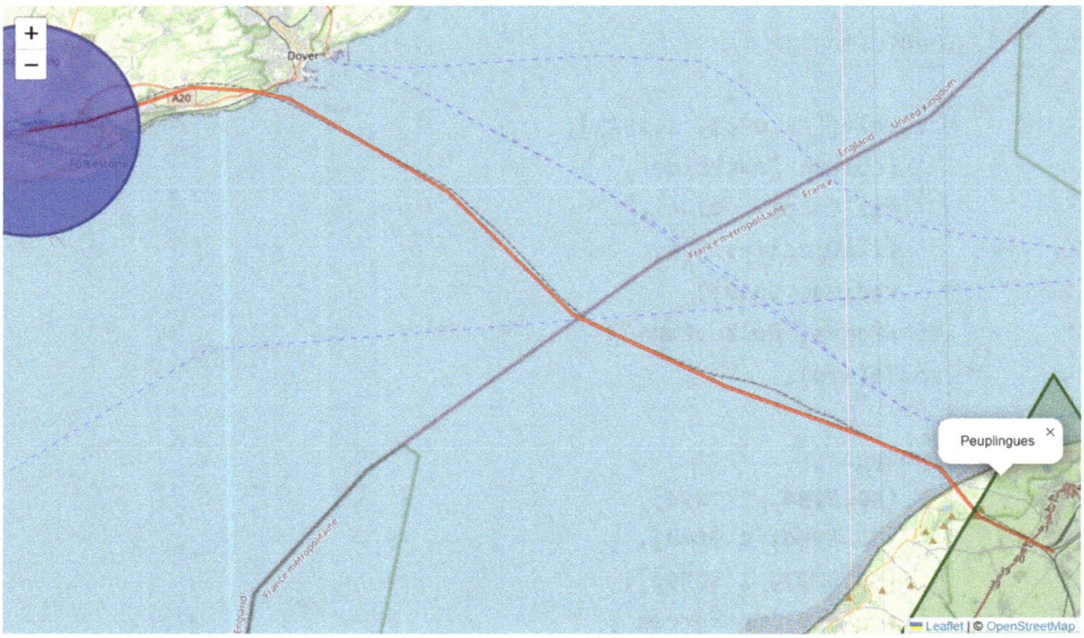

Figure 4-14. *Map created with Leaflet using OpenStreetMap tiles. Data © OpenStreetMap contributors, licensed under the ODbL*

The circle and triangle are truncated, as the map zoom level is adjusted on the polyline (due to the `map.fitBounds(polyline.getBounds());` statement on line 70).

To correct this issue, we will modify the code so that the map adjusts to fit the polyline, the circle, and the triangle. Polylines and polygons are features in Leaflet and can be grouped using a feature group. We will create such a group and then adjust the map's zoom level to fit this feature group:

```
    ...

47      const polyline = L.polyline(latlngs, {
48          color: 'red'
49      })
50      .bindTooltip('Tunnel sous la Manche')
```

```
        .addTo(map);

        const circle = L.circle([51.0933, 1.1343],
            {color: 'darkblue',
            fillColor: 'blue',
            fillOpacity: 0.5,
            radius: 5000})
        .bindPopup('Folkestone')
        .addTo(map);

        const triangle = L.polygon([
            [50.8794, 1.7070],
            [50.8794, 1.9148],
            [50.9897, 1.8059]],
            {color: 'darkgreen',
            fillColor: 'green'})
        .bindPopup('Peuplingues')
        .addTo(map);

        const group = L.featureGroup([polyline, circle, triangle]);

        map.fitBounds(group.getBounds());

    } catch(e) {

    ...
```

The three objects have been grouped together in an object of type `L.featureGroup`, and the map adjustment performed on it.

Opening the file again in a browser and clicking on the circle produces the following page:

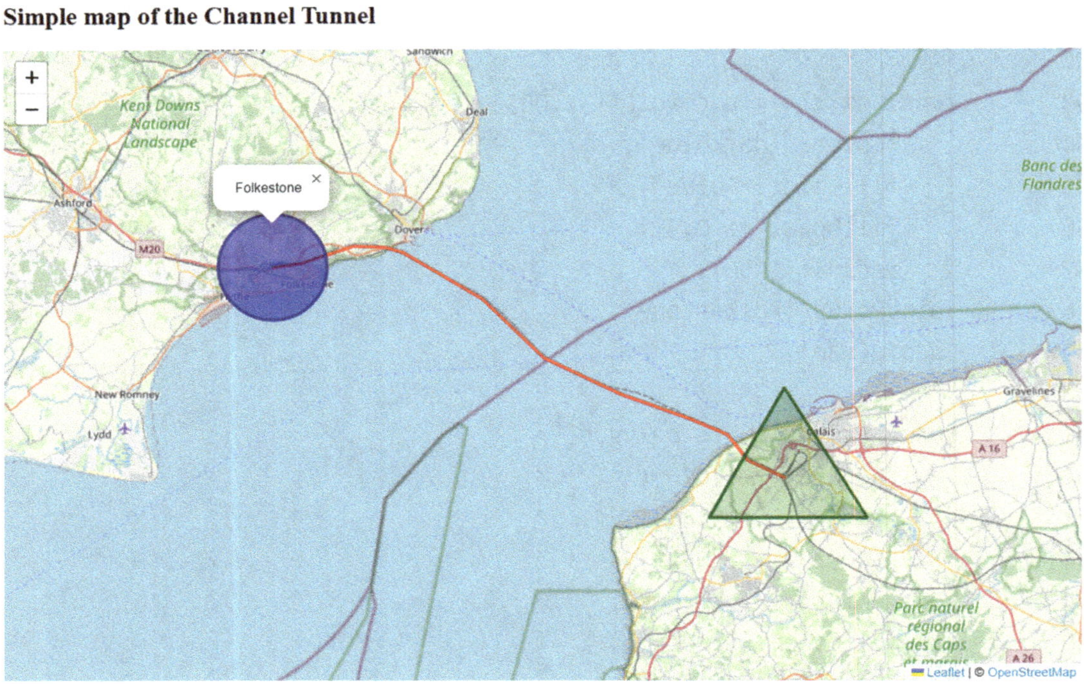

Figure 4-15. *Map created with Leaflet using OpenStreetMap tiles. Data © OpenStreetMap contributors, licensed under the ODbL*

We have switched to next lower zoom level, and this allows us to see all three map elements in the group. However, this zoom level is a little too low and not well adjusted to the three elements. This is because, by default, only 19 zoom levels are available (from 0 to 18). But we can change this behavior by allowing non-integer zoom levels. To do this, we will modify the statement on line 25 of the previous example:

```
25      map = L.map('map', {zoomSnap: 0.1}).setView([51.065, 1.408], 14);
```

- The `zoomSnap: 0.1` option allows fractional zoom steps in increments of 0.1, instead of restricting zoom levels to whole numbers between 0 and 18.

Let's also add a scale to the map (statement in **bold**):

```
        ...
72      map.fitBounds(group.getBounds());
73
```

```
        L.control.scale().addTo(map);

    } catch(e) {

    ...
```

And also a layer control to show or hide the three map elements at will (statements in **bold**):

```
    ...

        L.control.scale().addTo(map);

        const layers = {
            "Tunnel": polyline,
            "Folkestone": circle,
            "Peuplingues": triangle
        };

        L.control.layers(null, layers).addTo(map);

    } catch(e) {

    ...
```

With these modifications, our map fits much more precisely around the three map elements in the group, a scale is displayed in the bottom left-hand corner, and a layer control is present in the top right-hand corner.

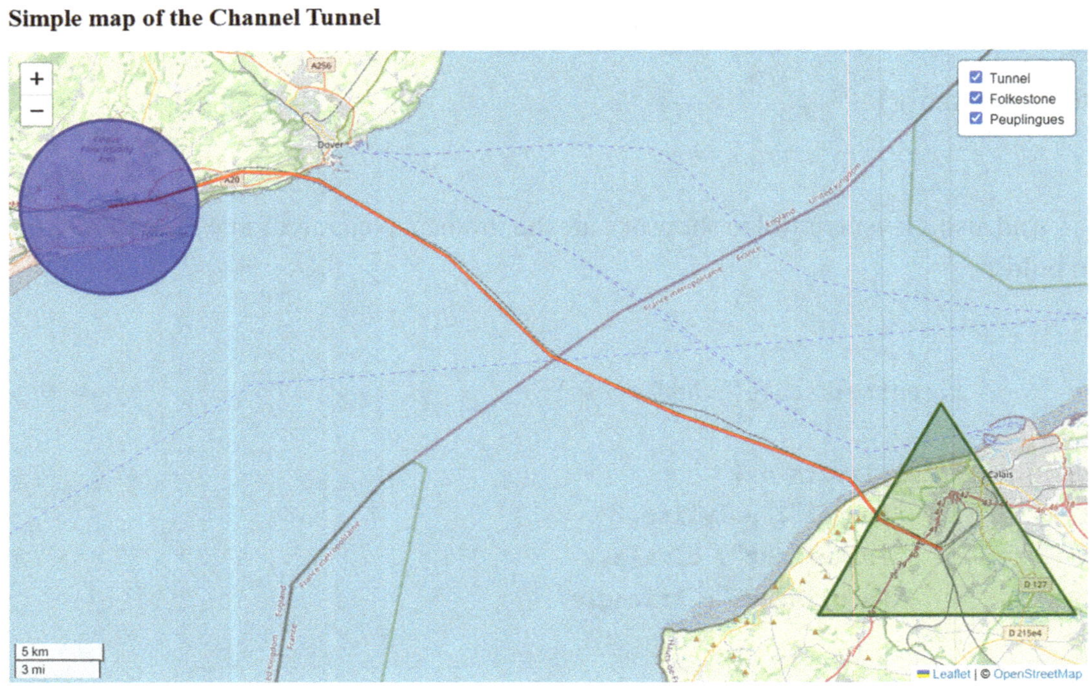

Figure 4-16. *Map created with Leaflet using OpenStreetMap tiles. Data © OpenStreetMap contributors, licensed under the ODbL*

Adding GeoJSON and TopoJSON Data

The **GeoJSON** format is a particular form of JSON format, respecting a specific structure designed for the exchange of geographic data. The same applies to **TopoJSON**, which is an extension of the GeoJSON format designed to be lighter and avoid redundancy. GeoJSON and TopoJSON are standards for data exchange between mapping tools.

GeoJSON

Leaflet allows you to add data in **GeoJSON format** to maps. The features defined in the GeoJSON file are then overlaid on the map as markers (for specific points), polylines (for roads, power lines, borders, etc.), and polygons (for specific areas such as regions or countries). You can customize these elements by specifying marker styles, line colors and thicknesses, fill colors and opacities for areas, etc. These elements are added to the map as an additional layer, which can be shown or hidden using a layer control.

GeoJSON data can be added to a Leaflet map using an `L.geoJSON` object. For example:

```
L.geoJSON(geojsonData).addTo(map);
```

- An `L.geoJSON` object is created using the GeoJSON data stored in the geojsonData variable. It is then added to the map, resulting in markers, polylines, or polygons being overlaid.

Here is an extract of a GeoJSON file available on the Internet, describing the outlines of all the countries of the world:

```
{
    "type": "FeatureCollection",
    "features": [
        {
            "type": "Feature",
            "properties": {
                "name": "Afghanistan"
            },
            "geometry": {
                "type": "Polygon",
                "coordinates": [
                    [
                        [
                            61.210817,
                            35.650072
                        ],
                        [
                            62.230651,
                            35.270664
                        ],
                        [
                            62.984662,
                            35.404041
                        ],
```

```
                    [
                        63.193538,
                        35.857166
                    ],
                    [
                        63.982896,
                        36.007957
                    ],
                    ...

                ]
            ]
        },
        "id": "AFG"
    },
    {
        "type": "Feature",
        "properties": {
            "name": "Angola"
        },
        "geometry": {
            "type": "MultiPolygon",
            "coordinates": [
                [
                    [
                        [
                            16.326528,
                            -5.87747
                        ],
                        [
                            16.57318,
                            -6.622645
                        ],
```

```
                        [
                            16.860191,
                            -7.222298
                        ],
                        ...
                    ]
                ]
            ]
        },
        "id": "AGO"
    },
    ...
```

This file consists of a collection of features. Each feature represents a country and includes:

- The geometry of this country, of type `Polygon` or `MultiPolygon`, including a series of points (latitude and longitude). These points enable applications to draw the outline of the country.

  Other geometry types used in GeoJSON files are `Point`, `MutiPoint`, `LineString`, `MultiLineString`, and `GeometryCollection`.

- Some metadata as properties: here, the name of the country is provided.

Here is an HTML code using the preceding GeoJSON file: the GeoJSON data is downloaded from the Internet, then added to a map with Leaflet. Store this code in a new file `leaflet_maps/page_with_geojson.html` (in the `leaflet_maps` directory).

```
<html>
<head>
<link rel="stylesheet" href="https://unpkg.com/leaflet@1.9.4/
dist/leaflet.css" integrity="sha256-p4NxAoJBhIIN+hmNHrzRCf9tD/
miZyoHS5obTRR9BMY=" crossorigin=""/>
<title>Simple OSM map created with Leaflet</title>
</head>
```

```
<body>
<h2>Map with added GeoJSON data</h2>
<div id="map" style="width: 1030px; height: 620px;"></div>
<script src="https://unpkg.com/leaflet@1.9.4/dist/leaflet.js" integrity="sha256-20nQCchB9co0qIjJZRGuk2/Z9VM+kNiyxNV1lvTlZBo=" crossorigin=""></script>
<script type="text/javascript">
window.onerror = function(message, url, linenumber) {
    if (typeof map !== 'undefined' && map) {map.remove();}

    const div = document.getElementById("map")
    if (div) {div.remove();}

    const elemDiv = document.createElement('div');
    elemDiv.innerHTML = `<p>${message}, ligne ${linenumber}</p>`;
    document.body.appendChild(elemDiv);
}
</script>
<script type="text/javascript">
let map;
fetch("https://raw.githubusercontent.com/python-visualization/folium-example-data/main/world_countries.json")
.then(response => response.json())
.then(geojsonData => {
    map = L.map('map').setView([51.505, -0.09], 2);

    const tiles = L.tileLayer('https://tile.openstreetmap.org/{z}/{x}/{y}.png', {
        maxZoom: 19,
        attribution: '&copy; <a href="http://www.openstreetmap.org/copyright">OpenStreetMap</a>'
    }).addTo(map);

    const style = {
            'color': '#E31A1C',
        'fillColor': '#FED976',
```

```
37            'weight': 1,
38            'opacity': 1,
39            'fillOpacity': 0.7
40        };
41
42        L.geoJSON(geojsonData, {
43            style: style
44        }).addTo(map);
45    })
46    .catch(error => {
47        if (typeof map !== 'undefined' && map) {map.remove();}
48
49        const div = document.getElementById("map")
50        if (div) {div.remove();}
51
52        const elemDiv = document.createElement('div');
53        elemDiv.innerHTML = `<p>${error}</p>`;
54        document.body.appendChild(elemDiv);
55    });
56    </script>
57    </body>
58    </html>
```

- **Lines 24, 25, 26, and 46**: A GeoJSON file containing the contours of all the countries in the world is downloaded from the Internet using the `fetch()` function. This function is asynchronous, meaning it returns immediately and continues running in the background.

  In this case, error handling is performed using the `.then()` and `.catch()` methods rather than a synchronous try/catch block (unless the `fetch()` call is used inside an `async` function).

  It is associated with the two `.then()` blocks on lines 25 and 26 and with the `.catch()` block on line 46:

- The `fetch()` call invokes the URL asynchronously and immediately returns a Promise, which resolves when the server responds.
- The first `.then()` block (line 25) runs after the Promise resolves. It receives the response object and parses it.
- The second `.then()` block (lines 26 to 45) receives and processes the parsed data.
- The `.catch()` block (lines 46 to 55) handles any errors that occur during the `fetch()` call or in any of the preceding `.then()` blocks.

In case the GeoJSON file is no longer available on the Internet, it has also been downloaded and can be found at `https://github.com/Apress/Modern-Web-Cartography` (file: `leaflet_maps/World_countries.json`).

- **Lines 25 and 26**: The received data are parsed (using a JSON parser) and placed in a JSON-structured object called `geojsonData`.
- **Line 27**: A map is created (with a zoom level small enough to view all the countries in the world).
- **Lines 29 to 32**: A tile layer is added to the map.
- **Lines 34 to 40**: Style properties are defined for the JSON data to be displayed: outline color (red) and fill color (orange), line thickness, line and fill opacity levels.
- **Lines 42 to 44**: An object of type `L.geoJSON` is created with the object generated from the JSON data (downloaded from the Internet) and the style defined above, then it is added to the map.
- **Lines 46 to 55**: These lines handle any errors that may occur by displaying an error message on a web page.

Opening the file in a browser produces the following result:

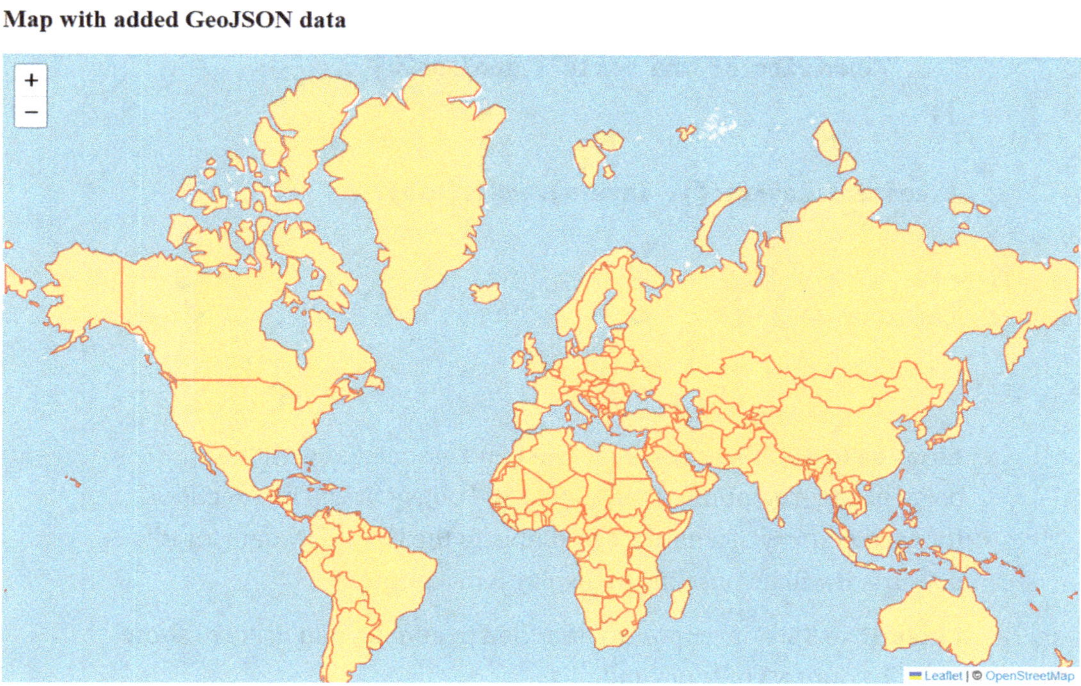

Figure 4-17. *Map created with Leaflet using OpenStreetMap tiles. Data © OpenStreetMap contributors, licensed under the ODbL. Country boundaries from the Folium example dataset (MIT License), source:* `https://github.com/python-visualization/folium-example-data`

We will now add the ability to click on a country to display its name in a popup. A layer control will also be added to show or hide the country contours and fills.

```
    ...

        }

        const geoJsonObj = L.geoJSON(geojsonData, {
            style: style,
            onEachFeature: function (feature, layer) {
                if (feature.properties.name) {
                    layer.bindPopup(feature.properties.name);
                }
            }
        }).addTo(map);

```

```
51        const layers = {
52            "Countries of the world": geoJsonObj
53        };
54
55        L.control.layers({}, layers).addTo(map);
56
57    })
58    .catch(error => {
      ...
```

- **Lines 44 to 48**: A function is defined and associated with the `onEachFeature` option of the `L.geoJSON` object. It will be executed for each country: if a name is available in the GeoJSON data for this country, it will be possible to display it in a popup.
- **Lines 51 to 55**: A layer control is added to enable map users to show or hide country contours and fills.

After refreshing the map, clicking on a country displays its name, and a layer control is available in the top right corner.

Map with added GeoJSON data

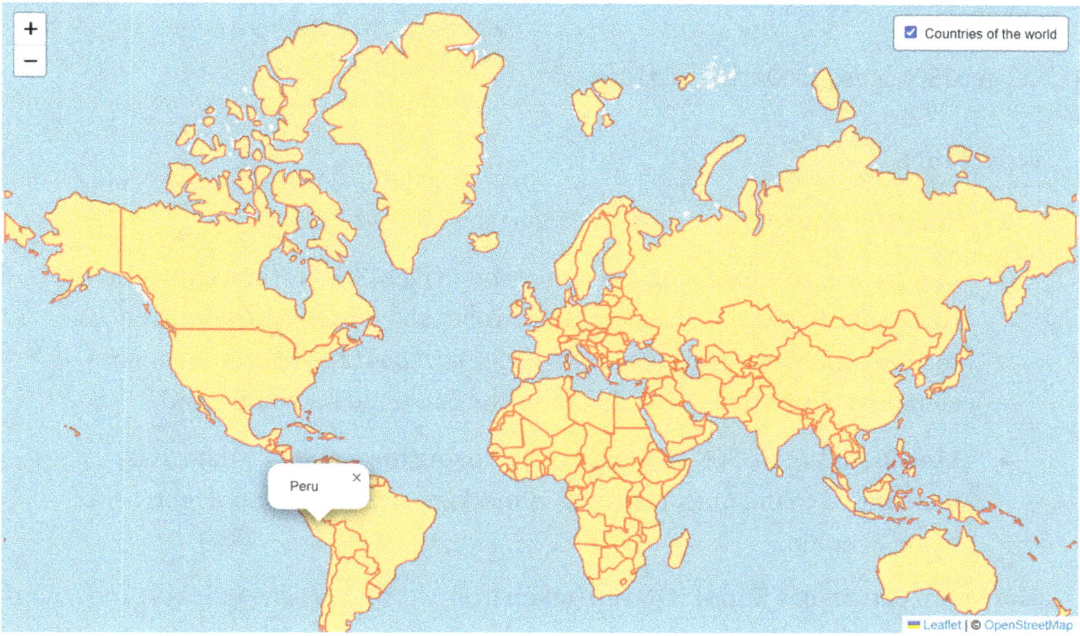

Figure 4-18. *Map created with Leaflet using OpenStreetMap tiles. Data © OpenStreetMap contributors, licensed under the ODbL. Country boundaries from the Folium example dataset (MIT License), source:* `https://github.com/python-visualization/folium-example-data`

TopoJSON

TopoJSON data are also supported, but indirectly: they must first be converted to GeoJSON format before Leaflet can use them. This conversion can be done using a JavaScript library such as `topojson-client` (see `https://github.com/topojson/topojson`). Once converted to GeoJSON, the data can be used to create an `L.geoJSON` object, which can then be added to a map (as shown in the example above). The following example illustrates this process:

```
    ...
1   <script src="https://unpkg.com/topojson-client@3"></script>
    ...
    <script type="text/javascript">
    ...
```

```
2  const geojsonData = topojson.feature(topojsonData, topojsonData.
   objects.counties);
3  L.geoJSON(geojsonData).addTo(map);
   ...
   </script>
```

- **Line 1**: A link to the code of the `topojson-client` library.
- **Line 2**: The TopoJSON data is converted to GeoJSON format using the `topojson.feature()` function of the `topojson-client` library. The second parameter `topojsonData.objects.counties` specifies which elements in the TopoJSON data must be extracted and converted.
- **Line 3**: An `L.geoJSON` object is created using the GeoJSON data. It is then added to the map, resulting in markers, polylines, or polygons being overlaid.

Here is an extract of a TopoJSON file (taken from: `https://cdn.jsdelivr.net/npm/us-atlas@3/counties-10m.json`) containing the outlines of the US counties, the US states, and the whole country:

```
{
    "type": "Topology",
    "bbox": [
        -179.14733999999999,
        -14.552548999999999,
        179.77847,
        71.352561
    ],
    "transform": {
        "scale": [
            0.003589293992939929,
            0.0008590596905969058
        ],
        "translate": [
            -179.14733999999999,
            -14.552548999999999
        ]
```

```
},
"objects": {
    "counties": {
        "type": "GeometryCollection",
        "geometries": [
            {
                "type": "Polygon",
                "arcs": [
                    [
                        0,
                        1,
                        2,
                        3,
                        4,
                        5,
                        6,
                        7,
                        8,
                        9,
                        10,
                        11
                    ]
                ],
                "id": "04015",
                "properties": {
                    "name": "Mohave"
                }
            },
            {
                "type": "Polygon",
                "arcs": [
                    [
                        12,
                        13,
                        14,
```

```
                        15,
                        16,
                        17,
                        18
                    ]
                ],
                "id": "22105",
                "properties": {
                    "name": "Tangipahoa"
                }
            },
            ...
    "states": {
        "type": "GeometryCollection",
        "geometries": [
            {
                "type": "MultiPolygon",
                "arcs": [
                    [
                        [
                            -1237,
                            -1236,
                            -6098,
                            -5228,
                            ...
                            725
                        ]
                    ]
                ],
                "id": "04",
                "properties": {
                    "name": "Arizona"
                }
            },
            ...
```

```
    "nation": {
        "type": "GeometryCollection",
        "geometries": [
            {
                "type": "MultiPolygon",
                "arcs": [
                    [
                        [
                            1626,
                            4919,
                            7192,
                            284,
                            4797,
                            ...
                        ]
                    ],
                    ...
"arcs": [
    [
        [
            18136,
            59828
        ],
        [
            0,
            183
        ]
    ],
    [
        [
            18136,
            60011
        ],
        ...
```

- The `objects.counties` section contains geometries for the US counties. Its structure is typically a collection of geometries with `type` equal to `Polygon` or `MultiPolygon`, one per county. These geometries are described using arcs defined in the last section of the file. These arcs are not referenced by an ID, but by their position in the array of arcs of the last section.
- The `objects.states` section contains geometries for the US states. Its structure is a collection of geometries with `type` equal to `MultiPolygon`, one per state. These geometries are described using arcs defined in the last section of the file.
- The `objects.nation` section contains a geometry for the whole country (the USA). Its structure is a single geometry with `type` equal to `MultiPolygon`. This geometry is described using arcs defined in the last section of the file.
- The `arcs` section contains the definitions of the arcs referenced by the three previous sections. An arc is a line described using a list of points. These points are not registered using their geographic coordinates (latitude and longitude), but with "delta encoding": the first point provides a start position, and the points next are defined by moves from the previous point. Moreover, the values of the `scale` and `translate` fields (in the last section) are also used to calculate the geometric coordinates from the delta encoded positions.

The use of arcs in TopoJSON helps eliminate redundancy in geographic data. Shared boundaries - such as borders between adjacent counties or states - can be defined once as an arc and then referenced multiple times by different geometries. This approach results in smaller file sizes and more efficient data representation compared to formats like GeoJSON, where such borders would typically be duplicated for each feature.

As already mentioned, the preceding TopoJSON file can be converted to GeoJSON format using the `topojson.feature()` function of the `topojson-client` JavaScript library, as shown here:

```
const geojsonData = topojson.feature(topojsonData, topojsonData.objects.
counties);
```

The TopoJSON data must have previously been set in the `topojsonData` variable.

The second parameter `topojsonData.objects.counties` indicates that only the geometries of the `counties` section must be extracted and converted.

This conversion produces the following GeoJSON data (truncated here):

```
{
    "type": "FeatureCollection",
    "features": [
        {
            "type": "Feature",
            "id": "04015",
            "properties": {
                "name": "Mohave"
            },
            "geometry": {
                "type": "Polygon",
                "coordinates": [
                    [
                        [
                            -114.05190414404143,
                            36.84327416903169
                        ],
                        [
                            -114.05190414404143,
                            37.00048209241092
                        ],
                        [
                            -113.33404534545345,
                            37.00048209241092
                        ],
                        ...
                        [
                            -114.05190414404143,
                            36.84327416903169
                        ]
                    ]
```

```
                ]
            }
        },
        {
            "type": "Feature",
            "id": "22105",
            "properties": {
                "name": "Tangipahoa"
            },
            "geometry": {
                "type": "Polygon",
                "coordinates": [
                    [
                        [
                            -90.56715354823548,
                            30.999950153591527
                        ],
                        [
                            -90.54920707827078,
                            30.999950153591527
                        ],
                        ...
```

The resulting GeoJSON file consists of a list of features of types `Polygon` or `MultiPolygon` with, for each one, a list of geographic coordinates (longitude and latitude) for the polygon's vertices.

Here is an example demonstrating the workflow for processing the previous TopoJSON data: downloading it from the Internet, converting it to GeoJSON, and then adding it to a map with Leaflet. Store this HTML code in a new file `leaflet_maps/page_with_topojson.html` (in the `leaflet_maps` directory):

```
<html>
<head>
<link rel="stylesheet" href="https://unpkg.com/leaflet@1.9.4/
dist/leaflet.css" integrity="sha256-p4NxAoJBhIIN+hmNHrzRCf9tD/
miZyoHS5obTRR9BMY=" crossorigin=""/>
```

```
<title>Simple OSM map created with Leaflet</title>
</head>
<body>
<h2>Map with added TopoJSON data</h2>
<div id="map" style="width: 1030px; height: 620px;"></div>
<script src="https://unpkg.com/leaflet@1.9.4/dist/leaflet.js" integrity="sha256-20nQCchB9co0qIjJZRGuk2/Z9VM+kNiyxNV1lvTlZBo=" crossorigin=""></script>
<script src="https://unpkg.com/topojson-client@3"></script>
<script type="text/javascript">
window.onerror = function(message, url, linenumber) {
    if (typeof map !== 'undefined' && map) {map.remove();}

    const div = document.getElementById("map")
    if (div) {div.remove();}

    const elemDiv = document.createElement('div');
    elemDiv.innerHTML = `<p>${message}, ligne ${linenumber}</p>`;
    document.body.appendChild(elemDiv);
}
</script>
<script type="text/javascript">
let map;
fetch("https://cdn.jsdelivr.net/npm/us-atlas@3/counties-10m.json")
.then(response => response.json())
.then(topojsonData => {
    const geojsonData = topojson.feature(topojsonData, topojsonData.objects.counties);

    map = L.map('map').setView([37.8, -96], 4);

    const tiles = L.tileLayer('https://tile.openstreetmap.org/{z}/{x}/{y}.png', {
        maxZoom: 19,
```

```
        attribution: '&copy; <a href="http://www.openstreetmap.org/copyright">OpenStreetMap</a>'
    }).addTo(map);

    const style = {
        'color': '#E34A4C',
        'fillColor': '#FED9A6',
        'weight': 1,
        'opacity': 1,
        'fillOpacity': 0.5
    };

    const geoJsonObj = L.geoJSON(geojsonData, {
        style: style,
        onEachFeature: function (feature, layer) {
            if (feature.properties.name) {
                layer.bindPopup(feature.properties.name);
            }
        }
    }).addTo(map);

    const layers = {
        "US counties ": geoJsonObj
    };

    L.control.layers({}, layers).addTo(map);
})
.catch(error => {
    if (typeof map !== 'undefined' && map) {map.remove();}

    const div = document.getElementById("map")
    if (div) {div.remove();}

    const elemDiv = document.createElement('div');
    elemDiv.innerHTML = `<p>${error}</p>`;
```

```
68        document.body.appendChild(elemDiv);
69    });
70    </script>
71    </body>
72    </html>
```

- **Line 10**: A link to the place where the code of the `topojson-client` library can be found.
- **Lines 25 to 27**: The TopoJSON data is downloaded from the Internet, then parsed and stored in a variable named `topojsonData`.

  In case this data is no longer available on the Internet, it has also been downloaded and can be found at `https://github.com/Apress/Modern-Web-Cartography` (file: `leaflet_maps/US_counties.json`).
- **Line 28**: The TopoJSON data is converted to GeoJSON format with the `topojson.feature()` function and stored in a variable named `geojsonData`. The `counties` section of the TopoJSON data is selected for extraction and conversion.
- The following statements are identical to those used to add GeoJSON data to a map in the preceding section:
  - A map is created and a tile layer is associated with it.
  - An object named `style` is defined with several layout properties.
  - An `L.geoJSON` object is created with the `geojsonData` object, the `style` object, and a function to display the county names in popups. Then it is added to the map.
  - A layer control is added to enable the user to show or hide the county outlines resulting from the TopoJSON data.

When opened in a browser, this HTML file produces the following result (after clicking on a county):

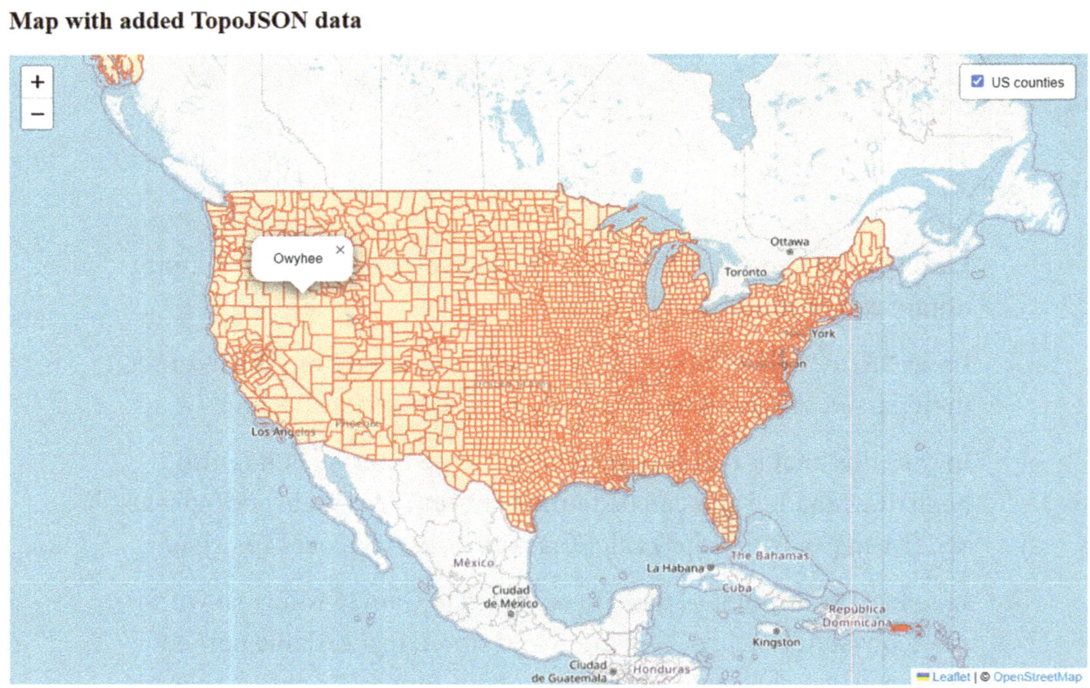

***Figure 4-19.** Map created with Leaflet using OpenStreetMap tiles. Data © OpenStreetMap contributors, licensed under the ODbL. County boundaries © Mike Bostock and contributors, MIT License*

Building Choropleth Maps

Choropleth maps are thematic maps in which areas are shaded or colored to represent an economic, statistical, or other quantitative value associated with each area. Leaflet allows you to build maps of this type. The elements that make up a choropleth map (zone outlines, color shading, associated popups, etc.) are overlaid on the map as an additional layer. Optionally, a layer control can be added to allow the user to show or hide this layer.

Leaflet does not provide specific objects and functions for building choropleth maps (unlike Folium). Instead, choropleth maps are created using `L.geoJSON` objects (as seen in the previous section), with the `style` option used to specify the fill color for each feature of the GeoJSON data. For example:

```
function style(feature) {
    return {
```

```
3           fillColor: getColor(<economic or statistical value for the
            feature>)
4       };
5   }
6
7   L.geoJSON(geojsonData, {style: style}).addTo(map);
```

- **Lines 1 to 5**: The `style()` function calls the `getColor()` function to determine the fill color for the polygon representing the feature in the GeoJSON data. The resulting color is assigned to the `fillColor` property.
- **Line 7**: An `L.geoJSON` object is created using the GeoJSON data. The `style` option refers to the `style()` function defined earlier (it is executed for each feature of the GeoJSON data).

Here is an example of an HTML code in which a choropleth map is built to show Donald Trump's scores by state in the US presidential election of November 2024. Store this HTML code in a new file `leaflet_maps/page_with_choropleth_map.html` (in the `leaflet_maps` directory).

```
<html>
<head>
<link rel="stylesheet" href="https://unpkg.com/leaflet@1.9.4/
dist/leaflet.css" integrity="sha256-p4NxAoJBhIIN+hmNHrzRCf9tD/
miZyoHS5obTRR9BMY=" crossorigin=""/>
<title>Simple OSM map created with Leaflet</title>
</head>
<body>
<h2>Choropleth map with percentages of votes for D. Trump in the
November 2024 presidential election</h2>
<div id="map" style="width: 1030px; height: 620px;"></div>
<script src="https://unpkg.com/leaflet@1.9.4/dist/leaflet.js" in
tegrity="sha256-20nQCchB9co0qIjJZRGuk2/Z9VM+kNiyxNV1lvTlZBo="
crossorigin=""></script>
<script type="text/javascript">
window.onerror = function(message, url, linenumber) {
```

```
    if (typeof map !== 'undefined' && map) {map.remove();}

    const div = document.getElementById("map")
    if (div) {div.remove();}

    const elemDiv = document.createElement('div');
    elemDiv.innerHTML = `<p>${message}, ligne ${linenumber}</p>`;
    document.body.appendChild(elemDiv);
}
</script>
<script type="text/javascript">
let map;
try {
    const statesData = {"type":"FeatureCollection","features":[
    {"type":"Feature","id":"01","properties":
    {"name":"Alabama","trump_votes_percent":64.6},
    "geometry":{"type":"Polygon","coordinates":
    [[[-87.359296,35.00118],[-85.606675,34.984749],
    [-85.431413,34.124869],[-85.184951,32.859696],
    [-85.069935,32.580372],[-84.960397,32.421541],
    [-85.004212,32.322956],[-84.889196,32.262709],
    [-85.058981,32.13674],[-85.053504,32.01077],
    [-85.141136,31.840985],[-85.042551,31.539753],
    [-85.113751,31.27686],[-85.004212,31.003013],
    [-85.497137,30.997536],[-87.600282,30.997536],
    [-87.633143,30.86609],[-87.408589,30.674397],
    [-87.446927,30.510088],[-87.37025,30.427934],
    [-87.518128,30.280057],[-87.655051,30.247195],
    [-87.90699,30.411504],[-87.934375,30.657966],
    [-88.011052,30.685351],[-88.10416,30.499135],
    [-88.137022,30.318396],[-88.394438,30.367688],
    [-88.471115,31.895754],[-88.241084,33.796253],
    [-88.098683,34.891641],[-88.202745,34.995703],
    [-87.359296,35.00118]]]}},

...
```

```
]};

map = L.map('map').setView([37.8, -96], 4);

L.tileLayer('https://tile.openstreetmap.org/{z}/{x}/{y}.png', {
    maxZoom: 19,
    attribution: '&copy; <a href="http://www.openstreetmap.org/copyright">OpenStreetMap</a>'
}).addTo(map);

function getColor(d) {
    return d > 70   ? '#800026' :
           d > 65   ? '#BD0026' :
           d > 60   ? '#E31A1C' :
           d > 55   ? '#FC4E2A' :
           d > 50   ? '#FD8D3C' :
           d > 45   ? '#FEB24C' :
           d > 40   ? '#FED976' :
           d > 35   ? '#FFEDA0' :
                      '#FFFEC5';
}

function style(feature) {
    return {
        fillColor: getColor(feature.properties.trump_votes_percent),
        weight: 1,
        opacity: 1,
        color: 'white',
        fillOpacity: 0.7
    };
}

L.geoJSON(statesData, {style: style}).addTo(map);

```

```
110    } catch(e) {
111    if (map) {map.remove();}
112
113        document.getElementById("map").remove();
114
115        const elemDiv = document.createElement('div');
116        elemDiv.innerHTML = `<p>${e}</p>`;
117        document.body.appendChild(elemDiv);
118    }
119    </script>
120    </body>
121    </html>
```

- **Lines 25 to 77**: An object is created with a structure that complies with the GeoJSON format. This content comprises a list of 50 states with, for each state:

  - The state name (`name` property).
  - The percentage of votes obtained by Donald Trump in this state (`trump_votes_percent` property).
  - Geometric data for outlining this state (`geometry` property).

  Above, only the data for the first state, Alabama, is shown. The data for all the states can be found at `https://github.com/Apress/Modern-Web-Cartography` (file: `leaflet_maps/November_2024_election_results_by_states.js`).

  These data were obtained:

  - By downloading from the Internet (from `https://raw.githubusercontent.com/python-visualization/folium-example-data/main/us_states.json`) the contours of the American states in GeoJSON format.
  - Then, by completing this GeoJSON file with a text editor: for each of the 50 states, the `trump_votes_percent` property was added with the correct value, found on the Internet site of CNN (this information is available on many sites).

In a more realistic context, these two types of data would not be placed in the same file:

- GeoJSON data describing state contours would be loaded from a local file or downloaded from the Internet via a URL and would contain nothing but contours.
- Another file, containing vote percentages by state, would be found from another source and used to complement the GeoJSON contour file.

- **Line 79**: A map is created and centered on the United States.
- **Lines 81 to 84**: A tile layer is associated with the map.
- **Lines 86 to 96**: A function named `getColor()` is defined. It determines the fill color to be used for a state based on the percentage of votes obtained by Donald Trump in that state (the higher the percentage, the darker the color used).
- **Lines 98 to 106**: A function named `style()` is defined. It returns an object containing style attributes in the form of key/value pairs. To determine the value to be given to the `fillColor` attribute (i.e., the state's fill color), the `getColor()` function defined above is used, passing the percentage of votes for this state found in the GeoJSON-structured object as an argument.
- **Line 108**: An `L.geoJSON` object is created using the GeoJSON-structured object (containing American state contours and voting percentages) and style settings defined by the `style()` function. Then it is added to the map.

Opening this file in a browser produces the following map:

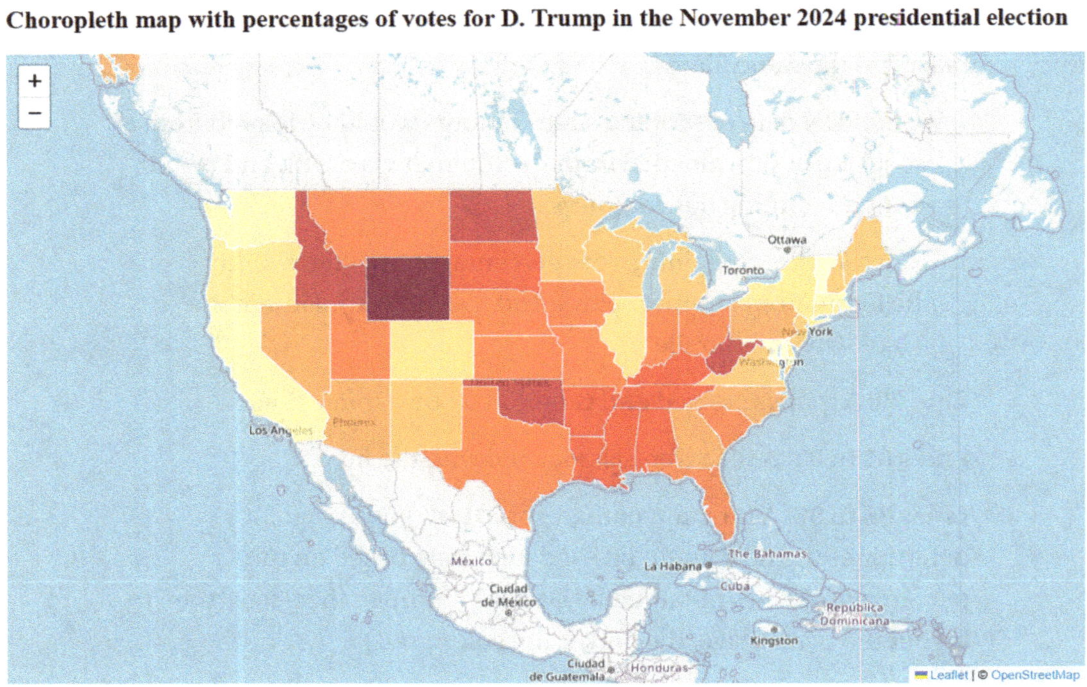

Figure 4-20. *Map created with Leaflet using OpenStreetMap tiles. Data © OpenStreetMap contributors, licensed under the ODbL. US state boundaries from the Folium example dataset (MIT License), source:* `https://github.com/python-visualization/folium-example-data`*. US 2024 presidential election vote percentages from CNN, source: CNN, "2024 Presidential Election Results," retrieved from* `https://edition.cnn.com/election/2024/results/president` *on 05/23/2025*

Leaflet allows you to go further with choropleth maps. In particular:

- In the example above, we use a variable containing GeoJSON-compliant data that includes both the data to be represented (the percentage of votes obtained by D. Trump in each state) and the geometric data of the state contours. As already explained, this is not realistic: most often, these two types of information are obtained from different sources, and two independent files are retrieved:
    - One in CSV (Comma-Separated Values) format, containing the economic or statistical data to be displayed.
    - The other one in GeoJSON format, containing the geometric coordinates of the zone contours.

Leaflet also lets you build choropleth maps from these two types of files, and we will see how in Chapter 8.

- We could have made the map more interactive, for example, by displaying the state name and the percentage of votes obtained by Donald Trump in a popup, when the user clicks on a state (or in a tooltip, when the mouse cursor hovers over a state). We will show how to implement this in Chapter 8.
- A legend should also be added to the map, indicating the percentage of votes corresponding to each color shade. This feature will also be demonstrated in Chapter 8.

Leveraging Leaflet's Interactive Capabilities

The preceding examples demonstrated the power and ease of use of Leaflet. But its other major advantage lies in its **interactive capabilities**. Indeed, Leaflet manages **events** on the map and the map elements, enabling the programmer to define **actions** to be executed when these events occur (for instance, when the user clicks anywhere on the map or places the mouse cursor over a map element, clicks it, or moves the cursor away from it).

For example, the code to handle a user's click on the map is very simple:

```
1   function onMapClick(e) {
2       alert("You clicked at position " + e.latlng);
3   }
4
5   map.on('click', onMapClick);
```

- **Lines 1 to 3**: The function named `onMapClick()` defines the action to be performed when the user clicks on the map: display an alert with the coordinates of the point that was clicked.
- **Line 5**: This statement binds the map's click event to the `onMapClick()` function, so that the function is executed each time the user clicks on the map.

We are going to implement an example with a map that reacts to user clicks: when the user clicks on any point on the map, it will immediately position itself at the antipode of that point and show its precise position.

Here is an HTML code to implement this functionality. Store it in a new file `leaflet_maps/page_with_event_processing.html`:

```
<html>
<head>
<link rel="stylesheet" href="https://unpkg.com/leaflet@1.9.4/dist/leaflet.css" integrity="sha256-p4NxAoJBhIIN+hmNHrzRCf9tD/miZyoHS5obTRR9BMY=" crossorigin=""/>
<title>Carte OSM simple créée avec Leaflet</title>
</head>
<body>
<h1>Map to find the antipode</h1>
<div id="map" style="width: 1030px; height: 620px;"></div>
<script src="https://unpkg.com/leaflet@1.9.4/dist/leaflet.js" integrity="sha256-20nQCchB9co0qIjJZRGuk2/Z9VM+kNiyxNV1lvTlZBo=" crossorigin=""></script>
<script type="text/javascript">
window.onerror = function(message, url, linenumber) {
    if (typeof map !== 'undefined' && map) {map.remove();}

    const div = document.getElementById("map")
    if (div) {div.remove();}

    const elemDiv = document.createElement('div');
    elemDiv.innerHTML = `<p>${message}, ligne ${linenumber}</p>`;
    document.body.appendChild(elemDiv);
}
</script>
<script type="text/javascript">
let map;
try {
    map = L.map('map').setView([-15.88632, -62.22656], 4);
```

```
    L.tileLayer('https://tile.openstreetmap.org/{z}/{x}/{y}.png', {
         maxZoom: 19,
         attribution: '&copy; <a href="http://www.openstreetmap.org/
         copyright">OpenStreetMap</a>'
    }).addTo(map);

    const popup = L.popup();

    function onMapClick(e) {
        const latlng = e.latlng;
        latlng.lat = -latlng.lat;
        if (latlng.lng > 0) {
            latlng.lng = latlng.lng - 180;
        } else {
            latlng.lng = latlng.lng + 180;
        }

        map.setView(latlng);

        popup
        .setLatLng(latlng)
        .setContent(`The antipode is there!<br>[${latlng.lat.
        toFixed(5)}, ${latlng.lng.toFixed(5)}]`)
        .openOn(map);
    }

    map.on('click', onMapClick);

} catch(e) {
    if (map) {map.remove();}

    document.getElementById("map").remove();

    const elemDiv = document.createElement('div');
    elemDiv.innerHTML = `<p>${e}</p>`;
```

```
60        document.body.appendChild(elemDiv);
61    }
62
63    </script>
64    </body>
65    </html>
```

- **Line 25**: A map is created (initially centered on South America and with an initial zoom level of 4).
- **Lines 27 to 30**: A tile layer is added to the map.
- **Line 32**: A popup is created (without adding it to the map yet).
- **Lines 34 to 49**: A function named `onMapClick()` is defined to handle the event of the user clicking on the map.
    - **Line 35**: The coordinates of the point where the user clicked are copied. This information is contained in the event received as an argument by the function.
    - **Lines 36 to 41**: The latitude and longitude of the antipode (of the point where the user clicked) are calculated.
    - **Line 43**: The map is repositioned and centered on the coordinates of the antipode.
    - **Lines 45 to 48**: The popup is positioned at the antipode point, assigned a text, and opened on the map.
- **Line 51**: An event listener is added to the map, associating the click event with the triggering of the `onMapClick()` function.

Opening this file in a browser produces the following map:

***Figure 4-21.** Map created with Leaflet using OpenStreetMap tiles. Data © OpenStreetMap contributors, licensed under the ODbL*

Click, for example, on the city of Santiago (Chile).

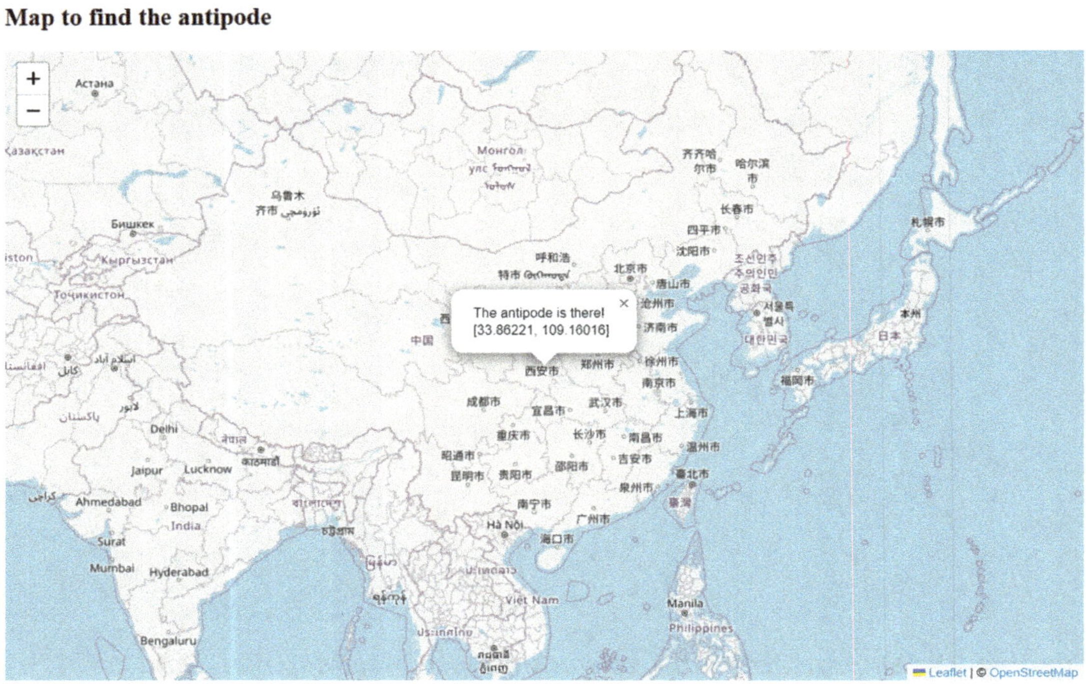

Figure 4-22. Map created with Leaflet using OpenStreetMap tiles. Data © OpenStreetMap contributors, licensed under the ODbL

The map is immediately positioned at the antipode of Santiago: in the province of Shaanxi (China).

You can keep clicking on the map to jump from antipode to antipode.

A variant is to use the `flyTo()` method instead of the `setView()` method on line 43. Instead of immediately positioning itself at the antipode of the point where you clicked, an animation takes place: the map moves away from the point where you clicked, makes a half-turn around the world, then approaches the antipode, and positions itself there.

```
    ...
41            }
42
43            map.flyTo(latlng);
44
45            popup
    ...
```

Note that during these operations, the map and popup objects are not recreated, but simply positioned at new coordinates.

Extending Leaflet with Plugins and Extensions

Leaflet is designed as a core set of basic mapping objects and functions that can be enhanced by partner-developed extensions in the form of **plugins**.

When you need functionality that isn't provided natively, so as not to spend energy on pointless development efforts ("Do not reinvent the wheel!"), you should first consult Leaflet plugin catalog.

Plugins

The **Leaflet plugin catalog** is available at `https://leafletjs.com/plugins.html`. It features a large number of plugins organized by category, extending Leaflet's capabilities in numerous areas.

For example, with the **Leaflet.fullscreen** plugin (developed by Bruno Bergot, with later versions maintained by Bernhard Pottler), you can get a control in a corner of the map to switch to full-screen mode or back to the normal view.

Figure 4-23. *Map created with Leaflet using OpenStreetMap tiles and the "Leaflet. fullscreen" plugin. Data © OpenStreetMap contributors, licensed under the ODbL*

When the map is displayed in full-screen mode, the same control, with a different icon, lets you return the map to its normal size.

Using this plugin is easy, you just need to add the following elements to the HTML file used to create the map:

- A link to a stylesheet:

```
<link href='https://api.mapbox.com/mapbox.js/plugins/leaflet-
fullscreen/v1.0.1/leaflet.fullscreen.css' rel='stylesheet' />
```

- A link to the plugin script:

```
<script src='https://api.mapbox.com/mapbox.js/plugins/leaflet-
fullscreen/v1.0.1/Leaflet.fullscreen.min.js'></script>
```

- A JavaScript statement to add the fullscreen control to the map:

```
...
map = L.map('map').setView([51.505, -0.09], 13);
...
```

```
map.addControl(new L.Control.Fullscreen());
...
```

The method for implementing new plugins is also documented in the Leaflet Plugin Guide: `https://github.com/Leaflet/Leaflet/blob/main/PLUGIN-GUIDE.md`. This topic is beyond the scope of this book.

Extensions

If no existing plugin meets your needs, you will have to develop a custom solution. Leaflet lets you create **extensions** to its classes, layers, event handlers, and controls:

- **Class extension** allows you to create new object types from object types that already exist in Leaflet. These new object types will inherit the functionality of their original object type and add new functionality: adding new methods and options, modifying methods and options inherited from the original object, etc. This technique is documented at `https://leafletjs.com/examples/extending/extending-1-classes.html`.
- **Layer extension** allows you to create new layer types from those that already exist in Leaflet. This technique is documented at `https://leafletjs.com/examples/extending/extending-2-layers.html`.
- **Event handler extension** lets you define new events to react to and actions to execute in response to these events. This technique is documented at `https://leafletjs.com/examples/extending/extending-3-controls.html`.
- Leaflet's controls are the zoom control at the top left of the map, the layer controls at the top right, the scale at the bottom left, etc. Leaflet lets you create **new controls** and place them on maps. This technique is documented at `https://leafletjs.com/examples/extending/extending-3-controls.html`.

These techniques are beyond the scope of this book.

CHAPTER 5

Creating Interactive Maps with Folium and Python

Folium is a Python library for building maps in the OpenStreetMap environment. Like Leaflet, it allows you to create maps and enhance them by adding various types of information: highlighting specific points or objects using markers, drawing polylines and geometric shapes, outlining and filling areas, displaying information through popups and tooltips, overlaying images on maps, adding data from GeoJSON and TopoJSON formats, creating choropleth maps, etc.

The main difference between Leaflet and Folium is that **Leaflet runs client-side in the browser using JavaScript**, while **Folium is a Python library that generates Leaflet maps server-side** – though Folium maps can also be displayed client-side in interactive environments such as Jupyter notebooks, where the map HTML and JavaScript are rendered in the browser.

The Folium library is a **Python-language wrapper for the Leaflet.js library** (designed for the JavaScript language and seen in the previous chapter). More precisely, Folium statements are placed in Python code used on a Web server supporting this language.

1. When the server executes Folium statements, they generate HTML and JavaScript code containing Leaflet statements, which is inserted into an HTML page before it is sent to the browser.

2. When the browser receives this page, it executes the JavaScript code and Leaflet statements it contains, resulting in a map being displayed.

The list of Folium functions described below is not exhaustive. In particular, the many options available for Folium objects are not detailed or explained. For a complete reference, please consult the Leaflet documentation: `https://python-visualization.github.io/folium/latest/index.html`.

P. Marie, *Modern Web Cartography*, https://doi.org/10.1007/979-8-8688-2335-0_5

At the time of writing this book, the current release of Folium is version 0.20.0, and it is the version used for all the examples. The version of Python used is 3.11.2, and the version of Flask is 3.1.2.

The code for the examples in this chapter can be found on the website `https://github.com/Apress/Modern-Web-Cartography` and in the `carto_osm` directory.

Basic Principles of Folium

Since the Folium library is a wrapper around the Leaflet.js library, their core principles are very similar.

Using Folium begins with creating a base map, which involves specifying the coordinates to center the map, a zoom level, and a tile set to render the map's visual style. This forms the map's base layer.

Additional map elements can then be added to highlight places, routes, regions, or other geographic features. These are overlay layers and can include:

- Markers, optionally enhanced with popups or tooltips.
- Polylines, polygons, and other geometric shapes.
- Images.
- Alternative tile sets.
- Etc.

Elements added to the map can be created through programming statements or generated from imported data in formats such as **GeoJSON**, **TopoJSON**, or **GPX**.

Choropleth maps can also be created.

These map elements can be grouped using layer groups or feature groups for better organization and control.

Controls allow users to interact with the map - for example, to zoom in or out or toggle the visibility of layers - or can be used to display additional information, such as a scale bar or an attribution notice. Some controls are included by default (but can be customized), while others must be explicitly added.

Because Folium runs server-side (in Python), it does not offer the same level of interactivity as Leaflet, which is client-side. While Folium supports basic built-in interactions such as panning and zooming, it does not natively support custom event handling (e.g., on-click behaviors). To achieve such behavior, developers must inject raw

JavaScript and Leaflet code directly into the map's HTML code. This is possible but less straightforward compared to working directly with Leaflet.

Environment Used to Build and Run the Following Examples

In this chapter, we will learn how to build maps with Folium and Python by creating a number of elementary maps. These examples will be implemented using the following Web application creation tools: the **VS Code** development workshop and the **Python** language, supplemented by the **Flask** framework and the **Jinja** model engine.

With Folium and Python running on a server and generating HTML, JavaScript, and Leaflet code for execution by a browser, the test architecture cannot be as simple as in the previous chapter related to Leaflet (where a simple HTML file was all that was needed to run a unit test).

The map examples below will be implemented in the Python/Flask Web environment, the setup of which is described in Appendix B, "Setting Up the Prototype Development and Test Environment." To be able to easily reproduce the examples following the instructions provided in this chapter, it is recommended to first prepare a VS Code environment as explained in Appendix B.

If you do not want to set up the environment as described in Appendix B, all you need to run the examples is a **Python** development environment with **Flask** and **Folium** installed, capable of executing Python applications and acting as a web server. The instructions provided below will not be directly applicable and may require a small amount of adaptation.

Although the environment described includes Flask, it should be pointed out that other web frameworks for Python exist (such as Django) and could be used just as well as Flask to build the Folium maps that follow.

The Flask package structure, described and created in Appendix B, will be used, and new routes and functions will be added to the `simple_maps` blueprint. The `page_with_map.html` template, created when setting up the environment (following the instructions in Appendix B), will be used for all the examples below. Here is a reminder of its contents:

```
{% extends 'base.html' %}

{% block title %}{{ name }}{% endblock %}
```

```
4
5   {% block header %}
6   <h2>{{ name }}</h2>
7   {% endblock %}
8
9   {% block content %}
10  {{ map_html | safe }}
11  {% endblock %}
```

- **Line 1**: This statement indicates that this template extends the page structure defined in the `base.html` base template (described in Appendix B) by inserting new elements.
- **Line 3**: A specific title will be received as an argument from the calling Python function when the construction of an HTML page from this template is triggered. While generating an HTML page, the Jinja engine will replace the `name` variable with the value received from Python.
- **Line 6**: A specific header will be received as an argument from the calling Python function when the construction of an HTML page is triggered (the Jinja engine will replace the `name` variable with the value received from Python).
- **Line 10**: The HTML and JavaScript code for the Folium map will be received as an argument from the calling Python function (this line will be processed by the Jinja engine on the server):
    - The `map_html` variable will be replaced by the actual HTML and JavaScript code (including Leaflet statements) generated by Folium to render the map.
    - The `safe` filter is necessary to prevent Jinja from escaping the HTML/JavaScript content. Without it, Jinja would convert characters such as "<" and ">" into "<" and ">", causing the browser to display raw code instead of rendering the map.

The architecture used by the examples of this chapter is the following:

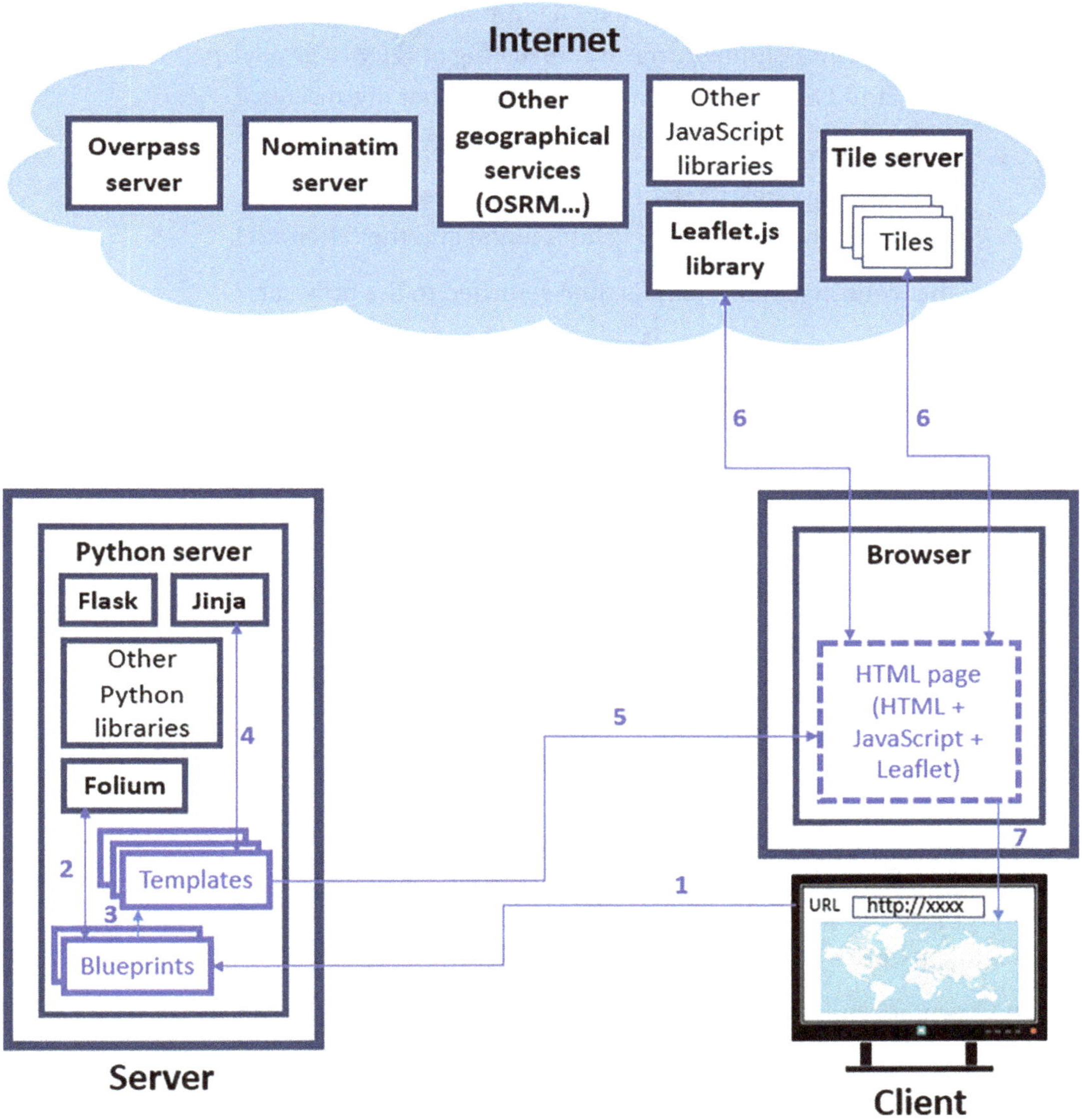

Figure 5-1. Architecture used for testing maps created with Folium

1. The user invokes a URL from the browser. This URL points to a route on the Python server.
2. The Python server retrieves the Python code associated with this route in a blueprint and executes it. When this Python code containing Folium statements is run, a map is generated: HTML and JavaScript code containing Leaflet statements is created.

3. The Python server then invokes a template, passing it the map created with Folium (currently consisting of HTML, JavaScript code, and Leaflet statements) and some other arguments.

4. The Jinja engine injects the dynamic data passed by the blueprint into the template: the map (currently consisting of HTML, JavaScript code, and Leaflet statements) and the other data.

5. The resulting HTML page is then returned to the browser.

6. The browser receives the HTML page, executes the embedded JavaScript code and the Leaflet statements, and downloads the tiles necessary to render the map.

7. The browser finally displays the resulting web page, including a map, on the screen.

Every time you test a new map, do not forget:

- to save the files that you have created or modified to produce it,
- to restart the VS Code test server (or start it if it is not running),

before opening its URL with a browser.

Beginning with a Simple Map

Let's start by building and displaying a simple Folium map.

Creating a map with Folium can be as simple as the following statement:

```
m = folium.Map(location=[36.103, -115.156], zoom_start=14)
```

This Python statement creates a `folium.Map` object (supplied by the Folium library). It includes a `location` parameter to specify the latitude (36.103) and longitude (-115.156) on which to center the map and a `zoom_start` parameter to specify the initial zoom level (14 is used, the default is 10). The resulting map is placed in a variable named `m`.

Note that we do not need to associate a tile layer with the map (as with Leaflet), as Folium associates one by default (OSM Standard Tiles - Mapnik).

At runtime, this map will be converted to HTML and JavaScript code including Leaflet statements, then inserted into an HTML page before being sent to the browser. The generated code does not include the map tiles themselves. These are loaded dynamically by the browser from the tile server as the map is displayed.

Please note that the tiles used by default above (`https://tile.openstreetmap.org/{z}/{x}/{y}.png` – OSM Standard Tiles – Mapnik), and all other tiles used in the following examples, are public tiles and intended only for demonstration, education, development, testing, and noncommercial web applications. For production or commercial applications, it is recommended to:

- Either set up your own tile server using OpenStreetMap data.
- Or use a commercial tile provider.

For more details, see `https://operations.osmfoundation.org/policies/tiles/`.

To implement a simple Web application that displays a map of Las Vegas, Nevada, in a browser when a URL is called up, add the following code to the `simple_maps.py` blueprint, at the end of the file, after the statements already there (the creation of the `simple_maps.py` blueprint is described in Appendix B):

```
1   @bp.route('/sm/map1', methods=('GET', 'POST'))
2   def map1():
3       try:
4           m = folium.Map(location=[36.10250, -115.15579], zoom_start=14)
5
6           return render_template('maps/page_with_map.html',
7               name="Simple map of Las Vegas",
8               map_html=m._repr_html_())
9
10      except Exception as e:
11          flash("An error has been encountered:<br/>" + str(e),
            category="error")
12          return render_template('maps/page_with_map.html', name="No map
            to display")
```

- As usual with Python, be careful with column alignments and indentations. The first statement (`@bp.route...`) must not be indented.
- **Line 1**: A route is defined and will be available for invoking the associated `map1()` view function using an HTTP request. This route determines the URL through which the `map1()` function can be called. The path `/sm/map1` should be added after the protocol, server

address, and server listening port (in the case of local execution with the VS Code test server: `http://127.0.0.1:5000`, therefore the complete URL is `http://127.0.0.1:5000/sm/map1`).

- **Line 2**: This line begins the definition of the `map1()` function.
- **Line 3**: This line begins the `try` part of a `try/except` block, used to intercept any errors during map construction (in the `try` part) and process them (in the `except` part). This is a Python feature, not something specific to Folium.
- **Line 4**: A map object is created and centered on geographic coordinates (latitude and longitude of a point south of the city of Las Vegas, Nevada). An initial zoom level that limits the map to the Strip district (actually part of the city of Paradise, south of Las Vegas) is also defined.

  The `folium` library, used here, is not imported in this snippet because it was already imported globally along with other libraries when the blueprint was created in Appendix B.
- **Lines 6 to 8**: A web page is rendered using the `page_with_map.html` template (created while setting up the environment in Appendix B), with the following arguments:
  - `name`: The name "Simple map of Las Vegas".
  - `map_html`: The map object, converted to HTML and JavaScript code (including Leaflet statements) using the `_repr_html_()` method.

  The `render_template` library, used here, is not imported in this snippet because it was already imported globally along with other libraries when the blueprint was created in Appendix B.

  The `m._repr_html_()` method returns an HTML string that includes both the necessary HTML and JavaScript code, which can easily be embedded in an HTML document using a Jinja template.

  An alternative option is to use `m.get_root().render()`, which generates a standalone HTML document containing the map. Although this approach is perfectly valid and convenient in some situations, we will not use it in our examples.

- **Line 10**: This line starts the `except` part of the `try/except` block. This part is responsible for handling errors. It is executed only if an error occurs during execution of the statements in the `try` part.
    - **Line 11**: An error message is prepared for display (using Flask's `flash()` function).

      The `flash` library, used here, is not imported in this snippet because it was already imported globally along with other libraries when the blueprint was created in Appendix B.
    - **Line 12**: A web page is rendered using the `page_with_map.html` template, with the `name` "No map to display" as an argument.

Launching (or restarting) the VS Code test server ("Run > Run Without Debugging") and opening the URL `http://127.0.0.1:5000/sm/map1` in a browser produces the following map:

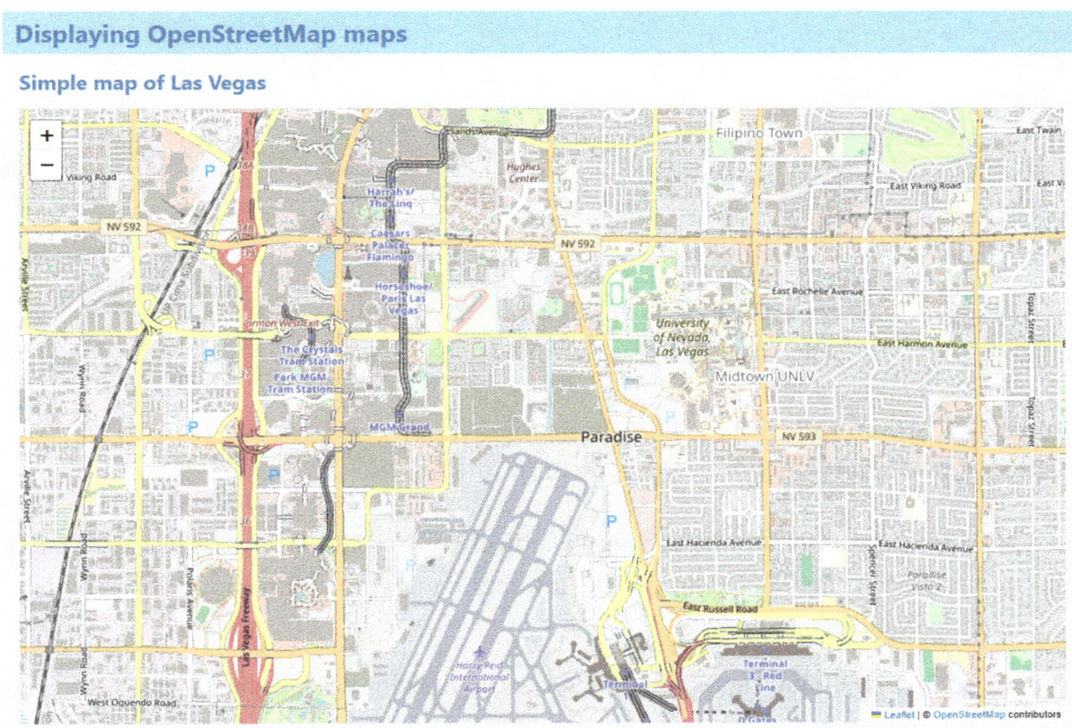

__Figure 5-2.__ Map created with Folium using OpenStreetMap tiles. Data © OpenStreetMap contributors, licensed under the ODbL

This map is not a static image, but an interactive map:

- Double-clicking with the mouse, using the "+" and "-" buttons (top left) or the mouse wheel, you can zoom in and out.
- Dragging the map (with the mouse), you can move it in any direction.

For example, by zooming out and shifting the map to the southwest:

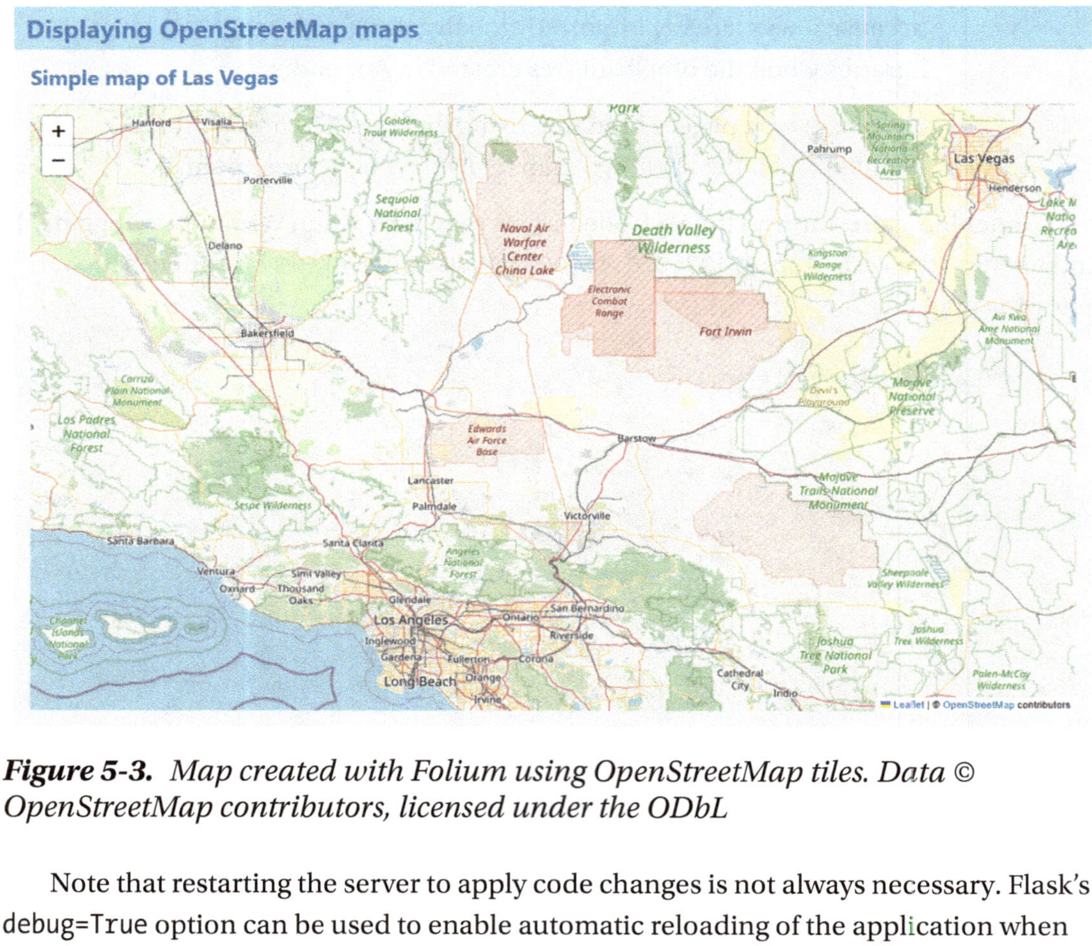

Figure 5-3. *Map created with Folium using OpenStreetMap tiles. Data © OpenStreetMap contributors, licensed under the ODbL*

Note that restarting the server to apply code changes is not always necessary. Flask's `debug=True` option can be used to enable automatic reloading of the application when source files change. It can be set:

- Either by including the following instructions in the Python blueprint:

```
if __name__ == "__main__":
    app.run(debug=True)
```

- Or by adding the `"FLASK_ENV": "development"` setting in the env stanza of the `launch.json` file, located in the `.vscode` directory of the workspace

We will not use this option for developing the examples below.

Using Tile Layers to Display Other Map Types

In the previous example, OpenStreetMap's default map style was used. However, other map styles referred to as **tiles** can be used, and, in this new example, we will use different tile sets.

The available tile sets are presented on the website `https://leaflet-extras.github.io/leaflet-providers/preview/`, which lets you visualize the rendering, check the area covered by each tile set, and choose the ones that suit you best. It also indicates the reference required to use a tile set and the attribution notice (equivalent to a copyright) to be inserted in the bottom right-hand corner of the map.

Here is an example of a statement for creating a map with specifying the tile set to be used:

```
1   m = folium.Map(location=[36.10250, -115.15579],
2         zoom_start=14,
3         tiles='https://{s}.tile-cyclosm.openstreetmap.fr/cyclosm/{z}/{x}/
          {y}.png',
4         attr='<a href="https://github.com/cyclosm/cyclosm-cartocss-
          style/releases" title="CyclOSM - Open Bicycle render">CyclOSM</
          a> | Map data: &copy; <a href="https://www.openstreetmap.org/
          copyright">OpenStreetMap</a> contributors')
```

- **Lines 1 and 2**: The `location` parameter defines the geographic coordinates on which to center the map, and the `zoom_start` parameter indicates the map's initial zoom level.
- **Line 3**: The `tiles` parameter defines the tile set (i.e., the map style) to be used. Here, the "CyclOSM" tile set has been chosen (the value to be used for this parameter can be found on `https://leaflet-extras.github.io/leaflet-providers/preview/`).

- **Line 4**: The `attr` parameter specifies the attribution notice to be inlaid in the bottom right-hand corner of the map for this tile set (the value to be used can also be found on the same site).

It is also possible to create the map with the default tile set, then to add another tile set as a new layer:

```
1  m = folium.Map(
2      location=(36.10250, -115.15579),
3      zoom_start=14
4  )
5
6  folium.TileLayer(
7      tiles='https://{s}.tile-cyclosm.openstreetmap.fr/cyclosm/{z}/{x}/{y}.png',
8      attr='<a href="https://github.com/cyclosm/cyclosm-cartocss-style/releases" title="CyclOSM - Open Bicycle render">CyclOSM</a> | Map data: &copy; <a href="https://www.openstreetmap.org/copyright">OpenStreetMap</a> contributors',
9      name="CyclOSM",
10     overlay=False,
11     control=True
12 ).add_to(m)
```

- **Line 6**: An object of type `folium.TileLayer` is created using the CycloOSM tile set.
- **Line 9**: The `name` parameter can be used in a layer control, enabling map users to show or hide the tile layer.
- **Line 10**: The `overlay` parameter specifies whether this tile layer is to be added:
    - **As a base layer (**`False`**)**: In this case, this new tile layer replaces the previous base layer, set by the `folium.Map()` function, and is the only one visible.
    - **Or as an overlay layer (**`True`**)**: In that case, several map layers can be visible at the same time, using transparency (controlled by another parameter named `opacity`).

- **Line 11**: The `control` parameter specifies whether this tile layer is included in a layer control (enabling it to be shown or hidden).
- **Line 12**: The `folium.TileLayer` object is added to the map. Because it is added as a base layer, it will be the only one visible on the map.

To obtain a new map with a new tile layer, we will build upon the code developed in the previous section. In the `simple_maps.py` blueprint:

1. Locate and duplicate the code for the `/sm/map1` route (this ensures that the original example remains unchanged).
2. In the duplicated code:
    1. Rename the route to `/sm/map2` and the view function to `map2()`.
    2. Insert or modify statements in the code of `/sm/map2` as shown in **bold** below:

```
@bp.route('/sm/map2')
def map2():
    try:
        m = folium.Map(
            location=(36.10250, -115.15579),
            zoom_start=14,
            tiles='https://{s}.tile-cyclosm.openstreetmap.fr/cyclosm/
            {z}/{x}/{y}.png',
            attr='<a href="https://github.com/cyclosm/cyclosm-
            cartocss-style/releases" title="CyclOSM - Open Bicycle
            render">CyclOSM</a> | Map data: &copy; <a href="https://
            www.openstreetmap.org/copyright">OpenStreetMap</a>
            contributors'
        )

        return render_template('maps/page_with_map.html',
            name="Map of Las Vegas with cycle lanes",
            map_html=m._repr_html_())

    except Exception as e:
```

```
16            flash("An error has been encountered:<br/>" + str(e),
              category="error")
17            return render_template('maps/page_with_map.html', name="No map
              to display")
```

- Again, be careful with column alignments and indentations. The first statement (`@bp.route...`) must not be indented.
- **Lines 4 to 9**: A map is built.
    - Centered on a geographical position (Las Vegas, Nevada).
    - With an initial zoom level of 14.
    - With a layer of "CyclOSM" tiles (cycle lanes and routes).
    - With the corresponding attribution (to be inserted in the bottom right-hand corner of the map).
- **Lines 11 to 13**: A web page is rendered using the `page_with_map.html` template. A name and the map object, converted to HTML and JavaScript code, are passed as arguments.
- **Lines 15 to 17**: If an error occurs while creating the map, a web page is rendered with the error message.

Launching the server and opening the URL `http://127.0.0.1:5000/sm/map2` in the browser produces the following map:

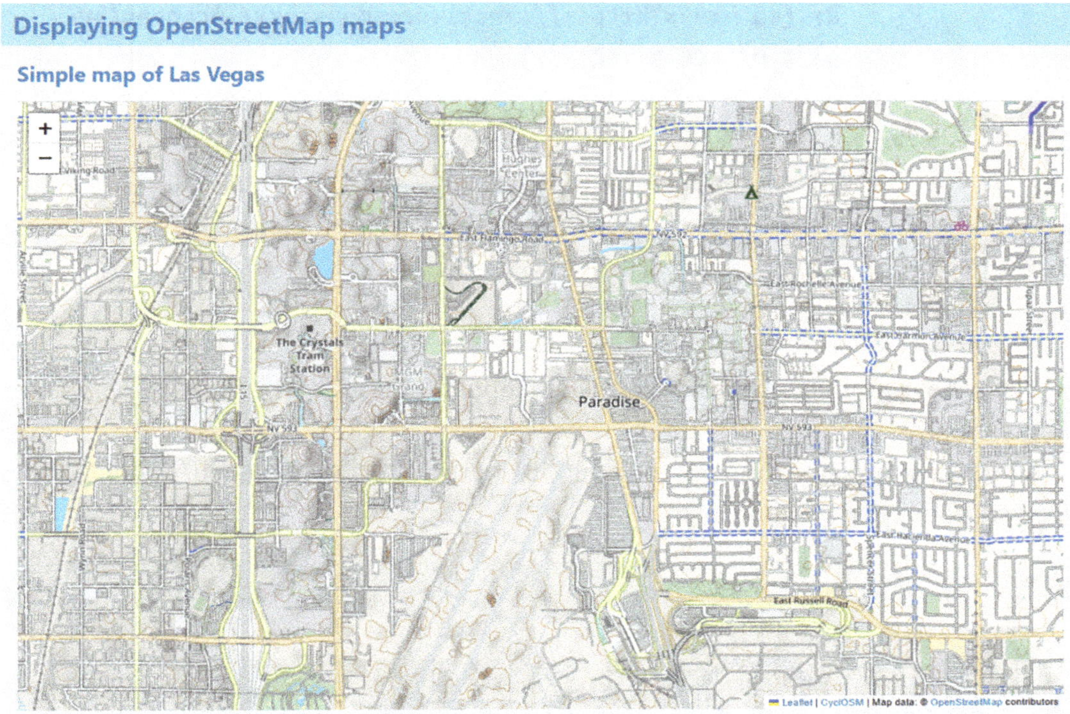

Figure 5-4. *Map created with Folium using CyclOSM tiles. Tiles © CyclOSM contributors. Data © OpenStreetMap contributors (ODbL). CyclOSM is an open source project. See* `https://cyclosm.org` *for details*

Cycle ways and routes are shown in blue (mostly limited to the lower right quadrant).

Let's now modify the statement on lines 4 to 9 to use a tile set with topographic features (rivers, mountains, etc.) and shaded relief:

```
...
4        m = folium.Map(
5            location=(36.10250, -115.15579),
6            zoom_start=14,
7            tiles='https://{s}.tile.opentopomap.org/{z}/{x}/{y}.png',
8            attr='Map data: &copy; <a href="https://www.openstreetmap.
             org/copyright">OpenStreetMap</a> contributors, <a
             href="http://viewfinderpanoramas.org">SRTM</a> | Map style:
             &copy; <a href="https://opentopomap.org">OpenTopoMap</
```

```
            a> (<a href="https://creativecommons.org/licenses/by-
            sa/3.0/">CC-BY-SA</a>)'
        )

        return render_template('maps/page_with_map.html',
            name="Map of Las Vegas with topographic features",
            map_html=m._repr_html_())

...
```

The `tiles` parameter specifies the "OpenTopoMap" tile set, and the `attr` parameter indicates the attribution notice for these new tiles.

Launching the server and opening the URL `http://127.0.0.1:5000/sm/map2` in the browser produces the following map:

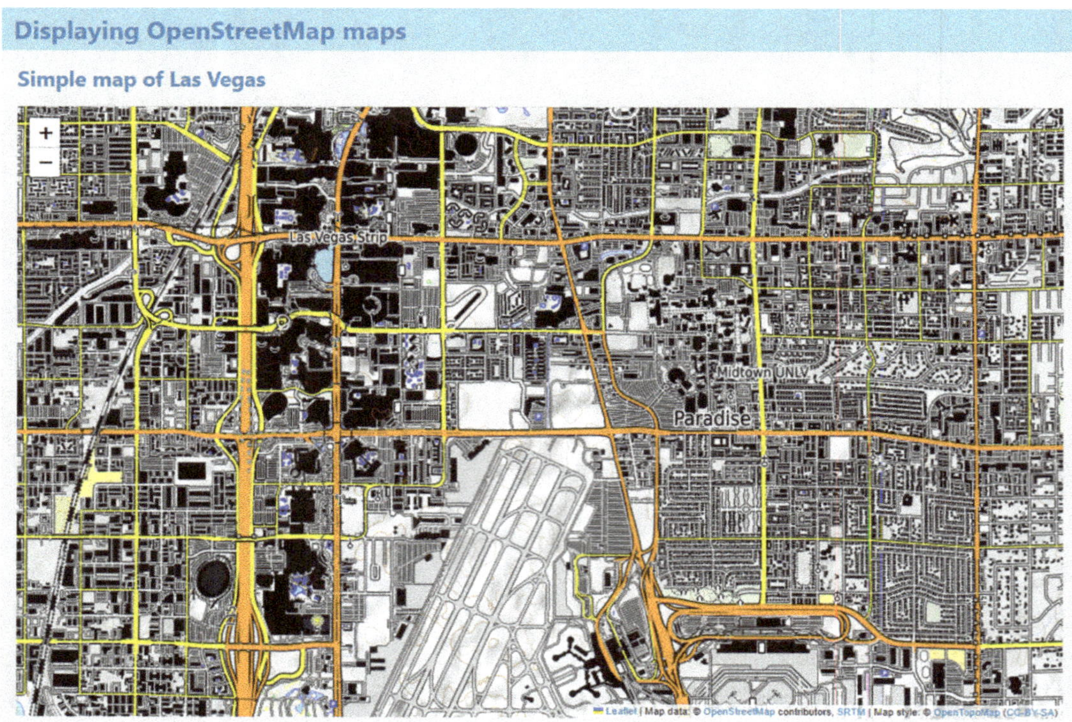

Figure 5-5. *Map created with Folium using OpenTopoMap tiles. Tiles © OpenTopoMap (CC BY-SA). Data © OpenStreetMap contributors (ODbL), SRTM*

Adding Markers, Popups, and Tooltips

Folium lets you add **markers** to maps in order to highlight specific locations. These markers can be of different shapes and colors, **icons** can be associated with them (either as an icon embedded in the marker symbol, or to replace the default marker symbol), text displayed in **popups** or **tooltips** can be associated with the markers, and an opacity level can also be defined (so that the underlying map is visible through transparency).

Here is an example of a statement creating a marker and adding it to a map:

```
1   folium.Marker(
2       location=[36.10250, -115.15579],
3       tooltip="Click me!",
4       popup="Caesars Palace"
5   ).add_to(m)
```

- **Lines 1 to 5**: An object of type `folium.Marker` is created:
    - **Line 2**: The `location` parameter specifies the geographical position at which to place the marker.
    - **Line 3**: The `tooltip` parameter specifies the text of a tooltip (to be displayed when the mouse cursor passes over the marker).
    - **Line 4**: The `popup` parameter indicates the text of a popup (to be displayed when the marker is clicked).
    - **Line 5**: The marker is eventually added to the map.

To obtain a new map with markers, we will again build upon the code created in the previous section. In the `simple_maps.py` blueprint, duplicate the code for the `/sm/map1` route again (to preserve the original). In the copy, rename the route to `/sm/map3` and the view function to `map3()`, and then insert the code shown in **bold** below:

```
@bp.route('/sm/map3')
def map3():
    try:
        m = folium.Map(location=[36.10250, -115.15579], zoom_start=14,
        control_scale=True)

        folium.Marker(
```

```
7               location=[36.11574, -115.17481],
8               tooltip="Click me!",
9               popup="Caesars Palace"
10          ).add_to(m)
11
12          folium.Marker(
13              location=[36.11236, -115.17206],
14              tooltip="Click me!",
15              popup="Paris Las Vegas"
16          ).add_to(m)
17
18          folium.Marker(
19              location=[36.10205, -115.17425],
20              tooltip="Click me!",
21              popup="New York-New York"
22          ).add_to(m)
23
24          folium.Marker(
25              location=[36.08421, -115.15333],
26              tooltip="Click me!",
27              popup="Harry Reid International Airport"
28          ).add_to(m)
29
30          return render_template('maps/page_with_map.html',
31              name="Map of Las Vegas with markers",
32              map_html=m._repr_html_())
33
34      except Exception as e:
35          flash("An error has been encountered:<br/>" + str(e),
            category="error")
36          return render_template('maps/page_with_map.html', name="No map
            to display")
```

- **Line 4**: A map is built and centered on the coordinates of Las Vegas, with an initial zoom level of 14. The `control_scale` parameter has been added to have a distance scale displayed in the bottom left-hand corner of the map (this modification is unrelated to the markers that are added next).

- **Lines 6 to 10**: A marker is created and added to the map. It is located at the coordinates of the Caesars Palace casino hotel, with a tooltip ("Click me!") and a popup ("Caesars Palace").
- **Lines 12 to 16, 18 to 22, and 24 to 28**: Three more markers are created for the Paris Las Vegas casino hotel, the New York-New York casino hotel, and the Harry Reid International Airport.
- **Lines 30 to 32**: A web page is rendered using the `page_with_map.html` template. A name and the previously constructed map, converted to HTML and JavaScript code, are passed as arguments.

Launch the server, open the URL `http://127.0.0.1:5000/sm/map3` in the browser, and click on one of the markers to obtain the following map:

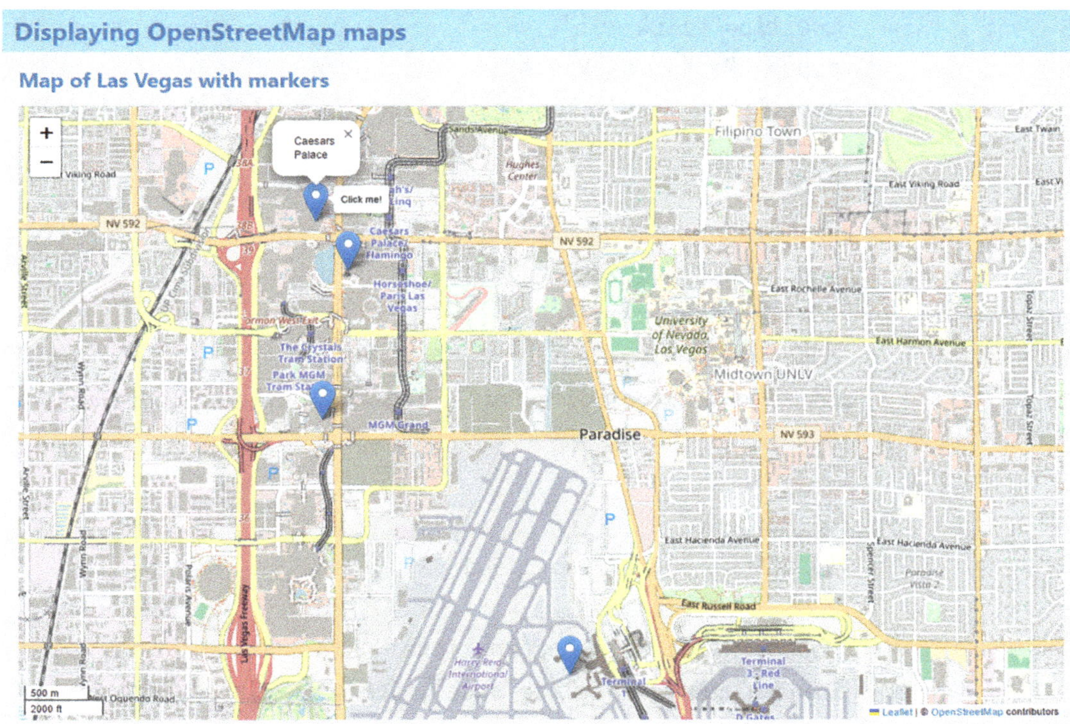

Figure 5-6. *Map created with Folium using OpenStreetMap tiles. Data © OpenStreetMap contributors, licensed under the ODbL*

Next, we are going to customize the markers to make them more relevant to the places they indicate. To do this, we will modify the code of the `map3()` function as follows (the modifications are shown in **bold** below):

```
...
        folium.Marker(
            location=[36.11574, -115.17481],
            tooltip="Click me!",
            popup="Caesars Palace",
            icon= folium.Icon(color="red", icon="fa-building",
            prefix="fa")
        ).add_to(m)

        folium.Marker(
            location=[36.11236, -115.17206],
            tooltip="Click me!",
            popup="Paris Las Vegas",
           icon= folium.Icon(color="red", icon="fa-building", prefix="fa")
        ).add_to(m)

        folium.Marker(
            location=[36.10205, -115.17425],
            tooltip="Click me!",
            popup="New York-New York",
           icon= folium.Icon(color="red", icon="fa-building", prefix="fa")
        ).add_to(m)

        folium.Marker(
            location=[36.08421, -115.15333],
            tooltip="Click me!",
            popup="Harry Reid International Airport",
            icon=folium.Icon(color="blue", icon="fa-plane", prefix="fa")
        ).add_to(m)
...
```

- **Lines 10, 17, and 24**: The `icon` parameter has been added to specify the marker color (red) and the icon to be placed in the marker (a building). The `prefix` parameter is needed to complete the icon's reference.
- **Line 31**: The `icon` parameter has been added to specify the marker color (blue) and the icon to be placed in the marker (a plane). The `prefix` parameter is needed to complete the icon's reference.

A list of available icons can be found at `https://fontawesome.com/v4/cheatsheet/`, and the supported colors are `red`, `blue`, `green`, `orange`, `yellow`, `purple`, `darkred`, `white`, `black`, and `cadetblue`.

Launch the server, open the URL `http://127.0.0.1:5000/sm/map3` in the browser, and click on one of the markers to obtain the following map:

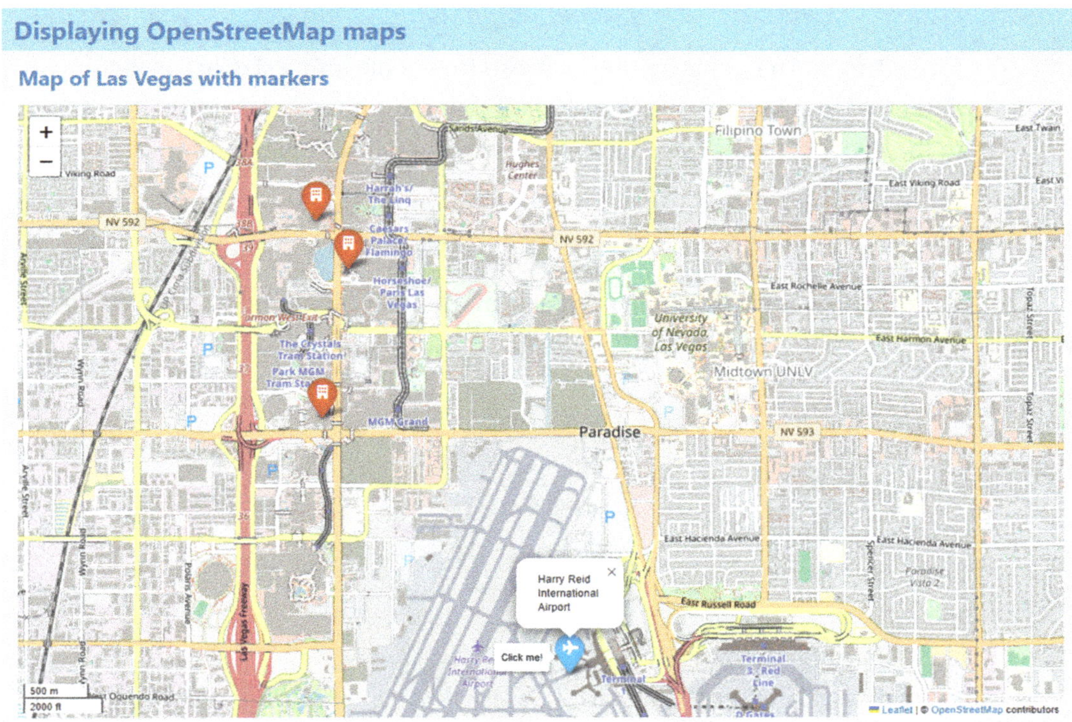

Figure 5-7. *Map created with Folium using OpenStreetMap tiles. Data © OpenStreetMap contributors, licensed under the ODbL*

We have just customized the markers by embedding an icon inside the marker symbol. It is also possible to replace the default marker symbol with an icon of your choice (e.g., use a train to indicate railway stations or coins for banks). For example:

```
1   train_icon = folium.CustomIcon(
2       icon_image='prototypes/static/nice_train.jpg',
3       icon_size=(40, 40)
4   )
5
6   folium.Marker(
7       location=[48.1355, 17.0062],
8       icon=train_icon,
9       popup="Railway station here!"
10  ).add_to(m)
```

- **Lines 1 to 4**: An icon is defined with an image file and a size in pixels. You must previously have found a suitable train image and stored it in the `prototypes/static` directory with the name "nice_train.jpg".
- **Lines 6 to 10**: A marker is created with the preceding icon, then added to the map.

Another possibility is to use circle markers (markers in the shape of a circle). Here is an example:

```
1   folium.CircleMarker(
2       location=[51.0520, 13.7415],
3       radius=10,
4       color='green',
5       fill=True,
6       opacity=1,
7       fill_opacity=0.5,
8       popup="Magnificient church!"
9   ).add_to(m)
```

- **Lines 1 to 9**: A circle marker is defined with a radius of ten pixels, a green stroke color, full stroke opacity, and medium fill opacity (meaning the background map will remain partially visible inside the circle).

Grouping Map Elements and Controlling Their Display

Elements added to maps (such as markers and other elements that will be presented in the following sections) can be associated within **groups**. A **layer control** can then be used to show or hide the elements of these groups. More precisely, elements added to the map (such as markers) are layers overlaid on the map, and a layer control lets you choose to show or hide them.

Here is an example of statements for creating and then grouping objects:

```
1   group1 = folium.FeatureGroup("First group").add_to(m)
2
3   folium.Marker(location=[36.11574, -115.17481], icon=folium.
    Icon(color="red")).add_to(group1)
4
5   folium.Marker(location=[36.11236, -115.17206], icon=folium.
    Icon(color="red")).add_to(group1)
6
7   folium.LayerControl().add_to(m)
```

- **Line 1**: A new feature group (object of type `folium.FeatureGroup`) named `First group` is created and added to the map.
- **Lines 3 and 5**: Markers are created and added to the previous feature group (and therefore also, indirectly, to the map).
- **Line 7**: Then, a layer control (object of type `folium.LayerControl`) is created. It will allow users to show or hide the map elements of the feature group at will.

Note In the preceding chapter with Leaflet, we used **layer groups** to group map elements and control their display. With Folium, however, we now use **feature groups** for this purpose. This is because, in Folium, feature groups are easier to use for grouping elements and controlling their display: they are automatically recognized by the layer control.

In contrast, using layer groups to control element display with Folium would require additional code, making the process more complex. Practically speaking, layer groups in Folium should be used only for grouping elements internally, not for display control.

To obtain a new map with feature groups and layer control, we will again build upon the code created in the previous section. In the `simple_maps.py` blueprint, duplicate the code of the `/sm/map3` route (to preserve the original). In the copy, rename the route to `/sm/map4` and the view function to `map4()`, then insert or modify the code as shown in **bold** below:

```
@bp.route('/sm/map4')
def map4():
    try:
        m = folium.Map(location=[36.10250, -115.15579], zoom_start=14, control_scale=True)

        casino_group = folium.FeatureGroup("Casinos hotels").add_to(m)

        folium.Marker(
            location=[36.11574, -115.17481],
            tooltip="Click me!",
            popup="Caesars Palace",
            icon= folium.Icon(color="red", icon="fa-building", prefix="fa")
        ).add_to(casino_group)

        folium.Marker(
            location=[36.11236, -115.17206],
            tooltip="Click me!",
            popup="Paris Las Vegas",
            icon= folium.Icon(color="red", icon="fa-building", prefix="fa")
        ).add_to(casino_group)

        folium.Marker(
```

```
23              location=[36.10205, -115.17425],
24              tooltip="Click me!",
25              popup="New York-New York",
26              icon= folium.Icon(color="red", icon="fa-building",
                prefix="fa")
27          ).add_to(casino_group)
28
29          airport_group = folium.FeatureGroup("Airports").add_to(m)
30
31          folium.Marker(
32              location=[36.08421, -115.15333],
33              tooltip="Click me!",
34              popup="Harry Reid International Airport",
35              icon=folium.Icon(color="blue", icon="fa-plane",
                prefix="fa")
36          ).add_to(airport_group)
37
38          folium.LayerControl().add_to(m)
39
40          return render_template('maps/page_with_map.html',
41              name="Map of Las Vegas with marker groups and layer
                control",
42              map_html=m._repr_html_())
43
44      except Exception as e:
45          flash("An error has been encountered:<br/>" + str(e),
            category="error")
46          return render_template('maps/page_with_map.html', name="No map
            to display")
```

- **Line 6**: A first feature group is created with the label "Casinos hotels" and added to the map.
- **Lines 13, 20, and 27**: The first three markers are added to the first group.
- **Line 29**: A second feature group is created with the label "Airports" and added to the map.

- **Line 36**: The last marker is added to the second group.
- **Line 38**: A layer control is created and added to the map. This allows the elements of the "Casinos hotels" and "Airports" feature groups to be shown or hidden at will by the user.
- **Lines 40 to 42**: A web page is rendered using the `page_with_map.html` template. A name and the previously constructed map, converted to HTML and JavaScript code, are passed as arguments.

Launching the server and opening the URL `http://127.0.0.1:5000/sm/map4` in a browser produces the following map:

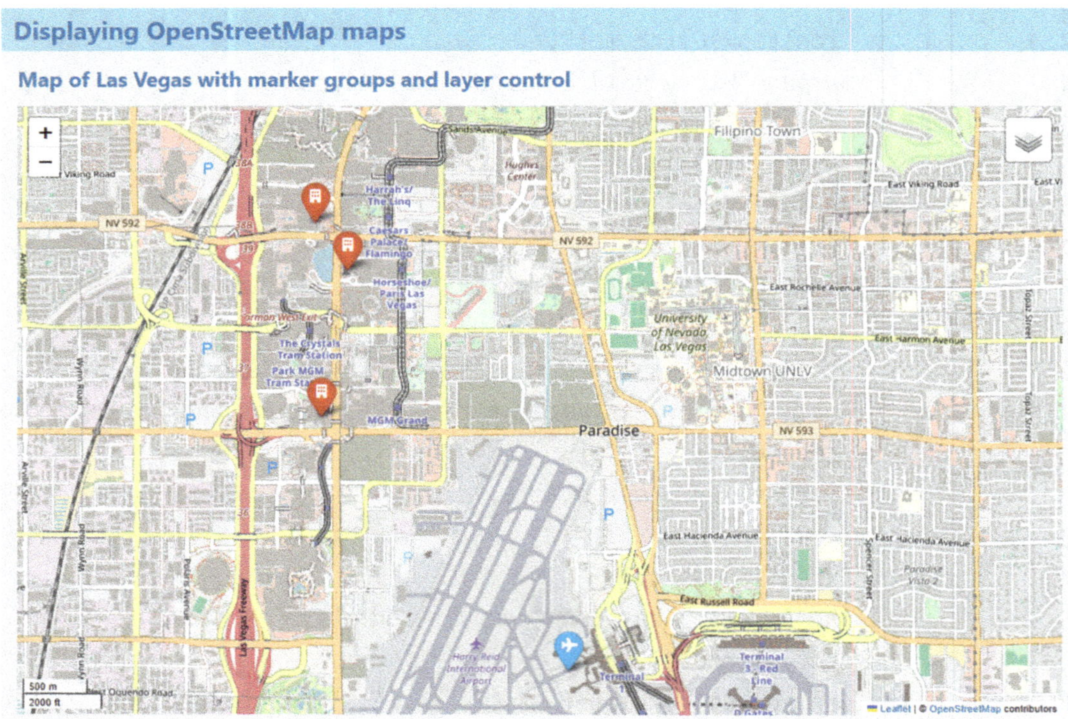

__Figure 5-8.__ Map created with Folium using OpenStreetMap tiles. Data © OpenStreetMap contributors, licensed under the ODbL

The layer control icon at the top right of the map controls whether or not the map elements of the "Casinos hotels" and "Airports" groups are displayed (a menu is displayed when the mouse cursor hovers over this icon). In the view below, the first group has been hidden:

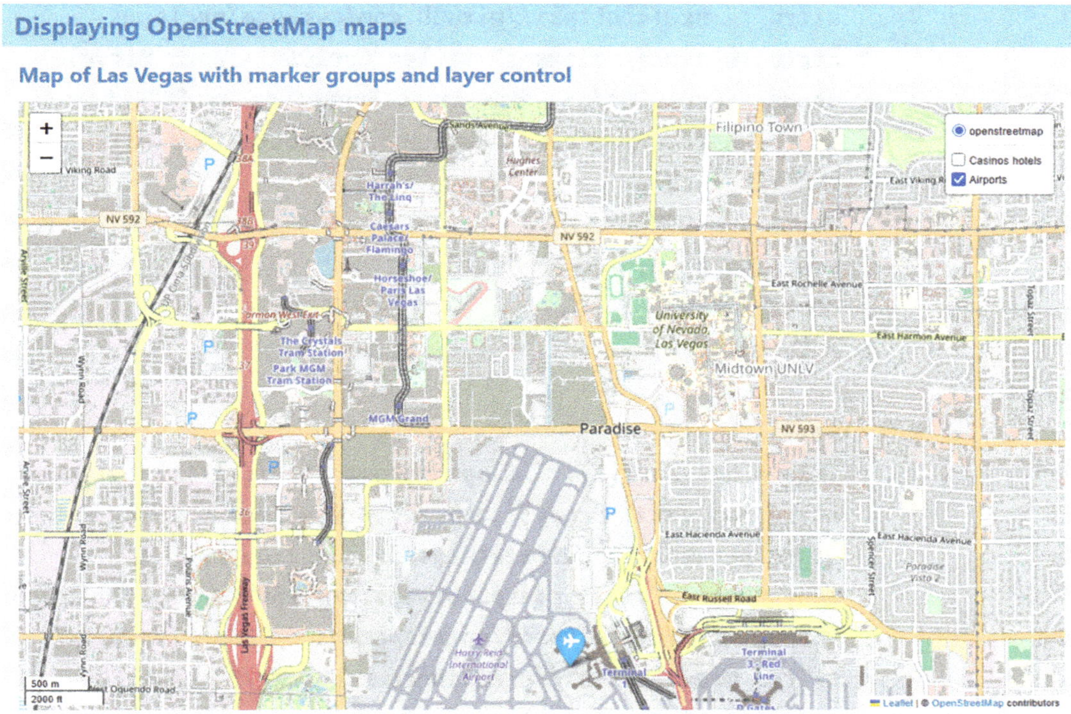

Figure 5-9. *Map created with Folium using OpenStreetMap tiles. Data © OpenStreetMap contributors, licensed under the ODbL*

Tile layers added to the map can also be shown or hidden using a layer control. Now we will insert the following code (in **bold**) into the preceding example, so that two additional tile layers are added to the map:

```
@bp.route('/sm/map4')
def map4():

...

        ).add_to(airport_group)

        folium.TileLayer(
            tiles='https://{s}.tile-cyclosm.openstreetmap.fr/cyclosm/
            {z}/{x}/{y}.png',
```

```
40              attr='<a href="https://github.com/cyclosm/cyclosm-cartocss-style/releases" title="CyclOSM - Open Bicycle render">CyclOSM</a> | Map data: &copy; <a href="https://www.openstreetmap.org/copyright">OpenStreetMap</a> contributors',
41              name="CyclOSM",
42              overlay=False,
43              control=True
44          ).add_to(m)
45
46          folium.TileLayer(
47              tiles='https://{s}.tile.opentopomap.org/{z}/{x}/{y}.png',
48              attr='Map data: &copy; <a href="https://www.openstreetmap.org/copyright">OpenStreetMap</a> contributors, <a href="http://viewfinderpanoramas.org">SRTM</a> | Map style: &copy; <a href="https://opentopomap.org">OpenTopoMap</a> (<a href="https://creativecommons.org/licenses/by-sa/3.0/">CC-BY-SA</a>)',
49              name="OpenTopoMap",
50              overlay=False,
51              control=True
52          ).add_to(m)
53
54          folium.LayerControl().add_to(m)
...
```

- **Lines 38 to 44**: A tile layer is created with a tile set dedicated to bicycle lanes (CyclOSM) and added to the map. The `overlay=False` parameter indicates that this new layer is added as a base layer, and the `control=True` parameter specifies that it is taken into account by the layer control on line 55 (this parameter could have been omitted, as its default value is `True`).
- **Lines 46 to 52**: Another tile layer is created with a tile set including natural features (OpenTopoMap) and added to the map. It is also added as a base layer and taken into account by the layer control.

This modification produces the following map, with the layer control icon at top right, allowing the user to choose which of the three tile layers to display (and also to specify which marker groups should be visible or not):

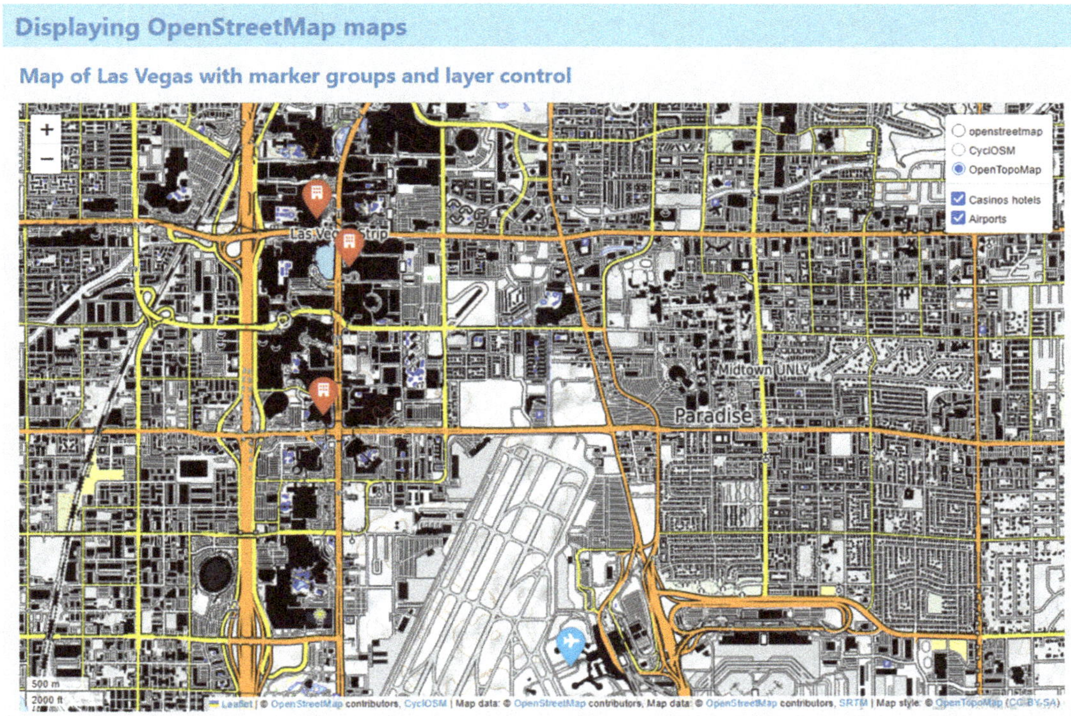

Figure 5-10. *Map created with Folium using OpenTopoMap tiles. Tiles © OpenTopoMap (CC BY-SA). Data © OpenStreetMap contributors (ODbL), SRTM*

By selecting another tile layer, the following map is obtained:

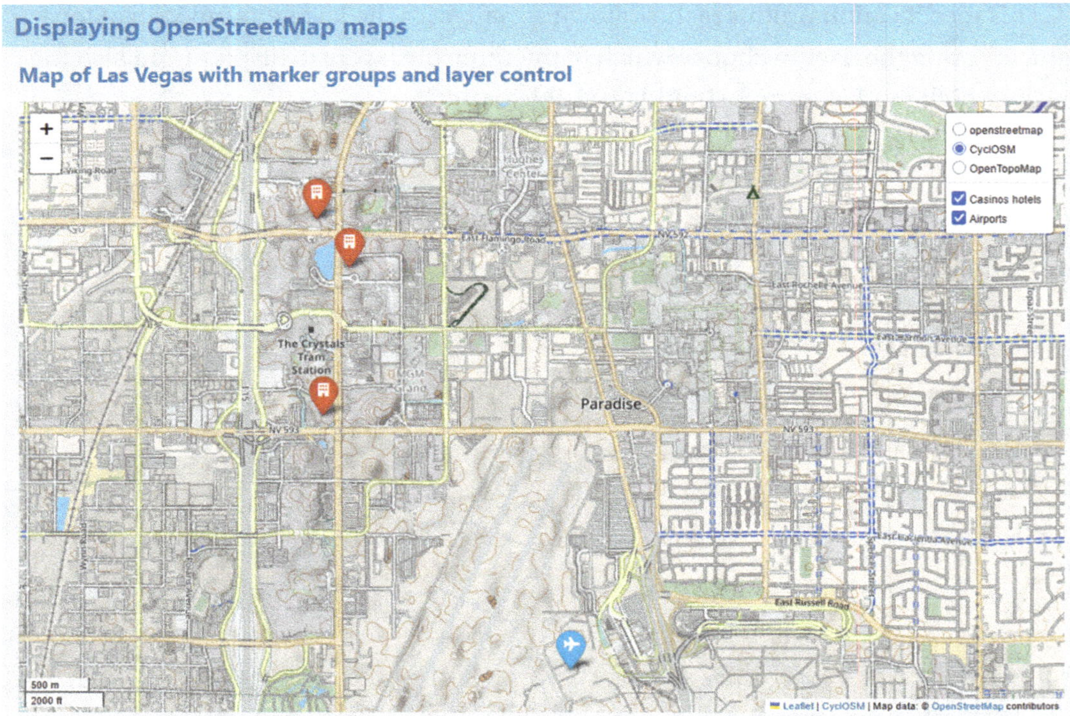

Figure 5-11. *Map created with Folium using CyclOSM tiles. Tiles © CyclOSM contributors. Data © OpenStreetMap contributors (ODbL). CyclOSM is an open source project. See* `https://cyclosm.org` *for details*

Note When you create a layer control with Folium, you cannot specify the groups or tile layers with which it is associated. Consequently, only one layer control can be created for the map, and it is associated with all the feature groups and tile layers defined (with Leaflet, on the other hand, you can create several layer controls and choose which groups or tile layers to associate with each of them). To work around this limitation, a plugin called **groupedlayercontrol** is available (see `https://python-visualization.github.io/folium/latest/user_guide/plugins/grouped_layer_control.html`).

Overlaying Images

Folium also allows you to overlay images on maps. The position, size, and opacity of the image can be specified using parameters.

Here is an example of a statement creating an image overlay and adding it to a map:

```
1   folium.raster_layers.ImageOverlay(
2       image="prototypes/static/Sphere.jpg",
3       bounds=[[36.08, -115.1318], [36.12, -115.1118]]
4   ).add_to(m)
```

- **Lines 1 to 4**: An object of type `folium.raster_layers.ImageOverlay` is created. The `image` parameter specifies the path to the image file (in this case, a local file), and the `bounds` parameter defines the geographic position and size of the image on the map: a list containing the coordinates of the bottom-left and top-right corners of a rectangle into which the image is adjusted (the image may be stretched or shrunk).
- **Line 4**: The image overlay object is then added to the map.

To obtain a new map with an image of the Stratosphere Tower overlaid, we will again build upon the code created in the previous section. In the `simple_maps.py` blueprint, duplicate the code for the `/sm/map4` route (to preserve the original). In the copy, rename the route to `/sm/map5` and the view function to `map5()`, and then insert or modify the code as shown in **bold** below:

```
@bp.route('/sm/map5')
def map5():

...

        ).add_to(m)

        folium.raster_layers.ImageOverlay(
            image="prototypes/static/Stratosphere.jpg",
            name="Stratosphere Tower",
            bounds=[ [36.083, -115.1330], [36.113, -115.1131]],
            opacity=0.5,
```

```
59              alt="Stratosphere Tower",
60              control=True
61          ).add_to(m)
62
63          folium.LayerControl().add_to(m)
64
65          return render_template('maps/page_with_map.html',
66               name="Map of Las Vegas with image overlay",
67               map_html=m._repr_html_())
68
69      except Exception as e:
70          flash("An error has been encountered:<br/>" + str(e), category="error")
71          return render_template('maps/page_with_map.html', name="No map to display")
```

- **Lines 54 to 61**: These lines create an object of type `folium.raster_layers.ImageOverlay` and add it to the map.
    - The `image` parameter defines where to find the image to be superimposed on the map. Here, a local file is used, but URLs are also supported (to download images dynamically from the Internet).

      This image can be found at `https://github.com/Apress/Modern-Web-Cartography` (file: `carto_osm/prototypes/static/Stratosphere.jpg`).

      Retrieve it and download it into the `prototypes/static` directory of the workspace.
    - The `name` parameter specifies the name of the ImageOverlay, as it will appear in the layer control panel (allowing the user to show or hide show the image at will).
    - The `bounds` parameter defines the rectangle (by the coordinates of its bottom left and top right corners) in which the image is placed and adjusted.

- The `opacity` parameter defines the opacity level of the image.
- The `alt` parameter defines the HTML "alt" attribute, i.e., a text to be displayed instead of the image in the event that the browser is unable to display images.
- The `control` parameter specifies whether or not the image should appear in the layer control, allowing the user to toggle its visibility on the map.

Launching the server and opening the address `http://127.0.0.1:5000/sm/map5` in a browser and hovering over the layer control produces the following map:

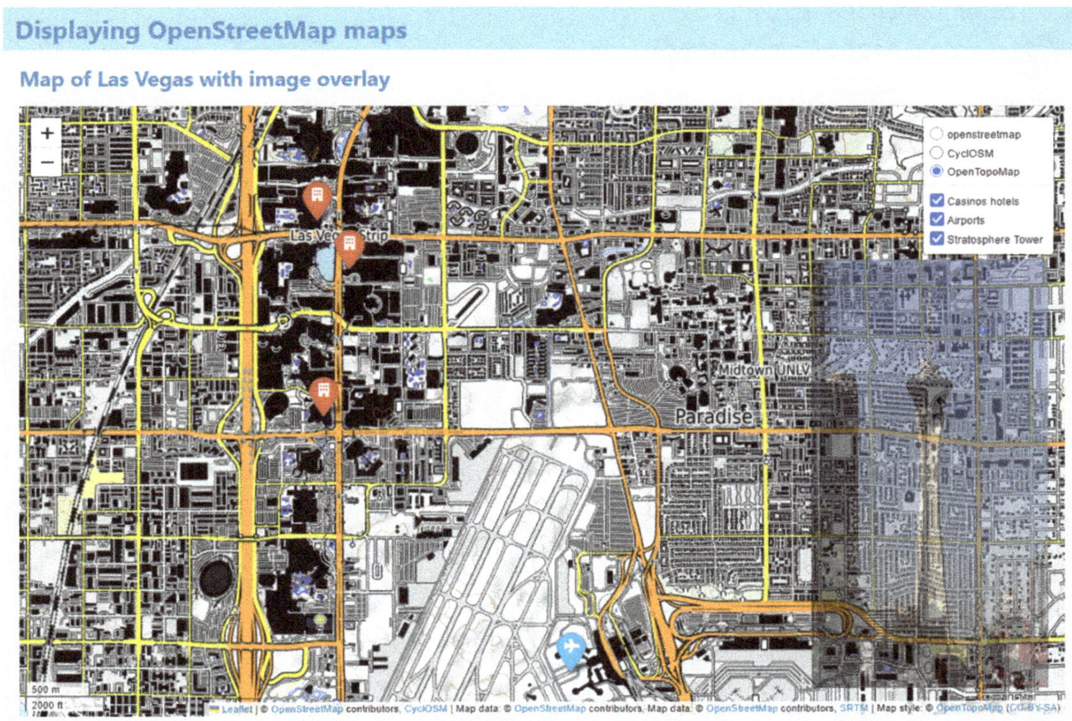

Figure 5-12. *Map created with Folium using OpenTopoMap tiles. Tiles © OpenTopoMap (CC BY-SA). Data © OpenStreetMap contributors (ODbL), SRTM. Image: "Stratosphere, Las Vegas" by Clément Bardot, licensed under CC BY-SA 4.0 (`https://creativecommons.org/`)*

A new entry "Stratosphere Tower" is available in the layer control panel to show or hide the image.

Drawing Polylines and Geometric Figures

Folium lets you add **polylines** (lines made up of several contiguous segments) to maps. These lines are defined by the coordinates of the points forming the vertices and the extremities of the line. A color, line thickness, smoothing level, opacity level, and text to be displayed in a popup or tooltip can be associated with a polyline.

Geometric figures such as circles, rectangles, or polygons can also be added to the maps. Line color, thickness and opacity, area fill color and opacity, and text to be displayed in a popup or tooltip can be specified for these figures.

Here is an example of how to create a polyline and add it to a map:

```
1   line_coordinates = [
2       (43.312071, 5.364504),
    ...
10      (41.922220, 8.740826)
11  ]
12
13  folium.PolyLine(line_coordinates, tooltip="Marseille-Ajaccio").add_to(m)
```

- **Lines 1 to 11**: A list of geographic coordinates (in the form of latitude-longitude pairs) is created.
- **Line 13**: A `folium.PolyLine` object is created:
    - The first parameter is the list of coordinates (for the polyline vertices and extremities).
    - The `tooltip` parameter specifies the text to be displayed in a tooltip (when the mouse cursor hovers over the polyline).
    - Finally, the polyline is added to the map.

To obtain a map with a polyline, the following code must be added to the `simple_maps.py` blueprint:

```
@bp.route('/sm/map6')
def map6():
    try:
```

```
4           m = folium.Map(location=[ 42.480656, 6.822510], zoom_start=8)
5
6           line_coordinates = [
7               (43.312071, 5.364504),
8               (43.297080, 5.357294),
9               (43.285398, 5.335579),
10              (43.201450, 5.299530),
11              (42.350882, 5.927124),
12              (41.816822, 8.640747),
13              (41.883423, 8.741683),
14              (41.906934, 8.756790),
15              (41.922220, 8.740825)
16          ]
17
18          folium.PolyLine(line_coordinates,
19              color='blue',
20              weight=4,
21              opacity=0.7,
22              tooltip="Marseille-Ajaccio"
23          ).add_to(m)
24
25          return render_template('maps/page_with_map.html',
26              name="Map of the Marseille-Ajaccio shipping route",
27              map_html=m._repr_html_())
28
29      except Exception as e:
30          flash("An error has been encountered:<br/>" + str(e),
            category="error")
31          return render_template('maps/page_with_map.html', name="No map
            to display")
```

- **Line 4**: A map is built and centered on a geographical position between Provence and Corsica (in France), with an initial zoom level of 8.

- **Lines 6 to 16**: A list is created with the geographic coordinates (latitude and longitude) of the vertices and extremities of the polyline to be drawn.
- **Lines 18 to 23**: A polyline is created using the previous coordinate list, then added to the map. Additional options also specify its color, line thickness and opacity, and an associated tooltip.
- **Lines 25 to 27**: A web page is rendered using the `page_with_map.html` template. A name and the previously constructed map, converted to HTML and JavaScript code, are passed as arguments.

Launching the server and opening the URL `http://127.0.0.1:5000/sm/map6` in the browser produces the following map, on which the polyline is drawn (the text "Marseille-Ajaccio" appears when the mouse cursor passes over the polyline):

***Figure 5-13.** Map created with Folium using OpenStreetMap tiles. Data © OpenStreetMap contributors, licensed under the ODbL*

Let's add a circle and a polygon to this map. To do this, we will insert the following statements (in **bold**) into the previous code:

```
...
          ).add_to(m)

          folium.Circle(
              location=[43.085840, 7.943115],
              radius=50000,
              stroke=False,
              fill=True,
              fill_color="cornflowerblue",
              fill_opacity=0.6,
              popup="Blue circle 100 km in diameter",
              tooltip="Click me!"
          ).add_to(m)

          locations = [
              [42.163861, 4.916382],
              [42.314335, 5.619507],
              [41.972203, 6.690674],
              [41.496698, 6.503906],
              [41.529605, 5.234985]
          ]

          folium.Polygon(
              locations=locations,
              color="orange",
              weight=6,
              opacity=1,
              fill=True,
              fill_color="yellow",
              fill_opacity=0.5,
              popup="Yellow polygon",
              tooltip="Click me!"
```

```
54          ).add_to(m)
55
56          return render_template('maps/page_with_map.html',
57              name="Map of the Marseille-Ajaccio shipping route",
58              map_html=m._repr_html_())
...
```

- **Lines 25 to 34**: A circle is created (using the `folium.Circle` object type) with a radius of 50,000 meters (50 km or 31 miles), with no outline but filled with blue, with an associated tooltip and popup. Then it is added to the map.
- **Lines 36 to 42**: A list of geographic coordinates is defined, in order to build a polygon in the next statement.
- **Lines 44 to 54**: A polygon is created (using the `folium.Polygon` object type) with the preceding list of coordinates. It is assigned an outline color, a fill color, line thickness (for the outline), opacity levels (for both the outline and fill), a popup, and a tooltip. Then it is added to the map.

Restarting the server and reopening the URL `http://127.0.0.1:5000/sm/map6` in the browser produces the following map, on which the circle and the polygon are drawn (click on the circle or the polygon to see their popups):

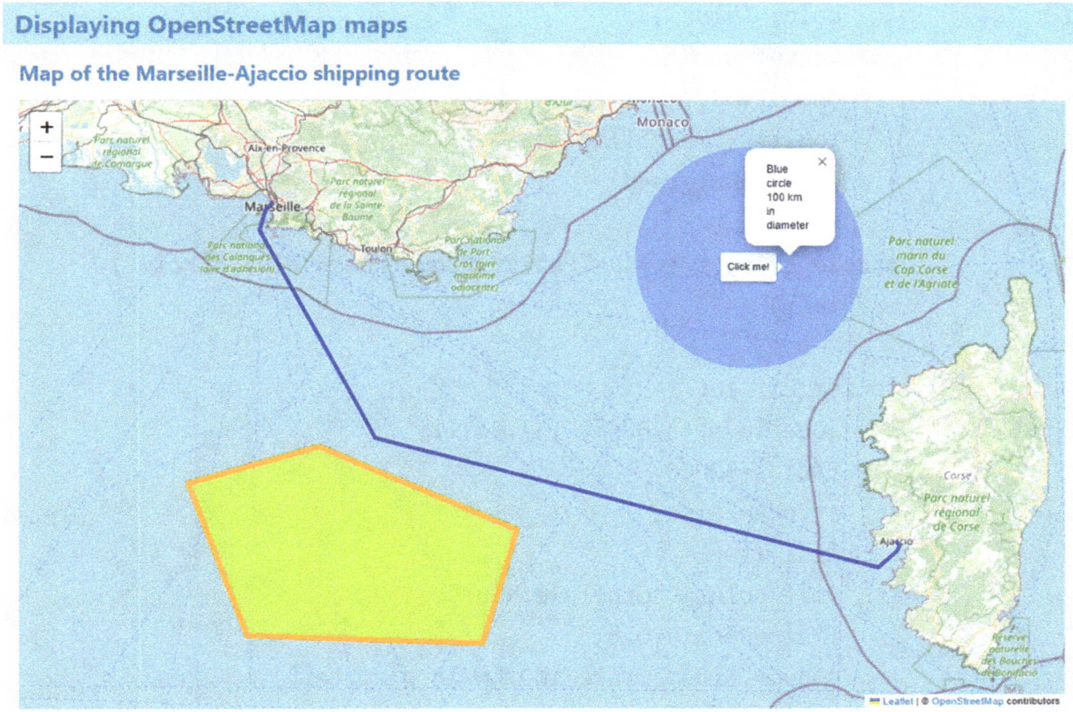

Figure 5-14. *Map created with Folium using OpenStreetMap tiles. Data © OpenStreetMap contributors, licensed under the ODbL*

Finally, we will add a scale to the map and a layer control to show or hide the three map elements (changes in **bold**):

```
...
    m = folium.Map(location=[42.480656, 6.822510], zoom_start=8,
    control_scale=True)
...
    ]

    line_group = folium.FeatureGroup(name="Marseille-Ajaccio
    route").add_to(m)

    folium.PolyLine(line_coordinates,
```

```
        color='blue',
        weight=4,
        opacity=0.7,
        tooltip="Marseille-Ajaccio"
    ).add_to(line_group)

    circle_group = folium.FeatureGroup(name="Blue circle").add_to(m)

    folium.Circle(
        location=[43.085840, 7.943115],
        radius=50000,
        stroke=False,
        fill=True,
        fill_color="cornflowerblue",
        fill_opacity=0.6,
        popup="Blue circle 100 km in diameter",
        tooltip="Click me!"
    ).add_to(circle_group)

    locations = [
        [42.163861, 4.916382],
        [42.314335, 5.619507],
        [41.972203, 6.690674],
        [41.496698, 6.503906],
        [41.529605, 5.234985]
    ]

    polygon_group = folium.FeatureGroup(name="Yellow polygon").add_to(m)

    folium.Polygon(
        locations=locations,
        color="orange",
        weight=6,
```

```
54              opacity=1,
55              fill=True,
56              fill_color="yellow",
57              fill_opacity=0.5,
58              popup="Yellow polygon",
59              tooltip="Click me!"
60          ).add_to(polygon_group)
61
62          folium.LayerControl().add_to(m)
63
64          return render_template('maps/page_with_map.html',
65              name="Map of the Marseille-Ajaccio shipping route",
66              map_html=m._repr_html_())
...
```

- **Line 4**: A scale is added to the map.
- **Line 18**: A feature group is created (to contain the polyline) and added to the map.
- **Line 25**: The polyline is added to the previous feature group.
- **Line 27**: A feature group is created (to contain the circle) and added to the map.
- **Line 38**: The circle is added to the previous feature group.
- **Line 48**: A feature group is created (to contain the polygon) and added to the map.
- **Line 60**: The polygon is added to the previous feature group.
- **Line 62**: A layer control is added to the map. This layer control enables the user to show or hide the polyline, circle, and polygon.

By refreshing the map and placing the mouse cursor over the layer control at the top right, we obtain the following map:

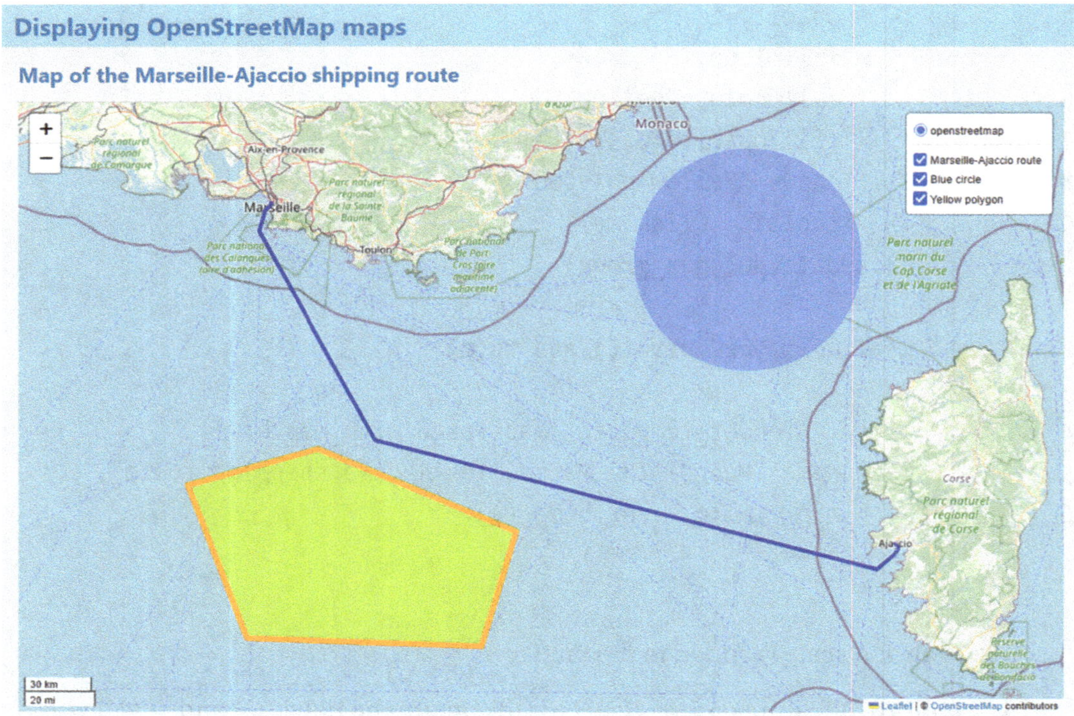

Figure 5-15. *Map created with Folium using OpenStreetMap tiles. Data © OpenStreetMap contributors, licensed under the ODbL*

The user can show or hide the three map elements using the layer control.

Adding GeoJSON and TopoJSON Data

Folium enables you to overlay maps with map elements generated from data supplied in **GeoJSON** or **TopoJSON** format. These are JSON-format datasets with specific structures designed to organize geographic data. GeoJSON and TopoJSON are standards for the exchange of geographic data between mapping tools.

Depending on the contents of the GeoJSON/TopoJSON data and the options used, markers, polylines, polygons (delimiting area contours), and more may be added to the map. These elements are overlaid as an extra layer that can be shown or hidden via a layer control.

GeoJSON

Using GeoJSON data, a `folium.GeoJson` object can be created; then this object can be added to the map, thus overlaying map elements (markers, polylines, polygons, etc.) on the map. Here is an example of a statement for enhancing a map with GeoJSON data:

```
folium.GeoJson(geojson_data).add_to(m)
```

- A `folium.GeoJson` object is created using GeoJSON dataset in a variable named `geojson_data` (the GeoJSON data must have previously been placed in this variable).

  Then the object is added to the map, thus generating map elements such as markers, polylines, or polygons on the map.

In the example below, the GeoJSON data to be overlaid on the map will be downloaded from the Internet (it is the same GeoJSON data that was used with Leaflet in the preceding chapter, containing the outlines of all the countries of the world). To do this, we first need to install the `requests` Python library, which enables HTTP requests to be submitted from Python code. To install it, run the following command from the VS Code terminal in the `carto_osm` directory:

```
pip install requests
```

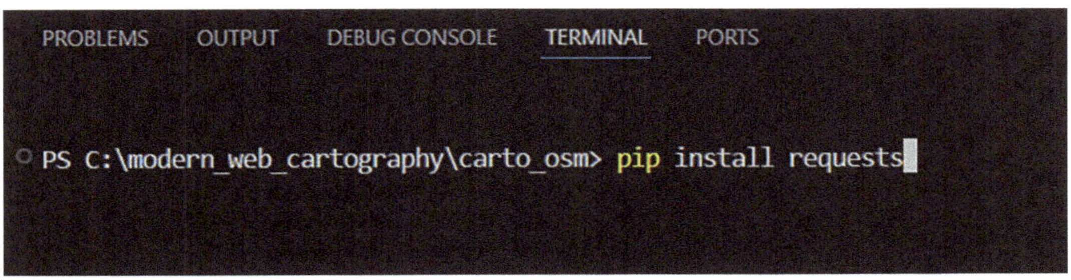

Figure 5-16. *Screenshot of Visual Studio Code. © Microsoft Corporation (`http://www.microsoft.com`). All rights reserved. Used with permission from Microsoft*

To obtain this new map, the following code must be added to the `simple_maps.py` blueprint:

```
@bp.route('/sm/map7')
def map7():
    import requests
```

```

    try:
        geojson_data = requests.get(
            "https://raw.githubusercontent.com/python-visualization/folium-example-data/main/world_countries.json"
        ).json()

        m = folium.Map(location=[15, 0],
            zoom_start=2)

        folium.GeoJson(geojson_data,
            name="Countries",
            style_function=lambda feature: {
                'color': 'darkgreen',
                'weight': 2,
                'fillColor': 'green',
                'fillOpacity': 0.5
            },
            tooltip= folium.GeoJsonTooltip(fields=['name'], aliases=['Country:'])
        ).add_to(m)

        folium.LayerControl().add_to(m)

        return render_template('maps/page_with_map.html',
            name="Map of the world with country outlines highlighted",
            map_html=m._repr_html_())

    except Exception as e:
        flash("An error has been encountered:<br/>" + str(e), category="error")
        return render_template('maps/page_with_map.html', name="No map to display")
```

- **Line 3**: The `requests` library is imported. It will be used to send an HTTP request to a website.
- **Lines 6 to 8**: An HTTP request is submitted to the `https://raw.githubusercontent.com/python-visualization/folium-example-data/main/world_countries.json` website to download GeoJSON data describing the contours of the countries of the world. The `json()` method is applied to the response, parsing it and placing its data into an object (containing dictionaries and lists reproducing the GeoJSON structure of the data).

  In case this data is no longer available on the Internet, it has also been downloaded and can be found at `https://github.com/Apress/Modern-Web-Cartography` (file: `carto_osm/prototypes/static/World_countries.json`).
- **Lines 10 and 11**: A map of the world is built.
- **Lines 13 to 22**: With the object created with GeoJSON data at line 8, a `folium.GeoJson` object is created and added to the map.
  - The `name` attribute specifies the name to be used in the layer control panel to show or hide the map elements generated from the GeoJSON data.
  - The `style_function` attribute defines the layout properties of the map elements generated from the GeoJSON data: contour color and weight, fill color, and opacity.
  - A tooltip will display the country name when the mouse cursor hovers over it.
- **Line 24**: A layer control is added to the map, so that the layer containing country contours and fills can be shown or hidden at will.
- **Lines 26 to 28**: A web page is rendered using the `page_with_map.html` template. A name and the previously constructed map, converted to HTML and JavaScript code, are passed as arguments.

Launch the server, open the URL `http://127.0.0.1:5000/sm/map7` in the browser, and place the mouse cursor over a country on the map to obtain the following map:

Figure 5-17. *Map created with Folium using OpenStreetMap tiles. Data © OpenStreetMap contributors, licensed under the ODbL. Country boundaries from the Folium example dataset (MIT License), source:* `https://github.com/python-visualization/folium-example-data`

Country contours and fills can be shown or hidden using the layer control icon at the top right-hand corner.

TopoJSON

TopoJSON data can also be used in a way very similar to GeoJSON:

- Unlike Leaflet, Folium provides a function named `folium.TopoJson()` that accepts TopoJSON data as an input - therefore, no conversion to GeoJSON is needed before adding data to a map. Additional options can be used to specify how the data must be rendered on a map.

- The resulting TopoJSON object can then be added to a map to enrich it. Depending on the content of the TopoJSON data and the options used with `folium.TopoJson()`, various map elements such as markers, polylines, and polygons may be added to the map.

Here is an example:

```
folium.TopoJson(topojson_data,
    object_path="objects.counties"
).add_to(m)
```

- A `folium.TopoJson` object is created using TopoJSON dataset in a variable named `topojson_data` (the TopoJSON data must have previously been placed in this variable).

  The `object_path` parameter specifies which section of the TopoJSON data is selected: only the geometries of this section will be extracted and included in the TopoJSON object.

  Then the object is added to the map object, thus generating map elements such as markers, polylines, or polygons on the map.

In the example below, the TopoJSON data to be overlaid on the map will be downloaded from the Internet (it is the same data that was used with Leaflet in the preceding chapter, available at `https://cdn.jsdelivr.net/npm/us-atlas@3/counties-10m.json` and containing the outlines of the US counties). To download the data from the Internet, the `requests` Python library must have been installed.

The following code must be added to the `simple_maps.py` blueprint for this example:

```
@bp.route('/sm/map8')
def map8():
    import requests

    try:
        topojson_data = requests.get(
            "https://cdn.jsdelivr.net/npm/us-atlas@3/counties-10m.json"
        ).json()

        m = folium.Map(location=[37.8, -96],
```

```
11              zoom_start=4)
12
13          folium.TopoJson(topojson_data,
14              object_path="objects.counties",
15              name="US counties",
16              style_function=lambda feature: {
17                  'color': 'green',
18                  'weight': 2,
19                  'fillColor': 'lightgreen',
20                  'fillOpacity': 0.5
21              },
22              tooltip=folium.GeoJsonTooltip(fields=['name'],
                aliases=['County:'])
23          ).add_to(m)
24
25          folium.LayerControl().add_to(m)
26
27          return render_template('maps/page_with_map.html',
28              name = "Map of the US with county outlines",
29              map_html=m._repr_html_())
30
31      except Exception as e:
32          flash("An error has been encountered:<br/>" + str(e),
            category="error")
33          return render_template('maps/page_with_map.html', name = "No
            map to display")
```

- **Lines 6 to 8**: The TopoJSON data is downloaded from the Internet, parsed, and set in a variable.

  In case this data is no longer available on the Internet, it has also been downloaded and can be found at `https://github.com/Apress/Modern-Web-Cartography` (file: `carto_osm/prototypes/static/US_counties.json`).

- **Lines 10 and 11**: A map is built.
- **Lines 13 to 23**: A `folium.TopoJson` object is created with the TopoJSON data, then added to the map.
    - The `object_path` parameter specifies the path of the desired objects in the TopoJSON structure: only the geometries of the `counties` section will be retained.
    - The `name` parameter will be used in the layer control panel.
    - The `style_function` parameter specifies various layout features (outline and fill colors, line thickness, fill opacity) for the map elements to be added to the map.
    - A tooltip will display the county name when the mouse hovers over it.
- **Line 25**: A layer control is created and added to the map.
- **Lines 27 to 29**: An HTML page is rendered from the `page_with_map.html` template. A name and the map, converted to HTML and JavaScript code, are passed as arguments.

Launch the server and open the URL `http://127.0.0.1:5000/sm/map8` in the browser to get the following map:

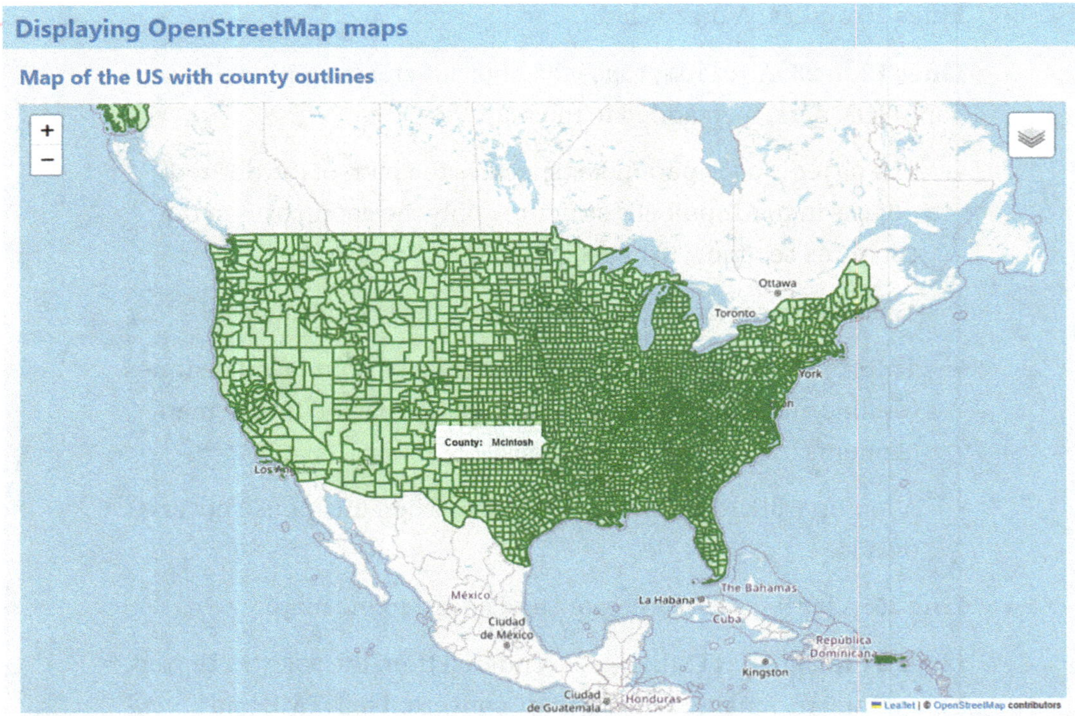

Figure 5-18. *Map created with Folium using OpenStreetMap tiles. Data © OpenStreetMap contributors, licensed under the ODbL. County boundaries © Mike Bostock and contributors, MIT License*

Building Choropleth Maps

Choropleth maps are thematic maps in which areas are shaded or colored to represent an economic, statistical, or other quantitative value associated with each area. Folium allows you to build maps of this type. The elements that make up a choropleth map (area outlines, color shading, associated popups, etc.) are added to the map as an additional layer, which a layer control can enable the user to show or hide.

Unlike Leaflet, Folium provides a `folium.Choropleth` object that facilitates the creation of choropleth maps. Therefore, this is easier to do with Folium than with Leaflet. Here is an example of a statement for creating a choropleth map:

```
folium.Choropleth(
    geo_data=state_geo,
    data=state_data,
```

```
    columns=["Country", "2022"],
    key_on="feature.properties.NAME",
    fill_color="BuGn",
    bins=16
).add_to(m)
```

- These statements define an object of type `folium.Choropleth` and add it to a map:
    - The `geo_data` parameter specifies the object containing the outlines of the areas to be displayed on the map in GeoJSON format (this parameter can also be set with a URL or a file path). This outline data must have previously been placed in the `state_geo` variable.

      It is also possible to use TopoJSON data with this parameter (in that case, an additional parameter `topojson` must be supplied to specify the reference of the TopoJSON object to use). Internally, Folium will convert the TopoJSON data to GeoJSON before adding it to the map.
    - The `data` parameter specifies the object (a multidimensional array of type Pandas DataFrame) containing the economic or statistical data to be represented on the map as shades of color. Data must have previously been placed in the `state_data` variable.
    - The `columns` parameter is used to specify which two columns of the preceding multidimensional array object must be used:
        - The first contains the area names to be correlated with the area names in the object specified with the `geo_data` parameter.
        - The second contains the data (economic or statistical) to be represented in the form of color shades on the map.

      The two columns are specified using their headers (first row contents).

- The `key_on` parameter is used to indicate which attribute, present in the object specified with the `geo_data` parameter, contains the area names (to be correlated with the area names in the first column (specified by the `columns` parameter) of the multidimensional array object specified with the `data` parameter).
- The `fill_color` parameter defines the color palette used to fill the areas based on their associated economic/statistical data (the higher the data value, the darker the area fill shade).
- The `bins` parameter defines the number of color shades to be used to fill map areas (this parameter can also receive a sequence of numeric thresholds that determine the exact breakpoints between color shades).
- On line 8, the choropleth object is added to the map object.

Please note that the country names in the country outline data object (`geo_data` parameter) and in the economic or statistical data object (`data` parameter) must match exactly; otherwise correlation is not possible (e.g., if you have "Germany" on one side and "Deutschland" on the other, Folium will fail to assign the correct color shade to that country).

Now, we will create an example of a choropleth map representing the median living standards of the countries of the European Union.

First, we need to install the additional `pandas` library, which will enable us to load CSV format data from a local file into a Pandas DataFrame object (a multidimensional array). To do this, run the following command from the VS Code terminal in the `carto_osm` directory:

```
pip install pandas
```

To build this new map, the following code must be added to the `simple_maps.py` blueprint:

```
@bp.route('/sm/map9')
def map9():
    import json, pandas

```

```
5       try:
6           with open("prototypes/static/europe.geojson") as f:
7               state_geo = json.loads(f.read())
8
9           state_data = pandas.read_csv("prototypes/static/EU_median_living_standards.csv")
10
11          m = folium.Map(location=[53, 12], zoom_start=4)
12
13          folium.Choropleth(
14              geo_data= state_geo,
15              data=state_data,
16              columns=["Country", "2022"],
17              key_on="feature.properties.NAME",
18              fill_color="YlGn",
19              bins=16,
20              fill_opacity=0.7,
21              line_opacity=0.2,
22              name="choropleth",
23              legend_name="Median living standards (€)"
24          ).add_to(m)
25
26          return render_template('maps/page_with_map.html',
27              name="Choropleth map of median living standards in the European Union",
28              map_html=m._repr_html_())
29
30      except Exception as e:
31          flash("An error has been encountered:<br/>" + str(e), category="error")
32          return render_template('maps/page_with_map.html', name="No map to display")
```

- **Line 3**: The json library (used to parse the GeoJSON file of country contours) and pandas library (used to read and parse a CSV format file of the median living standards) are imported.

- **Lines 6 and 7**: A local GeoJSON file, containing the contours of the European Union countries, is imported (these data could also have been imported from the Internet using a URL). The data is parsed and placed in a JSON-structured object (using the `json.loads()` function).

  GeoJSON outlines of the states of the European Union are available at `https://github.com/leakyMirror/map-of-europe/blob/master/GeoJSON/europe.geojson?short_path=e6b7b33`. The file was downloaded, reworked with a text editor (to keep only the contours of the countries belonging to the European Union). The resulting file can be found at `https://github.com/Apress/Modern-Web-Cartography` (file: `carto_osm/prototypes/static/europe.geojson`). Retrieve it and download it into the `prototypes/static` directory of the workspace.

- **Line 9**: The local CSV (Comma-Separated Values) format file, containing median living standards by European Union country is read (these data could also have been imported from the Internet using a URL). The data is parsed and placed in a Pandas DataFrame object (i.e., a multidimensional array).

  Data on median living standards by country in the European Union are available on the INSEE website (`https://www.insee.fr/fr/statistiques/2830158`) and are for the year 2022. These values were downloaded; their format was reworked with a text editor (to make them easier to use). The structure of the resulting file is as follows:

```
Country,2022
Austria,31443
Belgium,28997
Bulgaria,6523
Cyprus,18873
Croatia,9873
Czech Republic,13656
...
Sweden,27250
```

This file can be found at `https://github.com/Apress/Modern-Web-Cartography` (file: `carto_osm/prototypes/static/EU_median_living_standards.csv`). Retrieve it and download it into the `prototypes/static` directory of the workspace.

- **Line 11**: A map of Europe is created.
- **Lines 13 to 24**: A `folium.Choropleth` object is created with the country contours and median living standards per country obtained previously, then added to the map.

  The meaning of most of the parameters has already been explained just above. For the others:

  - The `fill_opacity` and `line_opacity` options define the opacity levels of the area outlines and fills on the map (depending on the opacity level used, the underlying map may remain more or less visible through transparency).
  - The `name` parameter defines the name assigned to the choropleth, which will be used in a layer control (to show or hide the choropleth elements).
  - The `legend_name` parameter specifies the name to be used for the legend.
- **Lines 26 to 28**: A web page is rendered using the `page_with_map.html` template. A name and the previously constructed map, converted to HTML and JSON code, are passed as arguments.

Launching the server and opening the URL `http://127.0.0.1:5000/sm/map9` in the browser produces the following map:

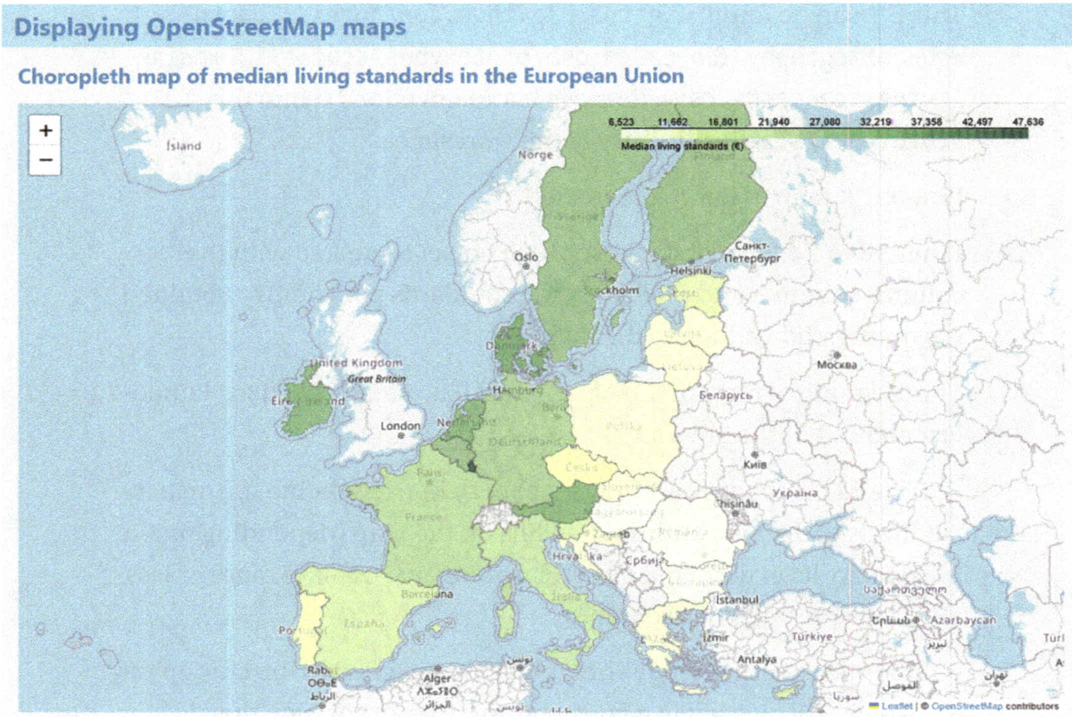

Figure 5-19. *Map created with Folium using OpenStreetMap tiles. Data © OpenStreetMap contributors, licensed under the ODbL. European country boundaries from the "Map of Europe" project by leakyMirror, licensed under the MIT License, source:* `https://github.com/leakyMirror/map-of-europe`*. European country unemployment rates from INSEE, Statistiques et Etudes, published under the Licence Ouverte v2.0, source:* `https://www.insee.fr/fr/statistiques/2830158`

In Chapter 8, we will learn how to add more interactivity to this type of map.

Injecting Leaflet Statements into Folium Maps to Add Interactivity

Unlike Leaflet, Folium does not natively support event handlers and therefore does not allow defining actions in response to user interactions. However, it is possible to achieve this level of interactivity by injecting custom JavaScript code, including Leaflet event handler statements, directly into Folium maps.

The following example demonstrates this technique:

```
1   leaflet_script = """
2       <script type="text/javascript">
3       window.onload = function() {
4           var map;
5           for (var key in window) {
6               if (window[key] instanceof L.Map) {
7                   map = window[key];
8                   break;
9               }
10          }
11
12          map.on('click', function(e) {
13              alert("You clicked at position " + e.latlng);
14          });
15      };
16      </script>
17  """
18
19  m.get_root().html.add_child(folium.Element(leaflet_script))
```

- **Lines 1 to 17**: A string containing JavaScript code (including Leaflet statements) is defined and set in a variable. This script will later be injected into the Folium map's HTML output.
    - **Line 3**: The `window.onload` function ensures that the code inside it (lines 4 to 14) is only executed after the entire web page, including the Folium map, has fully loaded. This is necessary to safely access the map object.
    - **Lines 4 to 10**: Since the map object is created by Folium in Python and its exact variable name isn't known in the JavaScript scope, a loop searches the global `window` object to find the map (i.e., the object that is an instance of `L.Map`).

- **Lines 12 to 14**: An event handler is added to the map: when the user clicks on the map, an alert box will display the latitude and longitude of the clicked position.

- **Line 19**: The JavaScript code is injected into the Folium map by adding it as a child element of the map's HTML structure.

We will now implement an interactive map using Folium:

- When the user clicks on the map, it will center and zoom in on the clicked point. Repeated clicks will continue to zoom in further.

- When the user presses the Esc key, the map will return to its initial zoom level.

To achieve this, add the following code at the end of the `simple_maps.py` blueprint:

```
@bp.route('/sm/map10')
def map10():

    leaflet_script = """
        <script type="text/javascript">
        window.onload = function() {
            var map;
            for (var key in window) {
                if (window[key] instanceof L.Map) {
                    map = window[key];
                    break;
                }
            }
            try {
                map.on('click', function(e) {
                    map.flyTo(e.latlng, map.getZoom() + 1);
                });

                map.on('keydown', function(e) {
                    if (e.originalEvent.key === 'Escape') {
                        map.setZoom(5);
                    }
```

```
23                  });
24              } catch(e) {
25                  if (typeof map !== 'undefined' && map) {map.remove();}
26
27                  const div = document.getElementById("map")
28                  if (div) {div.remove();}
29
30                  const elemDiv = document.createElement('div');
31                  elemDiv.innerHTML = `<p>${e}</p>`;
32                  document.body.appendChild(elemDiv);
33              }
34          };
35          </script>
36      """
37
38      try:
39          m = folium.Map(location=[21, 80], zoom_start=5)
40
41          m.get_root().html.add_child(folium.Element(leaflet_script))
42
43          return render_template('maps/page_with_map.html',
44              name="Map with interactivity added by injecting Leaflet
                statements",
45              map_html=m._repr_html_())
46
47      except Exception as e:
48          flash("An error has been encountered:<br/>" + str(e),
            category="error")
49          return render_template('maps/page_with_map.html', name = "No
            map to display")
```

- **Lines 4 to 36**: A string containing JavaScript code (including Leaflet statements) is defined and set in a variable. This script will later be injected into the Folium map's HTML output.

- **Line 6**: The `window.onload` function ensures that the code inside it (lines 7 to 33) is only executed after the entire web page, including the Folium map, has fully loaded.
- **Lines 7 to 13**: Because the map object is not created by Leaflet in JavaScript, but by Folium in Python, a loop searches the global `window` object to find the map.
- **Lines 15 to 17**: An event handler is added to the map: when the user clicks on the map, it centers on the clicked point and zooms in (the zoom level is increased by 1).
- **Lines 19 to 23**: Another event handler is added to the map: when the user presses the Esc key, the map returns to its initial zoom level (5).

- **Line 39**: A map is created (using Folium), centered on India.
- **Line 41**: The preceding JavaScript code (lines 5 to 35) is injected into the Folium map by adding it as a child element of the map's HTML structure.
- **Lines 43 to 45**: A web page is rendered using the `page_with_map.html` template. A name and the previously constructed map, converted to HTML and JSON code, are passed as arguments.

Launching the server and opening the URL `http://127.0.0.1:5000/sm/map10` in the browser produces a map of India.

Figure 5-20. Map created with Folium using OpenStreetMap tiles. Data © OpenStreetMap contributors, licensed under the ODbL

- When the user clicks on the map, it centers on the clicked point and zooms in. Repeated clicks continue to zoom in further.
- When the user presses the Esc key, the map returns to its initial zoom level.

Note The preceding code may seem complex, and some might be tempted to simply insert similar JavaScript code (with an event handler) into the HTML template. However, this approach won't work because Folium generates an <iframe> tag (used to embed one HTML document inside another) in which it places the Leaflet map.

If this JavaScript code were placed directly in the template (used to build the HTML page sent to the browser), it would not be able to access the map, since the map is located inside the inline frame dynamically created by Folium using the <iframe> tag.

Therefore, the JavaScript code must be inserted **inside the same iframe** as the map. This can be done by injecting the JavaScript code into the map's HTML structure from the Python code that creates the map with Folium.

Folium Extension Options

A number of Leaflet plugins have been packaged for use in Folium. The **Folium plugin catalog** is available at `https://python-visualization.github.io/folium/latest/user_guide/plugins.html`. These extensions already exist and are ready for use.

For example, with the **Leaflet.Terminator** plugin, you can overlay shaded night regions on the map, and they update continuously.

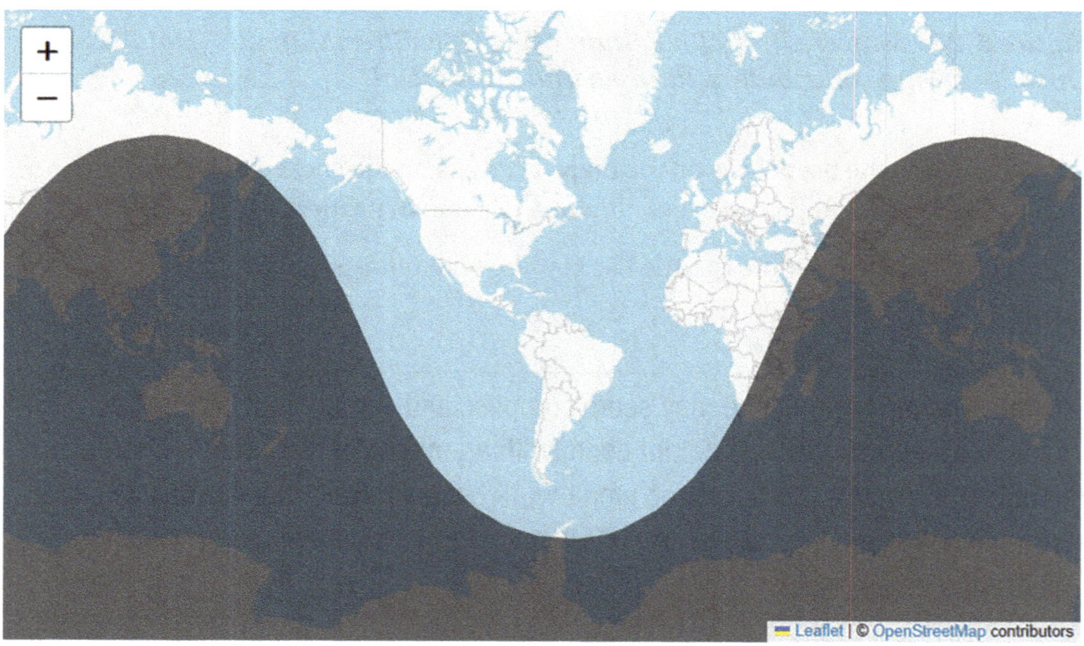

Figure 5-21. *Map created with Folium using OpenStreetMap tiles. Data © OpenStreetMap contributors, licensed under the ODbL*

The Python code to use it is very simple:

```
...
import folium.plugins
...
m = folium.Map(location=[45, 3], zoom_start=1)
    ...
folium.plugins.Terminator().add_to(m)
...
```

It is a little more complex, but also possible, to use Leaflet plugins that haven't been packaged for use in Folium. This can be accomplished by adding JavaScript code using the desired Leaflet plugin to the Folium map HTML code (`map.get_root().html`). This technique has been explained in the previous section.

Folium does not natively support creating new custom classes in the way Leaflet (JavaScript) does. However, you can still:

- Write JavaScript code involving any Leaflet feature you are interested in (including, if needed, custom Leaflet classes), and add it to the HTML code of your Folium maps, as explained in the previous section.
- Extend the `MacroElement` or `Element` classes from the Branca library (`https://python-visualization.github.io/branca/element.html#branca.element.MacroElement`) with your own JavaScript code and Leaflet statements, then add an instance of this class to the HTML section of the Folium map using `map.get_root().html`. This technique is beyond the scope of this book.

PART II

Building Advanced Maps with OpenStreetMap and Related Tools

In the preceding part, we looked separately at several of the tools that contribute to OpenStreetMap map-making. We are now going to use them together to create more ambitious map prototypes. In other words, it is no longer a matter of implementing basic examples function by function, but of rolling up our sleeves to create original, value-adding maps from a variety of sources.

Depending on the prototypes, the following tools will be used to build the maps:

- Leaflet with client-side JavaScript (in the browser).
- Folium with Python on the server side.

The aim is to learn how to use both techniques. However, whether the maps are created client-side with Leaflet or server-side with Folium, for the sake of realism, we will strive to build our prototypes as real web applications with Python and Flask on the server side.

The code for the prototypes in the following chapters can be found on the website `https://github.com/Apress/Modern-Web-Cartography` and in the `carto_osm` directory.

Preparing the Environment Used to Build and Display Advanced Maps

The prototypes will be implemented in a web environment based on **Python** and **Flask**, the setup of which is described in Appendix B, "Setting Up the Prototype Development and Test Environment." To be able to easily reproduce these prototypes following the instructions provided in this part, it is recommended to first carry out the setup steps explained in Appendix B.

The web applications for these prototypes will follow the Flask package structure, comprising:

- **Blueprints**: Python modules in which routes (i.e., URLs) and views (i.e., functions associated with routes and executed when the route is invoked by an HTTP request) are defined.
- **Templates**: HTML page models used to dynamically generate complete web pages by injecting data (via the **Jinja** engine) before sending them to the browser.

Depending on whether maps in prototypes are created with Leaflet or Folium:

- Either **Leaflet** map-building statements will be integrated in the JavaScript code of HTML templates.
- Or **Folium** map-building statements will be included in the Python code of blueprints (in the view functions).

A new blueprint, named `advanced_maps`, will be used to host the code used to build the new maps. The first step is to create it: just before the final `return app` statement in the `prototypes/__init__.py` file, add the two following statements (in **bold**) to import and declare the `advanced_maps` blueprint:

```
    ...

        app.register_blueprint(simple_maps.bp)

        from prototypes import advanced_maps
        app.register_blueprint(advanced_maps.bp)

    return app
```

Then, create an `advanced_maps.py` file in the `prototypes` directory of the workspace with the following contents, defining the new blueprint (after importing a few libraries):

```
1   import folium, folium.features
2   from flask import Blueprint, render_template, request, flash
3
4   bp = Blueprint('advanced_maps', __name__)
```

- As usual with Python, be careful with column alignments. The first statement (`import...`) must not be indented.
- Lines 1 and 2: Several libraries are imported globally here. They will be used multiple times in the following examples, but will not be imported again.

For most maps built with Folium, the versatile template `page_with_map.html`, already used in Chapter 5, "Creating Interactive Maps with Folium and Python," will be used. Its contents are as follows:

```
1   {% extends 'base.html' %}
2
3   {% block title %}{{ name }}{% endblock %}
4
5   {% block header %}
6   <h2>{{ name }}</h2>
7   {% endblock %}
8
9   {% block content %}
10  {{ map_html | safe }}
11  {% endblock %}
```

- Line 1 indicates that this template extends the page structure defined in the `base.html` base template (described in Appendix B, "Setting Up the Prototype Development and Test Environment"), by inserting new elements.
- Line 3: A specific title will be received as an argument from the calling Python function when the construction of an HTML page from this template is triggered. While generating an HTML page from

the template, the Jinja engine will replace the `name` variable with the value received from Python.

- Line 6: A specific header will be received as an argument from the calling Python function when the construction of an HTML page is triggered. The Jinja engine will replace the `name` variable with the value received from Python.
- Line 10: The HTML and JavaScript code for the Folium map will be received from the calling Python function. This line will be processed by the Jinja engine on the server:
    - The `map_html` variable will be replaced by the HTML and JavaScript code (including Leaflet statements) generated by Folium to render the map.
    - The `safe` filter is necessary to prevent Jinja from escaping the HTML/JavaScript content. Without it, Jinja would convert characters such as "<" and ">" into "<" and ">", causing the browser to display raw code instead of rendering the map.

For maps built with Leaflet, several templates containing specific Leaflet statements will be used, and a new base template named `base_ll.html` will be used to hold all the elements common to these templates. It will contain the same content as the `base.html` base template, with a few additional elements specific to Leaflet (in **bold**). Create it in the `prototypes/templates` directory with the following contents:

```
<!doctype html>
<html lang="en">
<head>
  <meta charset="utf-8">
  <title>{% block title %}{% endblock %} - Prototypes</title>
  <link rel="stylesheet" href="{{ url_for('static', filename='style.
  css') }}">
  <link rel="stylesheet" href="https://unpkg.com/leaflet@1.9.4/
  dist/leaflet.css" integrity="sha256-p4NxAoJBhIIN+hmNHrzRCf9tD/
  miZyoHS5obTRR9BMY=" crossorigin=""/>
  {% block links %}{% endblock %}
</head>
```

```
10  <body>
11    <nav>
12      <h1>Displaying OpenStreetMap maps</h1>
13    </nav>
14    <section class="content">
15      <header>
16        {% block header %}{% endblock %}
17      </header>
18      {% for message in get_flashed_messages() %}
19      <div class="flash">{{ message | safe }}</div>
20      {% endfor %}
21      <div id="map" style="width: 100%; height: 80vh;"></div>
22      {% block content %}{% endblock %}
23      </section>
24    <script src=https://unpkg.com/leaflet@1.9.4/dist/leaflet.js integrity="sha256-20nQCchB9co0qIjJZRGuk2/Z9VM+kNiyxNV1lvTlZBo=" crossorigin=""></script>
25    <script type="text/javascript">
26    window.onerror = function(message, url, linenumber) {
27        if (typeof map !== 'undefined' && map) map.remove();
28        const div = document.getElementById("map");
29        if (div) div.remove();
30        const elemDiv = document.createElement('div');
31        elemDiv.innerHTML = `<p>${message}, ligne ${linenumber}</p>`;
32        document.body.appendChild(elemDiv);
33      }
34    </script>
35    {% block script %}{% endblock %}
36  </body>
37  </html>
```

- **Line 7**: A link to a style sheet used by Leaflet. This statement is required to use Leaflet functions.
- **Line 21**: The division where the map built with Leaflet will be placed.

- **Line 24**: A link to the site where the code of the `leaflet.js` library can be found. This statement is required to use Leaflet functions.
- **Lines 25 to 34**: A JavaScript function for handling errors (particularly syntax errors) that have escaped the specific error handling defined in templates referencing `base_ll.html`.

Templates extending this model will mostly follow the structure below (those using asynchronous functions, such as `fetch()`, will use a different error-handling pattern that does not involve a `try/catch` block):

```
{% extends 'base_ll.html' %}

{% block title %}{{ name }}{% endblock %}

{% block header %}
<h2>{{ name }}</h2>
{% endblock %}

{% block script %}
<script type="text/javascript">
let map;
try {
    (Specific statements for building a map with Leaflet)
} catch(e) {
    if (typeof map !== 'undefined' && map) {map.remove();}

    const div = document.getElementById("map")
    if (div) {div.remove();}

    const elemDiv = document.createElement('div');
    elemDiv.innerHTML = `<p>${e}</p>`;
    document.body.appendChild(elemDiv);
}
</script>
{% endblock %}
```

- **Line 1**: This template extends the `base_ll.html` base template. It uses the structure of `base_ll.html` and inserts specific elements.
- **Line 3**: The name of the map, supplied by the calling Python function, will be included as the title by the Jinja engine.
- **Line 6**: The name of the map, supplied by the calling Python function, will be included as a level two header by the Jinja engine.
- **Line 11**: The variable that will contain the map is defined.
- **Lines 12 to 46**: These lines are a `try/catch` block. The `try` part contains statements for building the map with Leaflet.js. The `catch` part contains the error-handling statements and is triggered only if an error occurs during execution of the `try` part.

  The following line numbers are arbitrary, as the number of specific statements may vary. The line numbers below are only indicative, since the actual number of statements can differ.

  - **Line 38**: If the map exists, it is deleted.
  - **Lines 40 and 41**: If the `map` division exists, it is deleted.
  - **Lines 43 to 45**: A new division is added to the page, in which an error message is inserted.

To avoid repetition, only the statements specific to each prototype are provided below, except for the first prototype.

The main architectures used by the prototypes of this part are the following:

- **Browser- and Leaflet-Oriented Architecture**: Most of the processing executed by the browser on the client

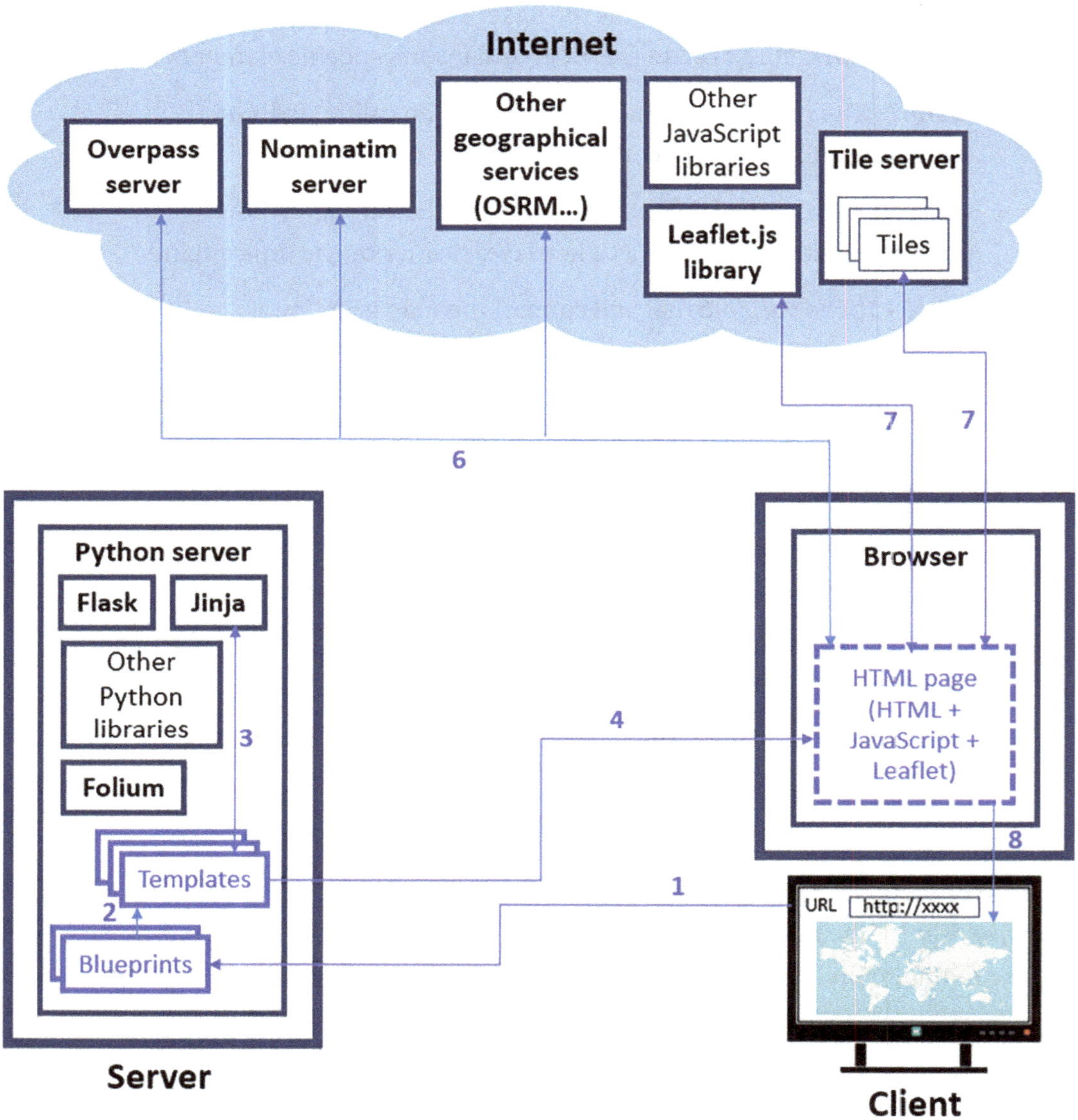

1. The user invokes a URL from the browser. This URL points to a route on the Python server.

2. The Python server retrieves the corresponding Python code for this route in a blueprint and executes it. This Python code is very simple and does not contain any Folium statement (no map is created yet).

3. The Python code invokes a template, passing it some arguments (such as a name). This template contains static HTML with embedded JavaScript code and Leaflet statements.

4. The Jinja engine injects the arguments passed by the Python code of the blueprint (such as a name) into the template.

5. The resulting HTML page is then returned to the browser.

6. The browser receives the HTML page, executes the embedded JavaScript code, and may query various external services (Overpass, Nominatim, OSRM, etc.) or fetch data from external sources to gather information to be displayed on a map.

7. It then executes the Leaflet statements, creating a map, enhancing it with the external data collected in step 6, and downloading the necessary tiles.

8. The browser finally displays the resulting web page, including a map, on the screen.

The example of Chapter 6, section "Retrieving Data and Creating a Map on the Client with Leaflet and JavaScript," typically uses this architecture.

- **Server and Folium-Oriented Architecture**: Most of the processing executed by the Python web server on the server platform

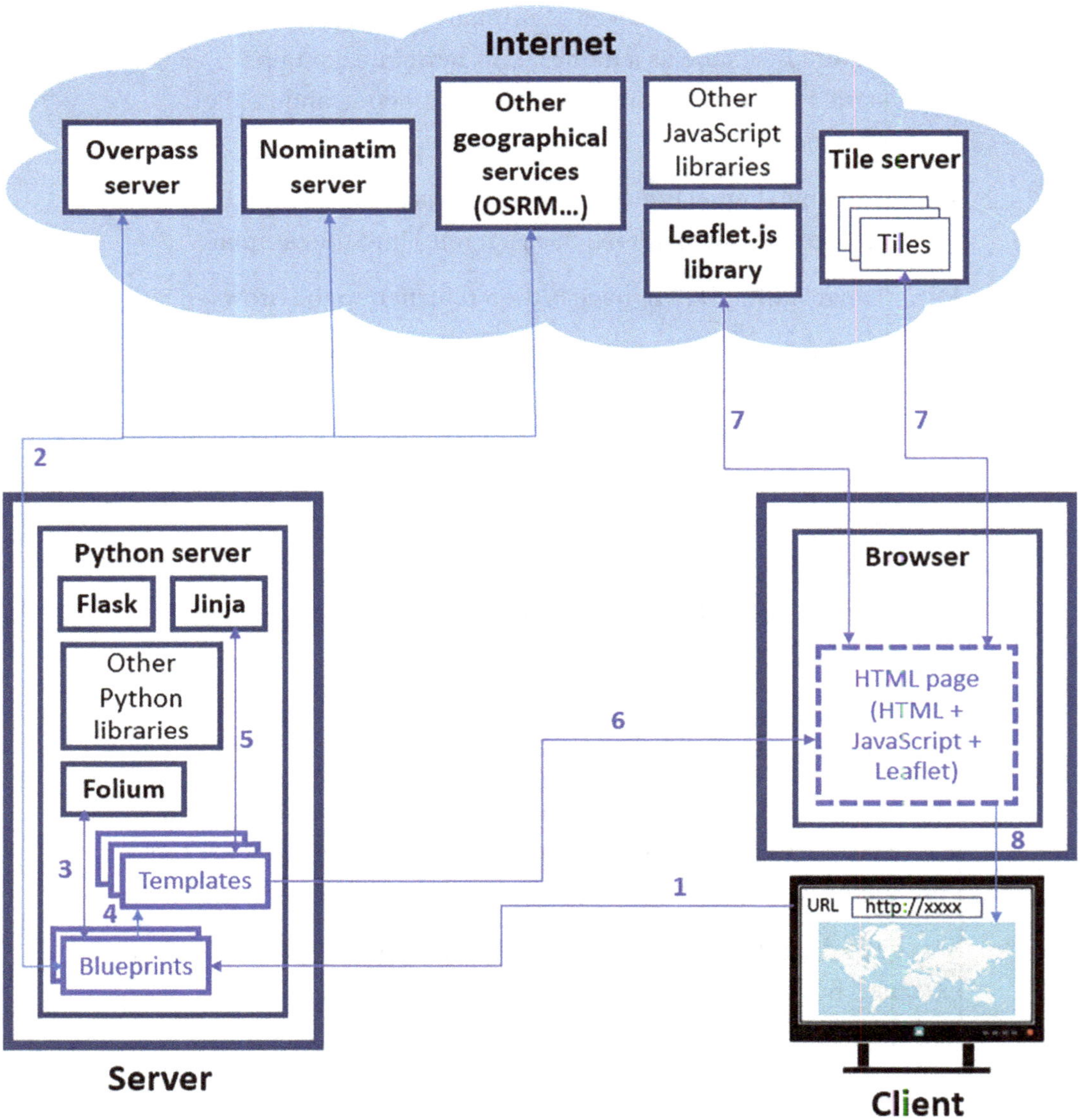

1. The user invokes a URL from the browser. This URL points to a route on the Python server.
2. The Python server retrieves the corresponding Python code for this route in a blueprint and executes it. This Python code may query various external services (Overpass,

Nominatim, OSRM, etc.) or fetch data from external sources to gather information to be displayed on a map.

3. The Python code then executes Folium statements, creating a map and enhancing it with the external data collected in step 2. The map is finally converted to HTML, JavaScript code, and Leaflet statements.

4. The Python code then invokes a template, passing it the Folium-generated map (currently consisting of HTML, JavaScript code, and Leaflet statements) and some other arguments (such as a name to be displayed as a header on the web page to be created).

5. The Jinja engine injects dynamic data passed by the Python code of the blueprint into the template: the map (currently consisting of HTML, JavaScript code, and Leaflet statements) and the other data (such as a name).

6. The resulting HTML page is then returned to the browser.

7. The browser receives the HTML page, executes the embedded JavaScript code and Leaflet statements, and downloads the tiles necessary to render the map.

8. The browser finally displays the resulting web page, including a map, on the screen.

The example of Chapter 6, section "Retrieving Data and Creating a Map on the Server with Folium and Python," typically uses this architecture.

- **Mixed Architecture**: Data prepared on the Python server and map built with Leaflet in the browser

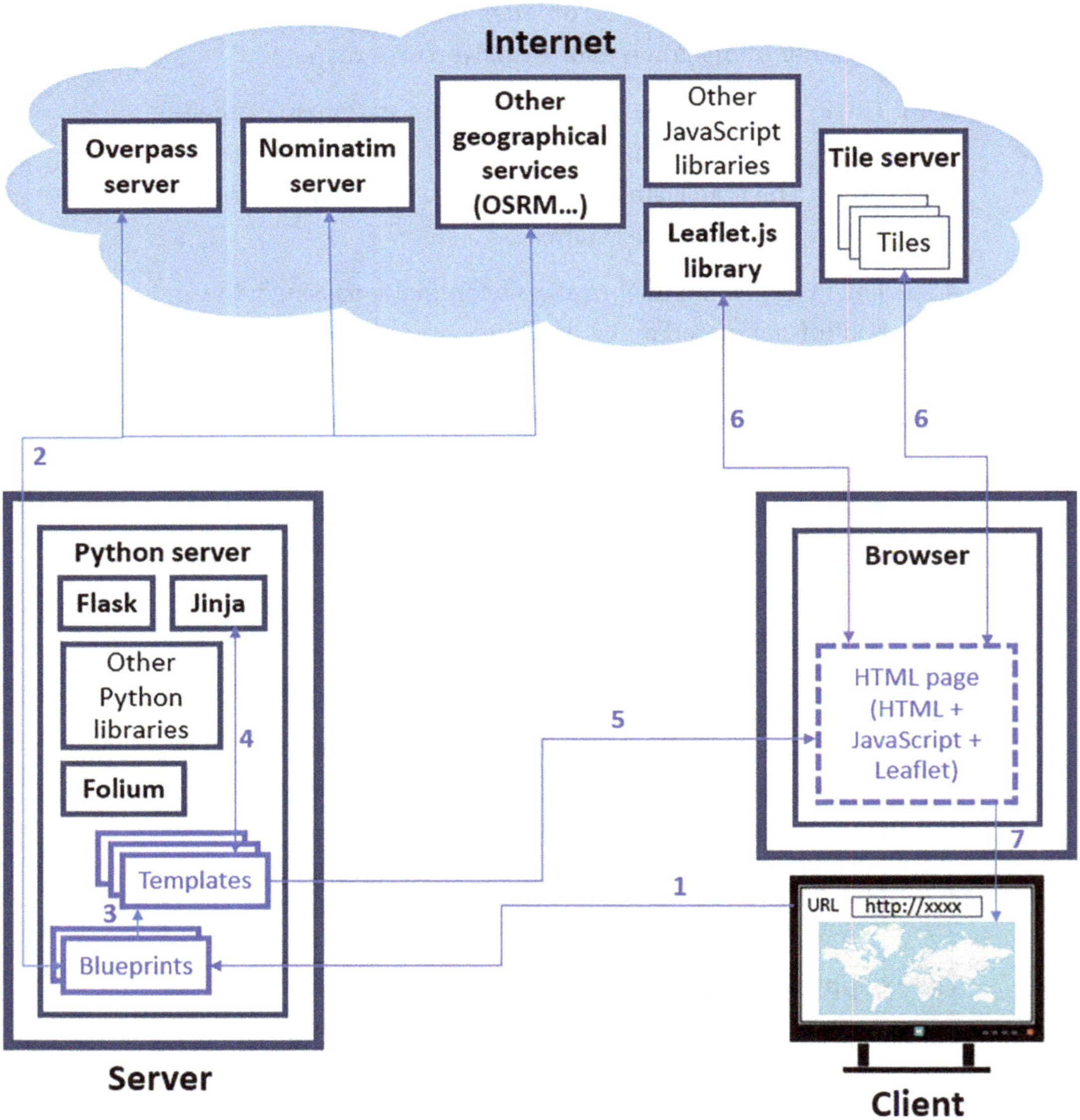

1. The user invokes a URL from the browser. This URL points to a route on the Python server.

2. The Python server retrieves the corresponding Python code for this route in a blueprint and executes it. This Python code may query various external services (Overpass, Nominatim, OSRM, etc.) or fetch data from external sources to gather information to be displayed on a map. It does not contain any Folium statement (no map is created yet).

3. The Python server then invokes a template, passing it the data gathered at step 2 and some other arguments (such as a name to be displayed as a header on the web page). This template contains static HTML with embedded JavaScript code and Leaflet statements.
4. The Jinja engine injects dynamic data passed by the blueprint into the template: the data gathered at step 2 and some others (such as a name).
5. The resulting HTML page is then returned to the browser.
6. The browser receives the HTML page and executes the embedded JavaScript code and Leaflet statements, creating a map, enhancing it with the external data collected in step 2, and downloading the necessary tiles.
7. The browser finally displays the resulting web page, including a map, on the screen.

The example of Chapter 6, section "Using a Mixed Approach: Data Retrieved on the Server and Map Created on the Client," typically uses this architecture.

CHAPTER 6

Enhancing Maps with Data from Overpass Queries

The goal of the following prototypes is to retrieve data from **OpenStreetMap** using the **Overpass API** and use it to enhance maps built with **Leaflet** (JavaScript) or **Folium** (Python).

The process involves two parts:

1. **Data retrieval**: Use the Overpass API to extract relevant geographic features from OSM.

   As already seen previously, Overpass is a general-purpose, high-level API associated with the OpenStreetMap mapping project. In Chapter 2, we looked at its query language, its HTTP API, and Overpass Turbo, its Web-based user interface for testing queries. We will now turn our attention to the API itself, focusing on how it can be used with Python or JavaScript in web applications.

2. **Map enhancement**: Convert the Overpass output into GeoJSON format, and then add this data to maps created with Leaflet or Folium. As already seen, both Leaflet and Folium can create a GeoJSON object from GeoJSON data, and then add it to a map, thereby overlaying it with additional elements such as markers, polylines, polygons, and popups.

While Overpass provides access to rich geographic data, the format of its output is typically not GeoJSON-compliant, requiring a transformation step before integration into mapping libraries.

P. Marie, *Modern Web Cartography*, https://doi.org/10.1007/979-8-8688-2335-0_6

Using the Overpass API to Enhance Maps Built with Leaflet and Folium

Overview

The Overpass API uses the custom, high-level query language already explored in Chapter 2 (this query language is fully documented on the web, for example, at https://wiki.openstreetmap.org/wiki/Overpass_API/Overpass_QL). With such queries, the API can be used to retrieve data useful for map enhancement.

Overpass queries can return data in JSON format (among other output formats). However, this JSON format differs from standard GeoJSON, which is required by most web mapping libraries such as Leaflet and Folium. Below is an example of an Overpass JSON response that includes one node, one way, and one relation:


```
{
  "version": 0.6,
  "generator": "Overpass API 0.7.62.5 1bd436f1",
  "osm3s": {
    "timestamp_osm_base": "2025-05-14T07:34:30Z",
    "copyright": "The data included in this document is from www.
    openstreetmap.org. The data is made available under ODbL."
  },
  "elements": [
    {
      "type": "node",
      "id": 6772305769,
      "lat": 50.2106895,
      "lon": 8.5857061,
      "tags": {
        "bicycle": "yes",
        "information": "guidepost",
        "ref": "OU.425.1",
        "tourism": "information"
      }
    },
```

```
{
  "type": "way",
  "id": 5123563,
  "bounds": {
    "minlat": 50.2106234,
    "minlon": 8.5856189,
    "maxlat": 50.2133772,
    "maxlon": 8.5873420
  },
  "nodes": [
    35352060,
    35352061,
    35352062,
    35352063,
    2056819178,
    35352001
  ],
  "geometry": [
    { "lat": 50.2133772, "lon": 8.5873420 },
    { "lat": 50.2129946, "lon": 8.5872135 },
    { "lat": 50.2126594, "lon": 8.5870583 },
    { "lat": 50.2121634, "lon": 8.5867429 },
    { "lat": 50.2111796, "lon": 8.5860363 },
    { "lat": 50.2106234, "lon": 8.5856189 }
  ],
  "tags": {
    "bicycle": "yes",
    "foot": "yes",
    "highway": "track",
    "horse": "yes",
    "incline": "down",
    "lcn": "yes",
    "motor_vehicle": "agricultural;forestry",
    "surface": "asphalt",
```

```
        "tracktype": "grade1",
        "traffic_sign": "DE:260,DE:1026-36"
      }
    },
    {
      "type": "relation",
      "id": 10000000,
      "bounds": {
        "minlat": 50.2106234,
        "minlon": 8.5856189,
        "maxlat": 50.2133772,
        "maxlon": 8.5873420
      },
      "members": [
        {
          "type": "node",
          "ref": 6772305769,
          "role": "sign",
          "lat": 50.2106895,
          "lon": 8.5857061
        },
        {
          "type": "node",
          "ref": 35352001,
          "role": "intersection",
          "lat": 50.2106234,
          "lon": 8.5856189
        },
        {
          "type": "way",
          "ref": 5123563,
          "role": "to",
          "geometry": [
             { "lat": 50.2133772, "lon": 8.5873420 },
             { "lat": 50.2129946, "lon": 8.5872135 },
```

```
            { "lat": 50.2126594, "lon": 8.5870583 },
            { "lat": 50.2121634, "lon": 8.5867429 },
            { "lat": 50.2111796, "lon": 8.5860363 },
            { "lat": 50.2106234, "lon": 8.5856189 }
          ]
        }
      ],
      "tags": {
        "bicycle": "yes",
        "colour:arrow": "green",
        "colour:back": "white",
        "colour:text": "green",
        "destination": "Friedrichsdorf;Bad Homburg",
        "destination:symbol": ";train_station",
        "distance": "8.0;4.6",
        "type": "destination_sign"
      }
    }
  ]
}
```

To be able to use this data with Leaflet or Folium, we must convert the Overpass JSON format into valid GeoJSON. The following is the data from the preceding example converted to GeoJSON format:

```
{
  'type': 'FeatureCollection',
  'features': [
    {
      'type': 'Feature',
      'properties': {
        'type': 'node',
        'id': 6772305769,
        'tags': {
          'bicycle': 'yes',
          'information': 'guidepost',
```

```
        'ref': 'OU.425.1',
        'tourism': 'information'
      }
    },
    'geometry': {
      'type': 'Point',
        'coordinates': [8.5857061, 50.2106895]
    }
  },
  {
    'type': 'Feature',
    'properties': {
      'type': 'way',
      'id': 5123563,
      'tags': {
        'bicycle': 'yes',
        'foot': 'yes',
        'highway': 'track',
        'horse': 'yes',
        'incline': 'down',
        'lcn': 'yes',
        'motor_vehicle': 'agricultural;forestry',
        'surface': 'asphalt',
        'tracktype': 'grade1',
        'traffic_sign': 'DE:260,DE:1026-36'
      },
      'nodes': [
        35352060,
        35352061,
        35352062,
        35352063,
        2056819178,
        35352001
      ]
    },
```

```
  'geometry': {
    'type': 'LineString',
      'coordinates': [
        [8.587342, 50.2133772],
        [8.5872135, 50.2129946],
        [8.5870583, 50.2126594],
        [8.5867429, 50.2121634],
        [8.5860363, 50.2111796],
        [8.5856189, 50.2106234]
    ]
  }
},
{
  'type': 'Feature',
  'properties': {
    'type': 'relation',
      'id': 10000000,
      'tags': {
      'bicycle': 'yes',
      'colour:arrow': 'green',
      'colour:back': 'white',
      'colour:text': 'green',
      'destination': 'Friedrichsdorf;Bad Homburg',
      'destination:symbol': ';train_station',
      'distance': '8.0;4.6',
      'type': 'destination_sign'
    }
  },
  'geometry': {
    'type': 'LineString',
      'coordinates': [
        [8.587342, 50.2133772],
        [8.5872135, 50.2129946],
        [8.5870583, 50.2126594],
        [8.5867429, 50.2121634],
```

```
                [8.5860363, 50.2111796],
                [8.5856189, 50.2106234]
            ]
        }
    }
  ]
}
```

This GeoJSON data consists of a collection of features. Each element (node, way, or relation) of the Overpass output is replaced by a feature of the same type, which includes:

- A `properties` field containing metadata (from the `tags` field of Overpass JSON).
- A `geometry` field with `Point` type (from the nodes of Overpass JSON) or `LineString` type (from the ways of Overpass JSON) and coordinates enabling to place the feature on a map.

Other geometry types used in GeoJSON are `MutiPoint`, `MultiLineString`, `Polygon`, `MultiPolygon`, and `GeometryCollection`.

Leaflet and Folium use this structure to render these features on a map:

- Points (as well as MultiPoints) are typically shown as markers.
- LineStrings (as well as MultiLineStrings) appear as polylines.
- Polygons (as well as MultiPolygons) are rendered as polygons.

The metadata located in the `properties` fields can also be used to display additional information in popups or tooltips.

What we still need, in order to query Overpass from JavaScript and Python and then to use its response to enrich maps, is:

- A way to invoke the Overpass API from within JavaScript and Python code.
- A way to convert the JSON data returned by Overpass into GeoJSON format compatible with Leaflet and Folium.

These points will be discussed in the next two sections.

As already explained in Chapter 2, there are public endpoints for using the Overpass API (such as `https://overpass-api.de/api/interpreter` and `https://overpass.kumi.systems/api/interpreter`), but they are intended for demonstration, education, and low-volume testing. For production use and high-throughput applications, it is recommended to self-host an Overpass server – either by setting up the infrastructure within your organization or by using a third-party Overpass service.

In the following examples, we will use the `https://overpass-api.de/api/interpreter` endpoint.

With Leaflet and JavaScript

Several options are possible for querying Overpass from a JavaScript program. One difficulty to overcome is that Leaflet, for which the data extracted from OSM is intended, expects this data in GeoJSON format, and Overpass does not allow data to be generated in this format (Overpass can provide its responses in JSON format, but with a different structure from that of GeoJSON).

The possible options are:

- Overpass provides an HTTP API (i.e., queries are submitted in the form of HTTP requests) that was already described at the end of Chapter 2. This API can be called from JavaScript code using the `fetch()` function.

  The response can be requested in JSON format. However, the structure of this JSON response differs from GeoJSON, and a conversion is necessary. This conversion can be achieved using the `osmtogeojson()` function of the `osmtogeojson` JavaScript library.

  This is the approach that we will use to query Overpass in our Leaflet prototypes.

- A JavaScript library named `overpass.js` is also available for querying Overpass from JavaScript code. It makes use of the Overpass HTTP API internally.

  However, it does not return raw JSON data (like the HTTP API described just above), but rather a series of JavaScript objects representing the nodes, ways, and relations returned by Overpass, along with their properties. To our knowledge, there is no JavaScript function for converting these objects to the GeoJSON format. To overcome this limitation, you could consider developing your own conversion program. This would be simple for nodes, somewhat more complex for ways, and significantly more complicated for relations (which can contain nodes, ways, and even other relations). Even if you made the effort of developing such a utility, questions of long-term maintenance and sustainability would still remain.

- Yet another option is to use the **Leaflet Overpass Layer** plugin, discussed in more detail at the end of this chapter. This plugin can both submit queries to Overpass (it uses the Overpass HTTP API internally) and convert the results into GeoJSON format, making the data ready for use by Leaflet.

Another possible solution is to use a mixed approach:

1. Query Overpass from Python code on the server, convert the response data into GeoJSON format, and then inject it into the HTML page built from a template and sent to the browser.
2. While the JavaScript code of the HTML page is executed by a browser, create a map with Leaflet.js, and enhance it using the GeoJSON data received from Python and included in the HTML page. The JavaScript and Leaflet code used to build the map must be placed in the template.

We will be using this approach in some of our prototypes (in particular in the section of this chapter entitled "Using a Mixed Approach: Data Retrieved on the Server and Map Created on the Client").

With Folium and Python

Overpass offers several ways of using its API from within a Python program. One difficulty to overcome is that Folium, for which the data extracted from OSM is intended, expects this data in GeoJSON format, and Overpass does not allow data to be generated in this format (Overpass can provide its responses in JSON format, but with a different structure from that of GeoJSON).

The possible options are:

- Overpass provides an HTTP API (i.e., queries are submitted in the form of HTTP requests) that was already described at the end of Chapter 2. This API can be used from Python code using the `requests` library.

The response can be requested in JSON format. However, the structure of this JSON response differs from GeoJSON, and a conversion is needed. This conversion can be achieved using the `json2geojson()` function of the `osm2geojson` Python library.

This is the approach that we will use to query Overpass in our Folium prototypes.

- A Python library named `overpass` is also available and perfectly meets the need. It allows you to query Overpass in a simple way (the Overpass HTTP API is used internally), and it can return the response in GeoJSON format, directly usable by Folium. However, this library is unfortunately no longer maintained, so we advise against using it: even if it works well at present, it would be unwise to commit to it in the long term.
- Another Python library named `overpy` is maintained and recommended for use. However, the data it returns, after sending requests to Overpass, is in a very crude JSON format (whose structure is inherited from the internal structure of the data in the OSM database). Moreover, there is (to our knowledge) no Python function for converting this JSON format to GeoJSON. To overcome this shortcoming, you could consider developing your own conversion program. This would be simple for nodes, a little less so for ways, and would become complex for relations (which can contain nodes, ways, and even other relations). Even if you made the effort to develop such a utility, questions of long-term maintenance and sustainability would still remain.

Retrieving Data and Creating a Map on the Client with Leaflet and JavaScript

This prototype will build a map of Paris (France) highlighting museums with markers. The museums and their geographic coordinates will be obtained by querying OSM data via the Overpass HTTP API from JavaScript code running in the browser. The map will then be created and enhanced with the Overpass data using Leaflet in the browser.

A suitable Overpass query to find all the museums in Paris is

```
1 [out:json][timeout:25];
2 area[name="Paris"][wikipedia="fr:Paris"]->.searchArea;
3 node[tourism="museum"](area.searchArea);
4 out;
```

- **Line 1**: The data returned by Overpass will be in JSON format, and a maximum time of 25 seconds is allowed to process the request (if this time is exceeded, the search is aborted, and an error returned).
- **Line 2**: A search area limited to the city of Paris (France) is defined.
    - The first filter `[name="Paris"]` limits the search area to cities with the name "Paris".
    - The second filter `[wikipedia="fr:Paris"]` must be added to restrict the search to Paris, France (and eliminate other cities named Paris and located in Texas or elsewhere).
    - The `->.searchArea` part of the statement places the selected area in the `searchArea` named set, used in the next statement.
- **Line 3**: The query searches for all nodes:
    - With a tag with key `tourism` and value "museum" (filter: `[tourism="museum"]`).
    - And located in the search area defined on the previous line (filter: `(area.searchArea)`).
- **Line 4**: This statement requests the nodes found by the previous statements to be included in the response. The `out;` statement (with no additional options) is sufficient to obtain the geographic coordinates of the nodes.

After testing the query with Overpass Turbo, the following simple code must be added to the `advanced_maps.py` blueprint:

```
@bp.route('/am/map_op1')
def map_op1():
    return render_template('maps/page_with_museums_paris.html',
```

```
4            name="Map with OSM data found with Overpass: museums in Paris",
5            city="Paris",
6            wikipedia_city="fr:Paris",
7            loc=[48.857481, 2.347425])
```

- As always with Python, be careful with column alignments and indentations. The first statement (`@bp.route...`) must not be indented.
- **Lines 3 to 7**: A web page is rendered using the `maps/page_with_museums_paris.html` template. These arguments are passed:
    - `name`: Will be used to create a title and a header in the resulting web page.
    - `city`: Will be used to build the Overpass query.
    - `wikipedia_city`: Will be used to build the Overpass query.
    - `loc`: Geographical coordinates of Paris, France, will be used to create the map.

  The `render_template` function (part of the Flask library), used here, is not imported in this snippet because it was already imported globally along with other libraries when the blueprint was created at the beginning of Part 2.

And the template `page_with_museums_paris.html` must be created in the `prototypes/templates/maps` directory with the following code:

```
{% extends 'base_ll.html' %}

{% block title %}{{ name }}{% endblock %}

{% block header %}
<h2>{{ name }}</h2>
{% endblock %}

{% block script %}
<script src="https://cdn.jsdelivr.net/npm/osmtogeojson@3.0.0-beta.5/
osmtogeojson.min.js"></script>
```

```

<script type="text/javascript">
let map;
const overpassQuery = `
    [out:json][timeout:25];
    area[name="{{ city }}"][wikipedia="{{ wikipedia_city }}"]->.searchArea;
    node[tourism="museum"](area.searchArea);
    out;
`;

fetch("https://overpass-api.de/api/interpreter?", {
    method: "POST",
    body: "data=" + encodeURIComponent(overpassQuery)})
.then(response => response.json())
.then(data => {
    const geojsonData = osmtogeojson(data);

    map = L.map('map').setView({{ loc }}, 13);

    L.tileLayer('https://tile.openstreetmap.org/{z}/{x}/{y}.png', {
        maxZoom: 19,
        attribution: '&copy; <a href="http://www.openstreetmap.org/copyright">OpenStreetMap</a>'
    }).addTo(map);

    L.geoJSON(geojsonData).addTo(map);
})
.catch(error => {
    if (typeof map !== 'undefined' && map) {map.remove();}

    const div = document.getElementById("map")
       if (div) {div.remove();}

    const elemDiv = document.createElement('div');
```

```
44        elemDiv.innerHTML = `<p>${error}</p>`;
45        document.body.appendChild(elemDiv);
46    });
47    </script>
48    {% endblock %}
```

- **Line 10**: This line defines where to find the code for the `osmtogeojson()` function on the Internet.

  This is a beta version not officially released as stable, but it works well and supports JSON. On the other hand, the stable version requires an XML input and is tricky to use.

- **Lines 14 to 19**: The Overpass query is prepared. This query contains a city name and a city Wikipedia name received from the calling Python function (variables substituted by their values by the Jinja engine on the server).

- **Lines 21 to 23**: The Overpass HTTP API is invoked via the `fetch()` function with the previously prepared Overpass query. The query is encoded with the `encodeURIComponent()` function to make sure it will be accepted by the Overpass server. This HTTP request is of POST type and cannot be tested individually with a browser, but can be with tools such as cURL, Postman, or SoapUI. For example, here is how to test it with cURL:

  ```
  curl -X POST https://overpass-api.de/api/interpreter -H
  "Content-Type: application/x-www-form-urlencoded" --data-
  urlencode "data=[out:json][timeout:25]; area[name=\"Paris\"]
  [wikipedia=\"fr:Paris\"]->.searchArea; node[tourism=\"museum\"]
  (area.searchArea); out;"
  ```

  Depending on the environment used, invoking Overpass from JavaScript in the browser may cause a CORS (cross-origin resource sharing) error. This is unlikely with the `https://overpass-api.de/api/interpreter` public endpoint, but may occur with other endpoints. In such cases, a common solution is to use a server-side proxy that adds a contact User-Agent (typically of the form:

MyApp/1.0 (me@example.com)) and appropriate headers. Appendix C explains how to do so.

Another possible solution is to route the request through a public CORS-proxy service (such as **corsproxy.io**, **CORS Anywhere** or **thingproxy**). For example, with **corsproxy.io**:

```
fetch("https://corsproxy.io/?" + encodeURIComponent(<your_
target_URL>));
```

However, be aware that public CORS-proxy services are generally intended for testing or non-commercial use only, and they may not be reliable or suitable for production environments.

The fetch() function is asynchronous, and the error catching and handling cannot be done with a try/catch block (as in the case of synchronous processing).

- **Lines 24 and 25**: The response is parsed (with a JSON parser) and placed in an object (with a structure similar to that of the JSON data included in the response).
- **Line 26**: This object (containing the data returned by Overpass) is converted to a GeoJSON-compliant object using the osmtogeojson() function.
- **Line 28**: A map is created and centered on the position received from the calling Python function (the coordinates of Paris).
- **Lines 30 to 33**: A tile layer is associated with the map.
- **Line 35**: An L.geoJSON object is built from the GeoJSON-compliant object (obtained at line 26 and containing the Overpass's response data), then added to the map.
- **Lines 37 to 46**: These lines handle any errors that may occur by displaying an error message on a web page.

Launching the server and opening the URL http://127.0.0.1:5000/am/map_op1 with a browser produces the following map, on which the museums returned by Overpass are represented by markers:

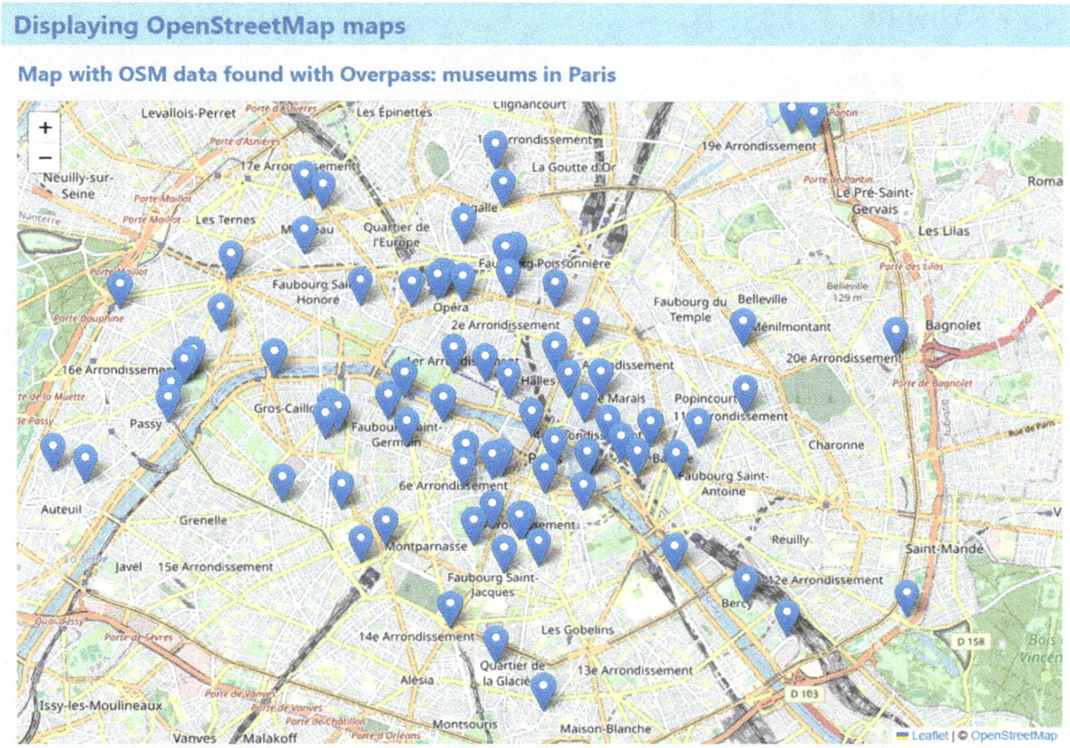

Figure 6-1. *Map created with Leaflet using OpenStreetMap tiles. Data © OpenStreetMap contributors, licensed under the ODbL. Overlaid data retrieved with Overpass*

If the map fails to display, it may be caused by a CORS error (as explained above). Verify this possibility and, if necessary, implement a server-side proxy as described in Appendix C or use a public CORS-proxy service (such as **corsproxy.io**).

It appears that some major Paris museums (notably the Louvre) have not been identified as museums on the map. After verification, it turns out that these museums, due to their large size, are not registered as nodes, but as ways or relations, in OSM. To take this into account, we must modify our query:

- **Line 17**:

  ```
  nwr[tourism="museum"](area.searchArea);
  ```

  `nwr` means nodes, ways, and relations. With this modification, museums are now queried not only as nodes but also as ways and relations.

- **Line 18**:

  ```
  out geom;
  ```

  The `geom` option must be added to the `out` statement to obtain the coordinates needed to represent the museums found as ways or relations on the map.

After restarting the server and reloading the map, we obtain the following map.

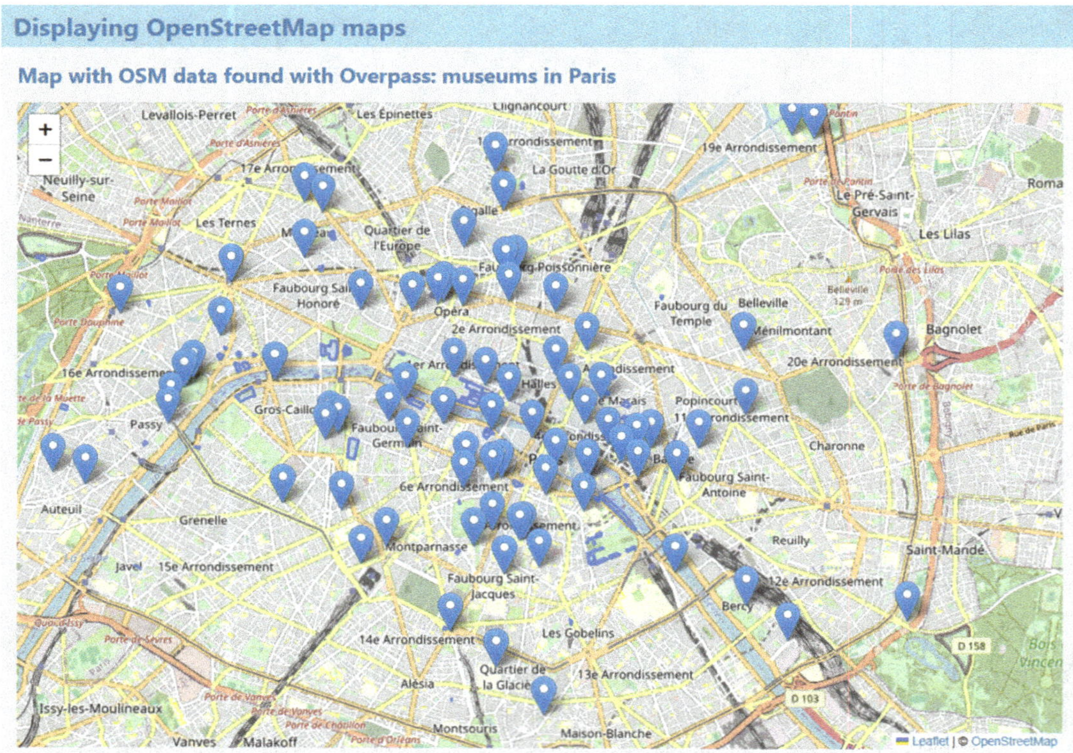

Figure 6-2. *Map created with Leaflet using OpenStreetMap tiles. Data © OpenStreetMap contributors, licensed under the ODbL. Overlaid data retrieved with Overpass*

Now, the Louvre Museum and a few other large museums are highlighted with blue polylines or polygons.

What is missing on our map is the ability to know which museums the markers, polylines, and polygons appearing on the map correspond to. We will fill this gap by modifying the code to associate a popup with the museum name to each marker, polyline, and polygon. We will also change the style of the markers, polylines, and polygons and the icons used for the markers. The **bold** lines below enable this enhancement:

```
{% extends 'base_ll.html' %}

{% block title %}{{ name }}{% endblock %}

{% block links %}
<link rel="stylesheet" href="https://cdnjs.cloudflare.com/ajax/libs/
font-awesome/4.7.0/css/font-awesome.min.css">
<link rel="stylesheet" href="https://cdn.jsdelivr.net/npm/leaflet-
extra-markers@1.2.1/dist/css/leaflet.extra-markers.min.css">
{% endblock %}

{% block header %}

...

{% block script %}
<script src="https://cdn.jsdelivr.net/npm/leaflet-extra-markers@1.2.1/
dist/js/leaflet.extra-markers.min.js"></script>
<script src="https://cdn.jsdelivr.net/npm/osmtogeojson@3.0.0-beta.5/
osmtogeojson.min.js"></script>

...

    }).addTo(map);

    L.geoJSON(geojsonData, {
        pointToLayer: function(feature, latlng) {
            const monumentMarker = L.ExtraMarkers.icon({
                icon: 'fa-university',
                markerColor: 'red',
                shape: 'square',
                prefix: 'fa'
            });

```

```
50                  return new L.Marker(latlng, {icon: monumentMarker});
51              },
52              style: function(feature) {
53                  return {
54                      color: "red",
55                      weight: 4,
56                      opacity: 0.7
57                  };
58              },
59              onEachFeature: function(feature, layer) {
60                  if (feature.properties.name) {
61                      layer.bindPopup(feature.properties.name, { permanent:
                        false, direction: "top" });
62                  }
63              }
64          }).addTo(map);
65
66      })
...
```

- **Lines 6 and 7**: These lines contain two links to style sheets that are needed to customize markers with icons.
- **Line 15**: This line contains a link to a JavaScript library needed to customize markers with icons.
- **Lines 41 to 64**: An `L.geoJSON` object is created with the data found with Overpass (and converted to GeoJSON).
    - **Lines 42 to 51**: These lines create a function that defines the object to be used to represent all the features of type `Point` in the GeoJSON data on the map. This function returns a marker, the style of which is defined using an `L.ExtraMarkers.icon` object defined on lines 43 to 48. This `L.ExtraMarkers.icon` object contains definitions for the icon to be inserted in the marker, the marker color, the marker shape, and a prefix (to complete the icon's reference).

A list of the icons that can be used can be found at `https://fontawesome.com/v4/cheatsheet/`, and the supported colors are `red`, `blue`, `green`, `orange`, `yellow`, `purple`, `darkred`, `white`, `black`, and `cadetblue`.

Museums will be represented by square red markers containing a colonnade icon.

- **Lines 52 to 58**: These lines create a function that defines the properties to be used to draw polylines and polygons for all features of type `LineString` and `Polygon` in the GeoJSON data on the map: color, line thickness, and opacity level.
- **Lines 59 to 63**: These lines create a function to be applied to each of the features in the GeoJSON data (i.e., of types `Point` as well as `LineString` and `Polygon`): if the feature has a `name` property, a popup containing its value is added to the map element (marker, polyline, or polygon) representing the feature on the map.

After restarting the server and reloading the map, we obtain the following map:

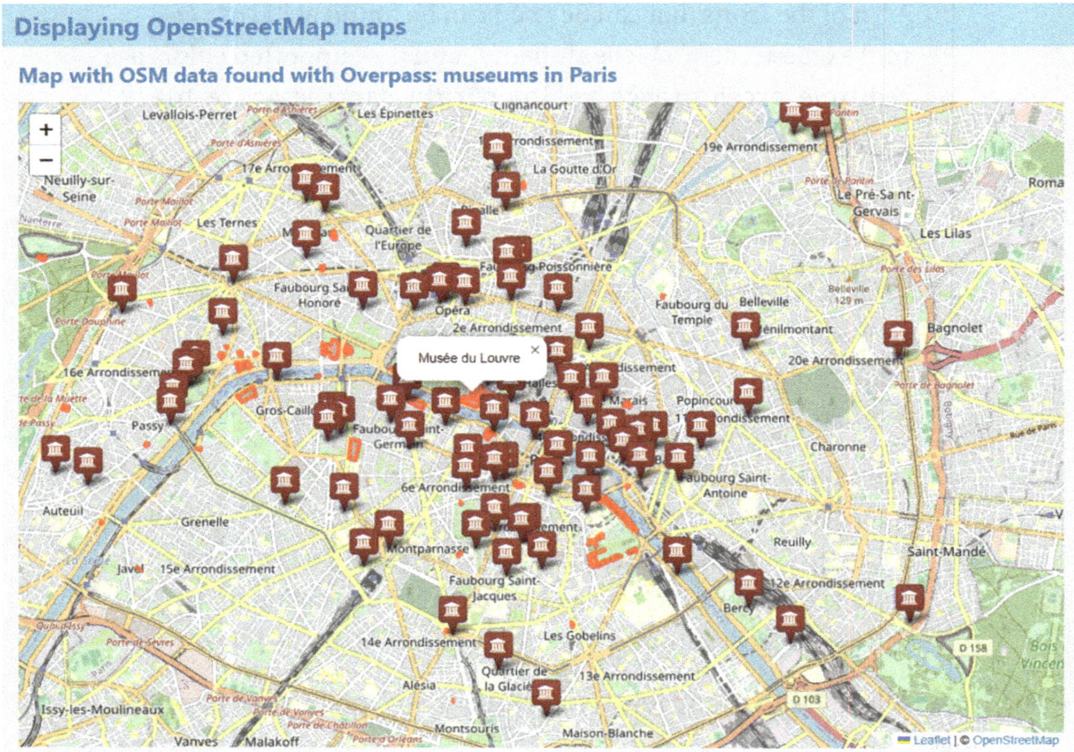

Figure 6-3. *Map created with Leaflet using OpenStreetMap tiles. Data © OpenStreetMap contributors, licensed under the ODbL. Overlaid data retrieved with Overpass*

The style of the markers and the color of the polylines and polygons have changed, and we can click on the markers, polylines, and polygons to obtain the name of the museum in a popup.

Let's add a last enhancement to our map: some museums offer full access to wheelchairs, while some others offer limited access, and others do not offer access or do not have this property set in OSM. We will display these four types of museums with different colors. Let's also add a layer control enabling the user to show or hide the markers, polylines, and polygons representing the museums.

These modifications are shown in **bold** below:

```
    ...
        }).addTo(map);

```

```
    function wheelchairColor(wheelchair) {
        if (!wheelchair) return 'purple';
        if (wheelchair === 'yes' || wheelchair === 'designated') return 'green';
        if (wheelchair === 'limited' || wheelchair === 'partial' || wheelchair === 'limited_access') return 'orange';
        if (wheelchair === 'no' || wheelchair === 'false') return 'red';
        return 'purple';
    }

    const geoJSON = L.geoJSON(geojsonData, {
        pointToLayer: function(feature, latlng) {
            const wheel = feature?.properties?.wheelchair;
            const color = wheelchairColor(wheel);
            const monumentMarker = L.ExtraMarkers.icon({
                icon: 'fa-university',
                markerColor: color,
                shape: 'square',
                prefix: 'fa'
            });


            return new L.Marker(latlng, {icon: monumentMarker});
        },
        style: function(feature) {
            const wheel = feature?.properties?.wheelchair;
            const color = wheelchairColor(wheel);
            const style = {
                color: color,
                weight: 4,
                opacity: 0.7
            };

            return style;
        },
```

```
74          onEachFeature: function(feature, layer) {
75              if (feature.properties.name) {
76                  layer.bindPopup(feature.properties.name, { permanent:
                    false, direction: "top" });
77              }
78          }
79      }).addTo(map);
80
81      const layer = {
82          "Museums": geoJSON
83      };
84
85      L.control.layers({}, layer).addTo(map);
86  })
    ...
```

- **Lines 41 to 47**: A `wheelchairColor()` function that selects a color (purple, green, orange, or red) from the value of an input string is defined.
- **Lines 51 and 52**: Using the value of the `wheelchair` property and the `wheelchairColor()` function, the right color is selected for each of the markers representing a museum.
- **Line 55**: This color is assigned to the `L.Extramarkers.icon` object that will be used to define the graphical properties of the marker.
- **Lines 64 and 65**: Using the value of the `wheelchair` property and the `wheelchairColor()` function, the right color is selected for each of the polylines and polygons representing a museum.
- **Line 67**: This color is assigned to the polyline or polygon.
- **Lines 81 to 85**: A layer control is created and will enable the user to show or hide the map elements representing museums.

After restarting the server and reloading the map, it now shows different colors for the museums, and a layer control is available to show or hide the museums' markers, polylines, and polygons.

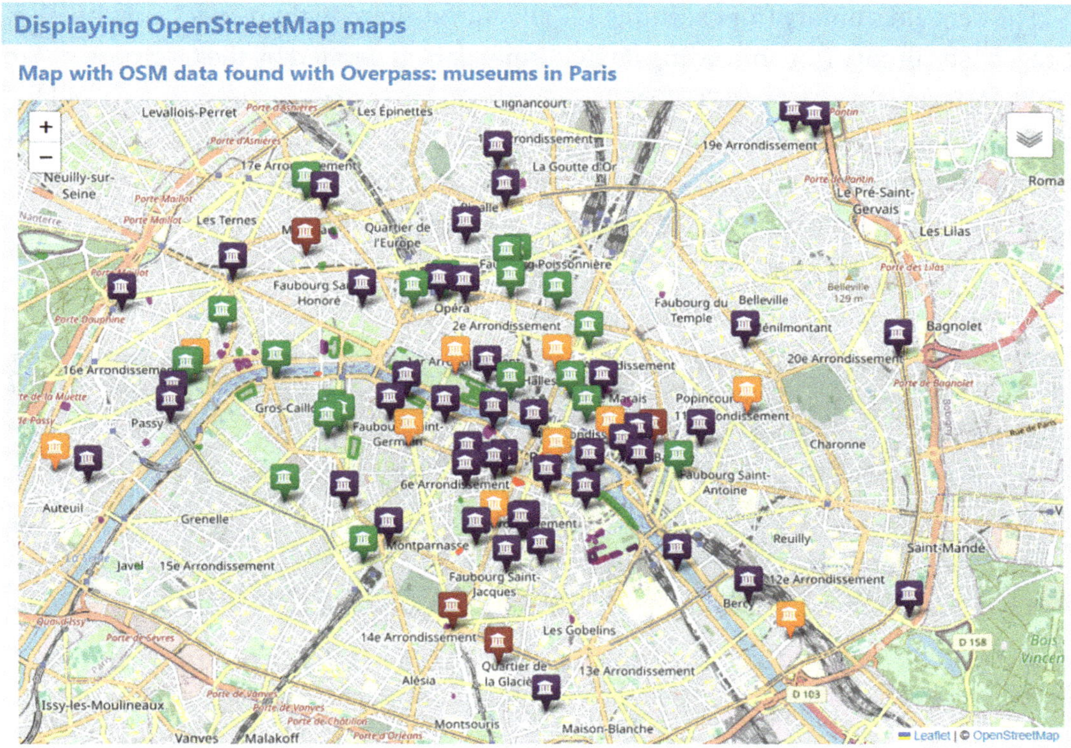

Figure 6-4. Map created with Leaflet using OpenStreetMap tiles. Data © OpenStreetMap contributors, licensed under the ODbL. Overlaid data retrieved with Overpass

Museums with full wheelchair access are highlighted in green, those with partial access in orange, those with no access in red, and those with unknown access ("wheelchair" tag not set in OSM) in purple. A layer control has been added in the top right corner.

Retrieving Data and Creating a Map on the Server with Folium and Python

This prototype will build a map of the city of Galway (Ireland) highlighting restaurants with markers. These restaurants and their coordinates will be obtained via the Overpass HTTP API from Python code running on the server. The map will then be created and enriched with the Overpass data using Folium, also on the server.

The `requests` library (for executing HTTP requests from Python code) and the `osm2geojson` library (for converting JSON-format data generated by the Overpass API to GeoJSON format) are needed. To install them, the following commands can be used in the application's virtual environment (using the VS Code terminal and executing these commands from the `carto_osm` directory of the workspace). If you already installed the `requests` library while running the examples of Chapter 5, do not install it again.

```
pip install requests
```

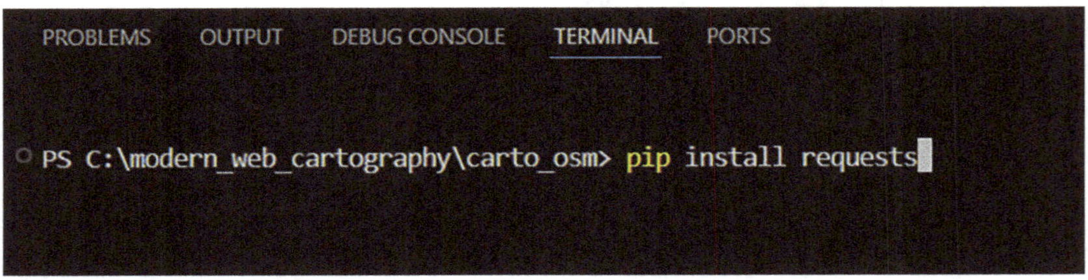

Figure 6-5. *Screenshot of Visual Studio Code. © Microsoft Corporation (`http://www.microsoft.com`). All rights reserved. Used with permission from Microsoft*

Then:

```
pip install osm2geojson
```

A suitable query to search for all restaurants in Galway is:

```
1  [out:json][timeout:25];
2  area[name="Galway"]->.searchArea;
3  node[amenity="restaurant"](area.searchArea);
4  out;
```

- **Line 1**: The data returned by Overpass will be in JSON format, and a maximum time of 25 seconds is allowed to process the request (if this time is exceeded, the search is aborted and an error returned).
- **Line 2**: A search zone limited to the city of Galway is defined. The `->.searchArea` part of the statement places the selected area in the `searchArea` named set, used in the next statement.

- **Line 3**: The query searches for all nodes:
  - With the `amenity` tag with the value "restaurant" (filter: `[amenity="restaurant"]`). This property indicates that the node corresponds to a restaurant.
  - And located in the search area (filter: `(area.searchArea)`).
- **Line 4**: This statement requests the nodes found by the previous statements to be included in the response. The `out;` statement (with no additional options) is sufficient to obtain the geographic coordinates of the nodes.

After testing the query with Overpass Turbo, the following code must be added to the `advanced_maps.py` blueprint:

```
@bp.route('/am/map_op2')
def map_op2():
    import requests
    from osm2geojson import json2geojson

    try:
        query = """
            [out:json][timeout:25];
            area[name="Galway"]->.searchArea;
            node[amenity="restaurant"](area.searchArea);
            out;
        """

        overpass_url = "https://overpass-api.de/api/interpreter"
        response = requests.post(overpass_url, data={'data': query})

        if response.status_code == 200:
            data = response.json()
        else:
            flash("Error retrieving data with Overpass API",
            category="error")
```

```
21              return render_template('maps/page_with_map.html', name="No
                map to display")
22
23          geojson_data = json2geojson(data)
24
25          m = folium.Map(location=[53.27340, -9.05080], zoom_start=17)
26
27          folium.GeoJson(
28              geojson_data
29          ).add_to(m)
30
31          return render_template('maps/page_with_map.html',
32              name="Map with OpenStreetMap data: restaurants in Galway",
33              map_html=m._repr_html_())
34
35      except Exception as e:
36          flash("An error has been encountered:<br/>" + str(e),
            category="error")
37            return render_template('maps/page_with_map.html', name="No
              map to display")
```

- Again, be careful with column alignments and indentations. The first statement (`@bp.route...`) must not be indented.
- **Lines 3 and 4**: The `requests` library and the `json2geojson()` function are imported.
- **Line 6**: This line starts a `try/except` block. If an error occurs in the `try` part of this block, it is caught, the processing of the `try` part is stopped, and the code in the `except` part is executed.
- **Lines 7 to 12**: The Overpass request is prepared.
- **Line 14**: The URL of the HTTP Overpass API is defined.
- **Line 15**: The HTTP request, containing the Overpass request previously prepared, is submitted. This HTTP request is of the POST type and cannot be tested individually with a browser, but could be

with tools such as cURL, Postman, or SoapUI. For example, here is how to test it with cURL:

```
curl -X POST https://overpass-api.de/api/interpreter -H "Content-
Type: application/x-www-form-urlencoded" --data-urlencode
"data=[out:json][timeout:25];area[name="Galway"]->.searchArea;node
[amenity="restaurant"](area.searchArea); out;"
```

- **Line 17**: The result of the HTTP request is checked:
  - **Line 18**: If the request was successful, the response is parsed and placed in a JSON-structured object.
  - **Lines 19 to 21**: These lines handle the case where the HTTP request has failed: processing is aborted, and an error message is displayed on a web page.
- **Line 23**: The JSON-structured object (containing the data returned by Overpass) is converted to a GeoJSON-structured object (by the `json2geojson()` function of the `osm2geojson` library).
- **Line 25**: A map is created and centered on the latitude and longitude coordinates of Galway, and an initial zoom level (17) is specified for the map.

  The `folium` library, used here, is not imported in this snippet because it was already imported globally along with other libraries when the blueprint was created at the beginning of Part 2.
- **Lines 27 to 29**: A `folium.GeoJson` object is created with the GeoJSON-structured object obtained previously. Then this new object is added to the map.
- **Lines 31 to 33**: A web page is rendered using the `page_with_map.html` template. A name ("Map with OpenStreetMap data: restaurants in Galway") and the map, converted to HTML and JavaScript code, are passed as arguments.

  The `render_template` function (part of the Flask library), used here, is not imported in this snippet because it was already imported globally along with other libraries when the blueprint was created at the beginning of Part 2.

- **Lines 35 to 37**: These lies make up the `except` part of the `try/except` block. These statements are executed only if an error occurs during execution of the statements in the `try` part. They prepare an error message for inclusion in an HTML page, then trigger its display.

  The `flash` function (part of the Flask library), used on line 37, is not imported in this snippet because it was already imported globally along with other libraries when the blueprint was created at the beginning of Part 2.

The `maps/page_with_map.html` template is a versatile template used to display most of the maps built with Folium. It has already been used in previous examples in Chapter 5 and is described in Appendix B.

Launching the server and opening the address `http://127.0.0.1:5000/am/map_op2` in the browser produces the following map, on which restaurants are indicated by markers:

Figure 6-6. *Map created with Folium using OpenStreetMap tiles. Data © OpenStreetMap contributors, licensed under the ODbL. Overlaid data retrieved with Overpass*

We will now enhance this map by using custom icons for the markers and associating them with tooltips containing the name of the restaurant and its style of food (available in the OSM data). Let's also add a layer control to enable the user to show or hide the restaurant markers.

To do this, we need to add or modify some statements (changes are shown in **bold** below):

```
...
        m = folium.Map(location=[53.27340, -9.05080], zoom_start=17)

        for feature in geojson_data['features']:
            tags = feature['properties'].get('tags', {})
            if 'name' in tags and 'cuisine' in tags:
                feature['properties']['desc'] = tags['name'] + ' (' +
                tags['cuisine'] + ')'
            elif 'name' in tags:
                feature['properties']['desc'] = tags['name']

        folium.GeoJson(
            geojson_data,
            name='Restaurants',
            marker = folium.Marker(icon=folium.Icon(icon='cutlery',
            prefix='fa', markerColor='orange')),
            tooltip=folium.GeoJsonTooltip(fields=['desc'],
            labels=False)
        ).add_to(m)

        folium.LayerControl().add_to(m)

        return render_template('maps/page_with_map.html',
            name="Map with OpenStreetMap data: restaurants in Galway",
            map_html=m._repr_html_())
...
```

- **Lines 27 to 32**: The content of the `geojson_data` GeoJSON-structured object is slightly changed. This change is necessary so that the `folium.GeoJson()` statement coming next can get access to the restaurant's name and food style (this statement cannot access data deeper than level 1 in the hierarchies of properties associated with features; therefore, this information must be brought to level 1).

  For that purpose, a loop is executed over the features (i.e., restaurants) contained in the `geojson_data` object. For each of them:

  - A list of its properties is obtained.
  - The existence of properties with names `tags/name` and `tags/cuisine` is checked.
  - Depending on their existence, a new property named `desc` is added just below the `properties` field. Its content is either the name and the food style of the restaurant or only the name (the new property is not added when none of these properties have been found).

  For example, here are the initial contents of one of the features (a restaurant):

```
{
    "type": "Feature",
    "properties": {
        "type": "node",
        "id": 1469414003,
        "tags": {
            "addr:city": "Galway",
            "addr:housenumber": "21",
            "addr:postcode": "H91 XR58",
            "addr:street": "Quay Street",
            "amenity": "restaurant",
            "check_date": "2024-11-19",
            "cuisine": "seafood",
            "name": "Martine's",
            "opening_hours": "We-Su 17:00-23:30",
            "phone": "+353 91 565 662",
```

```
                "start_date": "1986",
                "website": "https://martines.ie/",
                "wheelchair": "yes"
            }
        },
        "geometry": {
            "type": "Point",
            "coordinates": [
                -9.0541877,
                53.2708975
            ]
        }
    }
```

After the change, with the “desc” property added, these contents become

```
{
    "type": "Feature",
    "properties": {
        "type": "node",
        "id": 1469414003,
        "tags": {
            "addr:city": "Galway",
            "addr:housenumber": "21",
            "addr:postcode": "H91 XR58",
            "addr:street": "Quay Street",
            "amenity": "restaurant",
            "check_date": "2024-11-19",
            "cuisine": "seafood",
            "name": "Martine's",
            "opening_hours": "We-Su 17:00-23:30",
            "phone": "+353 91 565 662",
            "start_date": "1986",
            "website": "https://martines.ie/",
            "wheelchair": "yes"
        },
```

```
                "desc": "Martine's (seafood)"
            },
            "geometry": {
                "type": "Point",
                "coordinates": [
                    -9.0541877,
                    53.2708975
                ]
            }
        }
```

- **Lines 34 to 39**: A `Folium.GeoJson` object is created with the GeoJSON-structured object (completed with the `desc` property precedingly), then added to the map. Other parameters are specified:
    - A `name` parameter that will be used in the layer control panel to show or hide the restaurant markers.
    - A `marker` parameter to customize the markers' style: they will be orange and contain a cutlery icon (the icon can be found at `https://fontawesome.com/v4/cheatsheet/`, and the supported colors are `red`, `blue`, `green`, `orange`, `yellow`, `purple`, `darkred`, `white`, `black`, and `cadetblue`).
    - A `tooltip` parameter to associate a tooltip with each marker. The text of this tooltip will be the content of the `desc` attribute, set on lines 27 to 32.
- **Line 41**: A layer control is added to the map, so that the restaurant markers can be shown or hidden at will by the user.

Launching the server and opening the URL `http://127.0.0.1:5000/am/map_op2` with the browser produces the following map, after placing the mouse cursor on one of the restaurants:

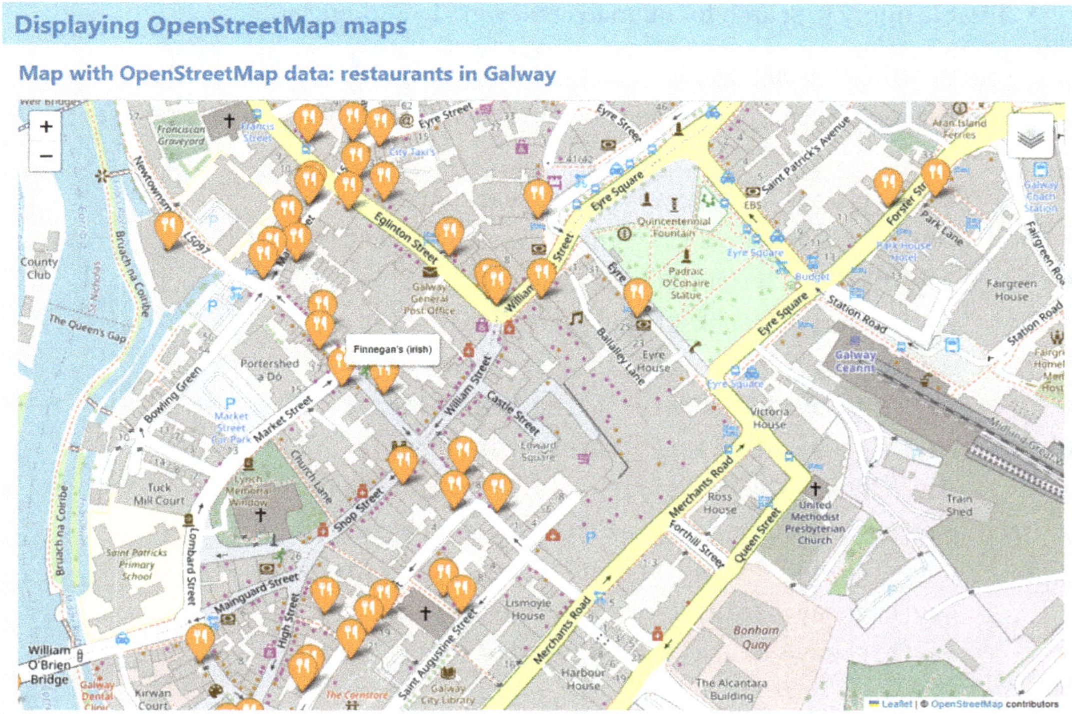

Figure 6-7. Map created with Folium using OpenStreetMap tiles. Data © OpenStreetMap contributors, licensed under the ODbL. Overlaid data retrieved with Overpass

Using a Mixed Approach: Data Retrieved on the Server and Map Created on the Client

This prototype will build a map of the city of Sedona, Arizona, highlighting the carriageways. A mixed processing scenario will be used:

1. On the server, Python code will query OpenStreetMap via the Overpass HTTP API. The response will then be converted to GeoJSON format and injected into an HTML page using a template.

2. When this HTML page is received and executed in the browser, JavaScript code including Leaflet statements (defined in the template) will create a map and enhance it with GeoJSON data returned by Overpass and provided by the server's Python code.

A suitable query to search for all roads and ways in Sedona is:

```
1   [out:json][timeout:25];
2   area[name="Sedona"][wikipedia="en:Sedona, Arizona"]->.searchArea;
3    (
4       way[highway="motorway"](area.searchArea);
5       way[highway="motorway_link"](area.searchArea);
6       way[highway="trunk"](area.searchArea);
7       way[highway="trunk_link"](area.searchArea);
8       way[highway="primary"](area.searchArea);
9       way[highway="secondary"](area.searchArea);
10      way[highway="tertiary"](area.searchArea);
11      way[highway="unclassified"](area.searchArea);
12      way[highway="residential"](area.searchArea);
13      way[highway="living_street"](area.searchArea);
14      way[highway="service"](area.searchArea);
15      way[highway="track"](area.searchArea);
16  );
17  out geom;
```

- **Line 1**: The data returned by Overpass is requested in JSON format, and a maximum processing time of 25 seconds is allowed.
- **Line 2**: A search zone limited to the area of Sedona, Arizona, is defined.
- **Lines 3 to 16**: Overpass is requested to search for all ways:
    - With the tag of key `highway` having one of the following values: “motorway”, “motorway_link”, “trunk”, “trunk_link”, “primary”, “secondary”, “tertiary”, “unclassified”, “residential”, “living_street”, “service”, or “track” (these are the values used by OSM to specify the different types of roads and ways).
    - And located in the search zone (filter: `(area.searchArea)`).

The 12 `way` statements in sequence, enclosed in parentheses, form a union: the ways selected by each of these `way` statements are combined to form the response.

Note that these 12 way statements can be replaced by a single one with a regular expression:

```
...
way[highway~"^(motorway|motorway_link|trunk|trunk_link
|primary|secondary|tertiary|unclassified|residential|living_
street|service|track)$"](area.searchArea);
...
```

This new version of the query is more efficient as it performs a single search pass instead of 12.

- **Line 17**: This statement requests the ways found by the previous statements to be output. The geom option is necessary to obtain the geographic coordinates needed to plot the ways on the map.

After testing this request with Overpass Turbo, the following code must be added to the advanced_maps.py blueprint:

```
@bp.route('/am/map_op3')
def map_op3():
    import requests, json
    from osm2geojson import json2geojson

    try:
        query = """
            [out:json][timeout:25];
            area[name="Sedona"][wikipedia="en:Sedona, Arizona"]->.
            searchArea;
            way[highway~"^(motorway|motorway_link|trunk|trunk_link|pr
            imary|secondary|tertiary|unclassified|residential|living_
            street|service|track)$"](area.searchArea);
            out geom;
        """

        overpass_url = "https://overpass-api.de/api/interpreter"
        response = requests.post(overpass_url, data={'data': query})
```

```
16
17          if response.status_code == 200:
18              data = response.json()
19          else:
20              flash("Error retrieving data with Overpass API", category="error")
21              return render_template('maps/page_with_map.html', name="No map to display")
22
23          geojson_data = json2geojson(data)
24
25          return render_template('maps/page_with_roads_sedona.html',
26              name="Map with OpenStreetMap data: carriageways in Sedona",
27              geojson_data=geojson_data,
28              loc=[34.8589, -111.7714])
29
30      except Exception as e:
31          flash("An error has been encountered:<br/>" + str(e), category="error")
32          return render_template('maps/page_with_map.html', name="No map to display")
```

- **Lines 3 and 4**: The `requests` and `json` libraries and the `osm2geojson()` function are imported.
- **Line 6**: This line starts a `try/except` block for proper error handling.
- **Lines 7 to 12**: The Overpass query is prepared. Its content has already been explained above.
- **Line 14**: The URL of the Overpass HTTP API is defined.
- **Line 15**: An HTTP request containing the query prepared earlier is submitted to Overpass.
- **Line 17**: The result of the HTTP request is checked:

- **Line 18**: If the request was successful, the response returned by Overpass is parsed and placed in a JSON-structured object.
- **Lines 19 to 21**: These lines deal with the case where the HTTP request has failed. Processing is aborted, and an error message is displayed on a web page.

- **Line 23**: The JSON-structured object (containing the data returned by Overpass) is converted to a GeoJSON-structured object.
- **Lines 25 to 28**: A web page is rendered using the `maps/page_with_roads_sedona.html` template. Several arguments are passed:
    - The name “Map with OpenStreetMap data: carriageways in Sedona”.
    - The data received from Overpass and converted to GeoJSON format.
    - The coordinates on which to center the map (those of Sedona).
- **Lines 30 to 32**: These lines constitute the `except` part of the `try/except` block: an error message is prepared and displayed on a web page.

The template `page_with_roads_sedona.html`, used by the previous Python function, must be created in the `prototypes/templates/maps` directory with the following contents:

```
{% extends 'base_ll.html' %}

{% block title %}{{ name }}{% endblock %}

{% block header %}
<h2>{{ name }}</h2>
{% endblock %}

{% block script %}
<script id="geojson-data" type="application/json">
{{ geojson_data | tojson }}
</script>

```

```
14  <script type="text/javascript">
15  let map;
16  try {
17      const geojsonData = JSON.parse(document.getElementById("geojson-data").textContent);
18
19      map = L.map('map').setView({{ loc }}, 14);
20
21      L.tileLayer('https://tile.openstreetmap.org/{z}/{x}/{y}.png', {
22          maxZoom: 19,
23          attribution: '&copy; <a href="http://www.openstreetmap.org/copyright">OpenStreetMap</a>'
24      }).addTo(map);
25
26      L.geoJSON(geojsonData).addTo(map);
27
28  } catch(e) {
29      if (typeof map !== 'undefined' && map) {map.remove();}
30
31      const div = document.getElementById("map")
32      if (div) {div.remove();}
33
34      const elemDiv = document.createElement('div');
35      elemDiv.innerHTML = `<p>${e}</p>`;
36      document.body.appendChild(elemDiv);
37  }
38  </script>
39  {% endblock %}
```

- **Lines 10 to 12**: These lines are a JSON script into which the Jinja engine will place the contents of the `geojon_data` argument passed by the calling Python function. This is the data extracted from Overpass and converted to GeoJSON format. The `tojson` filter is used to convert the received string into a JSON-formatted string.

This technique was made necessary by the amount of data that can be returned by Overpass depending on the requests (potentially, several million characters). If we tried to receive this data more simply, for example, in a JavaScript variable with a statement such as:

```
const geojson_data = {{ geojson_data | tojson }};
```

This would work fine for Overpass queries returning little data, but would typically cause a syntax error when the amount of data returned becomes larger (browsers usually impose limits on the length of JavaScript statements, and these limits may be exceeded when the content of `geojson_data` becomes too large).

- **Line 17**: The GeoJSON data mentioned above is retrieved, parsed, and placed in a GeoJSON-structured object.
- **Line 19**: A map is created and centered on the coordinates received from the calling Python function (those of Sedona).
- **Lines 21 to 24**: A tile layer is associated with the map.
- **Line 26**: An `L.geoJSON` object is created with the GeoJSON-structured object obtained on line 17, and it is added to the map.

Launching the server and opening the address `http://127.0.0.1:5000/am/map_op3` in the browser produces the following map, with the roads and ways highlighted with blue polylines or polygons:

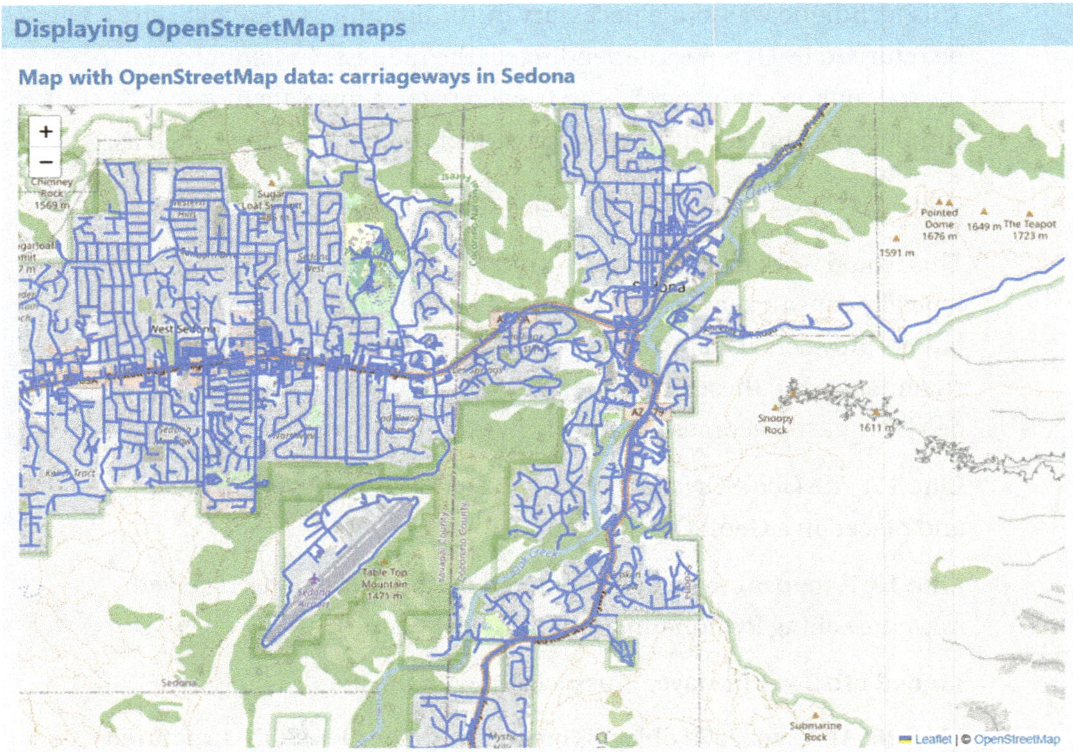

Figure 6-8. *Map created with Leaflet using OpenStreetMap tiles. Data © OpenStreetMap contributors, licensed under the ODbL. Overlaid data retrieved with Overpass*

Now we will modify the map so that road and street names can be displayed in tooltips. To do this, in the `page_with_roads_sedona.html` template, we must modify the statement adding GeoJSON data to the map as follows:

```
    ...
        }).addTo(map);

        L.geoJSON(geojsonData, {
            onEachFeature: function (feature, layer) {
                const tags = feature.properties?.tags;
                 if (tags && tags.name) {
                    layer.bindTooltip(feature.properties.tags.name,
                    {permanent: false, direction: "top"});
```

```
                }
            }
        }).addTo(map);

    } catch(e) {
    ...
```

A new option onEachFeature was added to the L.geoJSON() function. This option defines a function that, for all features present in the GeoJSON data (i.e., all ways returned by Overpass), checks whether it has a tag with the key name in its properties and, if so, adds a tooltip with the value of this name tag to the polyline or polygon resulting from this feature. The option permanent: false means that the tooltip will only be displayed when the mouse hovers over the polyline or polygon, and direction: "top" means that the tooltip will appear above the polyline or polygon (i.e., toward the top of the page).

After reloading the map, we can now display road and track names in tooltips by placing the mouse cursor over them.

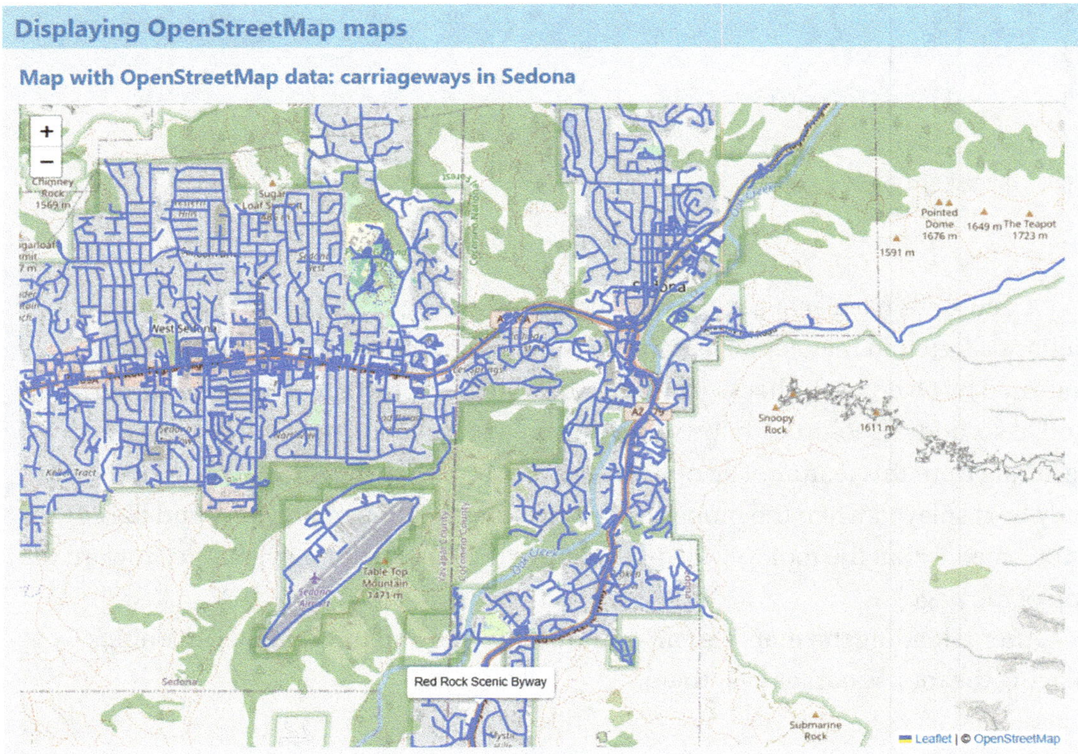

Figure 6-9. *Map created with Leaflet using OpenStreetMap tiles. Data © OpenStreetMap contributors, licensed under the ODbL. Overlaid data retrieved with Overpass*

Let's add an additional filter `[oneway!="yes"]` to the Overpass query in the `advanced_maps.py` blueprint, to select only roads and ways with two-way traffic. The Overpass query on lines 7 to 12 (in the blueprint) becomes:

```
...

7           query = """
8               [out:json][timeout:25];
9               area[name="Sedona"][wikipedia="en:Sedona, Arizona"]->.
                searchArea;
10              way[highway~"^(motorway|motorway_link|trunk|trunk_link|prima
                ry|secondary|tertiary|unclassified|residential|living_street|
                service|track)$"][oneway!="yes"](area.searchArea);
```

```
            out geom;
        """

    ...
```

The result is the following map:

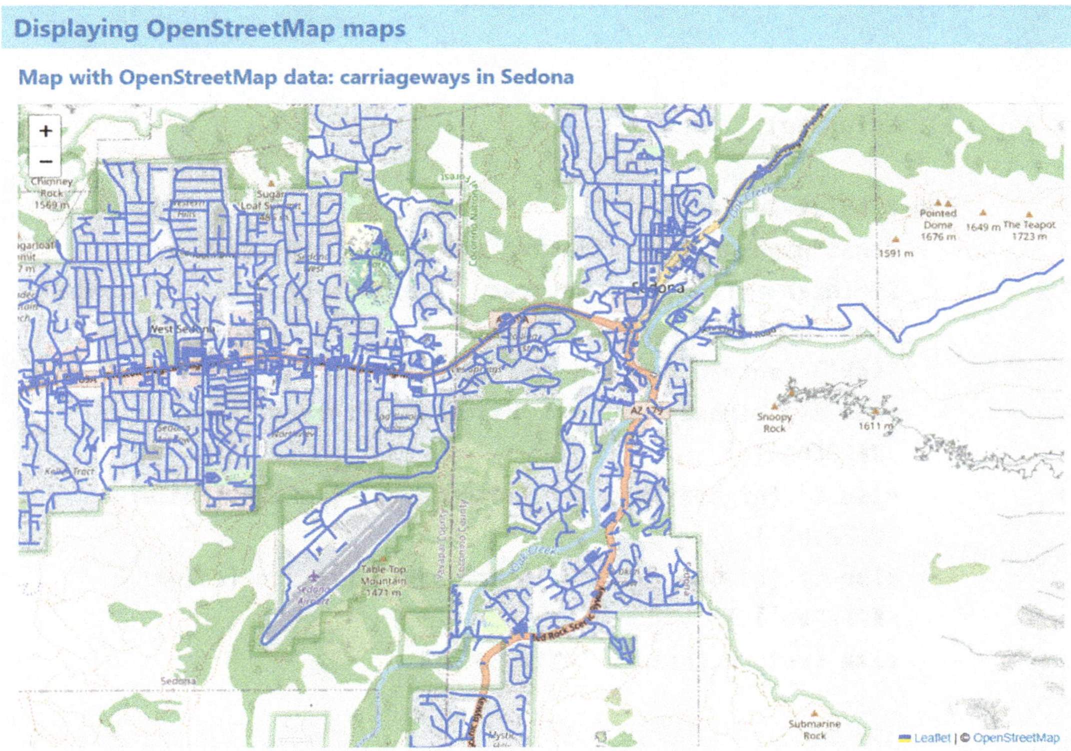

__Figure 6-10.__ Map created with Leaflet using OpenStreetMap tiles. Data © OpenStreetMap contributors, licensed under the ODbL. Overlaid data retrieved with Overpass

Compared with the previous map, some streets with one-way traffic have not been highlighted, especially the highway arriving from the south ("Red Rock Scenic Byway"), because it is registered in OSM as two one-way roads.

Let's remove the `[oneway!="yes"]` filter from the Overpass request now.

We will then add a final touch to our map: we are going to highlight the roads with different colors according to their type. In the following example, we will distinguish between freeways, primary, secondary, tertiary roads, and other streets and ways: roads of each of these types will be highlighted with different colors.

We will also add a layer control to make it possible to hide the polylines and polygons highlighting the ways.

The modifications to the `page_with_roads_sedona.html` template to implement these enhancements are as follows (shown in **bold**):

```
    ...
    }).addTo(map);

    function style(feature) {
        const highwayType = feature.properties.tags.highway;
        if (highwayType === 'motorway' || highwayType === 'motorway_link' || highwayType === 'trunk' || highwayType === 'trunk_link') {return {color: '#FF0000'};}
        else if (highwayType === 'primary') {return {color: '#FF8C00'};}
        else if (highwayType === 'secondary') {return {color: '#FFD700'};}
        else if (highwayType === 'tertiary') {return {color: '#FFFF00'};}
        else {return {color: '#32CD32'};}
    }

    const geoJSON = L.geoJSON(geojsonData, {
        style: style,
        onEachFeature: function (feature, layer) {
            const tags = feature.properties?.tags;
            if (tags && tags.name) {
                layer.bindTooltip(feature.properties.tags.name, { permanent: false, direction: "top" });
            }
        }
    }).addTo(map);

```

```
45        const layer = {
46            "Carriageways": geoJSON
47        };
48
49        L.control.layers({}, layer).addTo(map);
50
51    } catch(e) {
    ...
```

- **Lines 26 to 33**: A function named `style()` is defined for selecting and returning a color according to a road type:
    - For major roads (such as freeways), red is chosen.
    - For "primary" roads, orange is selected.
    - For "secondary" roads, golden yellow is chosen.
    - For "tertiary" roads, light yellow will be used.
    - Other roads and ways will be highlighted in green.
- **Line 36**: The `style()` function defined above is associated with the `style` option of the `L.geoJSON` object. This function will be called for each feature (i.e., carriageway) in the GeoJSON data to determine the color in which it should be represented on the map.
- **Lines 45 to 49**: A layer control is added to allow the map users to show or hide the polylines and polygons highlighting the ways.

After restarting the server and refreshing the map in the browser, the result is the following map:

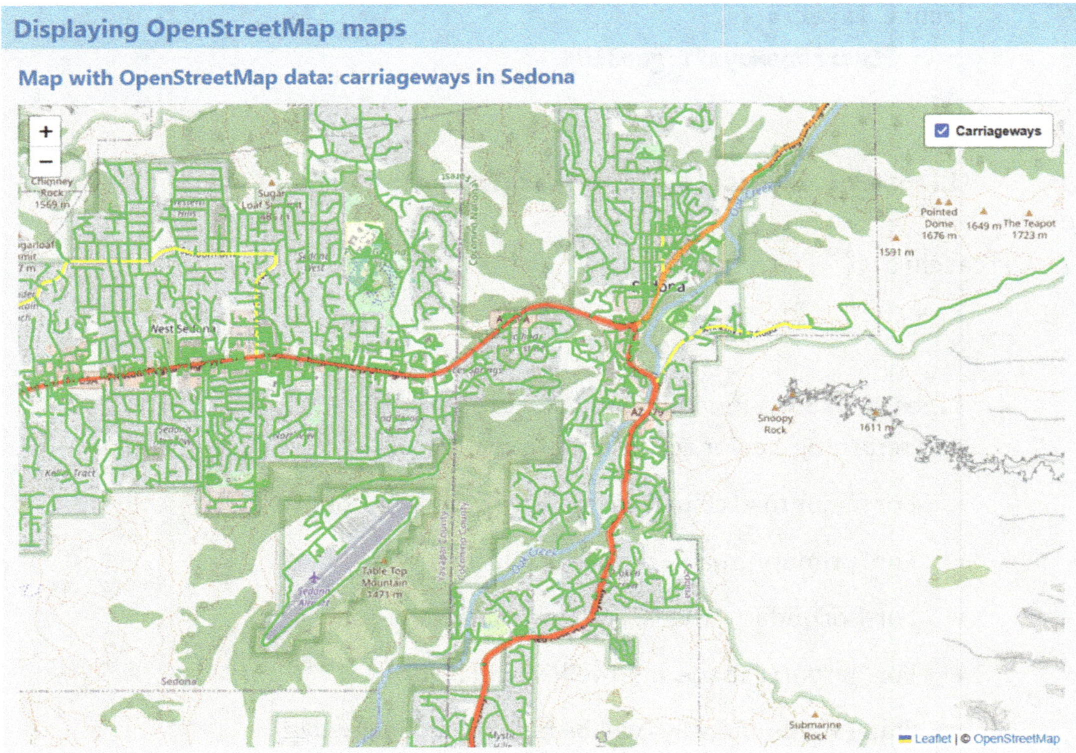

Figure 6-11. *Map created with Leaflet using OpenStreetMap tiles. Data © OpenStreetMap contributors, licensed under the ODbL. Overlaid data retrieved with Overpass*

Searching for Data and Creating Multiple Maps (on the Server with Folium and Python)

This prototype will search for all communities (cities, towns, and villages) with a given name and a population above a specified threshold and will build and display their maps. These communities and their geographic coordinates will be obtained by querying OSM data via the Overpass HTTP API from Python code running on the server. The maps will then be built using Folium, also on the server.

The same operations could also be performed entirely in the browser using JavaScript code with Leaflet or using a mixed approach: query Overpass with Python on the server, then build the maps with Leaflet and JavaScript in the browser.

Communities can be searched in OSM as nodes or relations:

- Looking for them as nodes, the `place` tag with the value "city", "town", or "village" must be used as a filter, and the coordinates of the point at the center of the community are returned.
- Looking for them as relations, you must filter by `[type=boundary]`. However, this filter alone will return all administrative boundaries (counties, states, countries, etc.), not just communities.

  To narrow the results to communities, you can add `[boundary=administrative][admin_level~"8|9|10"]`. Note, though, that some communities do not have an `admin_level` tag and will not be selected.

  Another option is to filter with `[type=boundary][border_type~"^(city|town|village)$"]`, but this will miss communities lacking the `border_type` tag.

  If you include `out geom;` in your query, the response will contain the community's boundary geometry (enabling you to represent it as a polygon on a map).

Below, we will be looking for communities as nodes.

A suitable Overpass query to search the world for all communities with the name "Paris" and a population greater than 5000 is as follows:

```
[out: json][timeout:25];
(
   node[name="Paris"][place="city"]
       (if:(is_number(t["population"]) &&
       number(t["population"])>= 5000));
   node[name="Paris"][place="town"]
       (if:(is_number(t["population"]) &&
       number(t["population"])>= 5000));
   node[name="Paris"][place="village"]
       (if:(is_number(t["population"]) &&
       number(t["population"])>= 5000));
);
out;
```

- **Line 1**: The data must be returned in JSON format, and a maximum processing time of 25 seconds is allowed.
- **Lines 2 to 12**: These lines form a union: The results returned by the three node statements are combined to form the query result:
  - **Lines 3 to 5**: All major community nodes (with the tag `place` equal to “city”), named “Paris”, and having a population greater than or equal to 5,000, are searched for.
  - **Lines 6 to 8**: All medium-sized community nodes (with the tag `place` equal to “town”), named “Paris”, and having a population greater than or equal to 5,000, are searched for.
  - **Lines 9 to 11**: All village nodes (with the tag `place` equal to “village”), named “Paris”, and having a population greater than or equal to 5,000, are searched for.

  This union with three node statements (lines 2 to 12) can be replaced by a single node statement using a regular expression to filter on the different community types:

  ```
  [out: json][timeout:25];
  node[name="Paris"][place~"^(city|town|village)$"]
      (if:(is_number(t["population"]) &&
      number(t["population"])>=5000));
  out;
  ```

  This new version of the query is more efficient as it performs a single search pass instead of three in the previous version.
- **Line 13**: This statement requests the nodes found by the previous statements to be included in the response. The `out;` statement is sufficient to output the geographic coordinates of the nodes.

After preparing and checking the query with Overpass Turbo, the following code must be added to the `advanced_maps.py` blueprint:

```
@bp.route('/am/map_op4')
def map_op4():
    import requests, re
```

```
from osm2geojson import json2geojson

city_raw = request.args.get('city', 'Paris')
city = re.sub(r'[^a-zA-Z0-9\s\-]', '', city_raw)
popmin = request.args.get('popmin', '0')

try:
    if isInt(popmin) != True:
        popmin = "0"

    query = '''
        [out:json][timeout:25];
        node[name="''' + city + '''"][place~"^(city|town|village)$"]
            (if:(is_number(t["population"]) &&
            number(t["population"])>=''' + popmin + '''));
        out;
    '''

    overpass_url = "https://overpass-api.de/api/interpreter"
    response = requests.post(overpass_url, data={'data': query})

    if response.status_code == 200:
        data = response.json()
    else:
        flash("Error retrieving data with Overpass API", category="error")
        return render_template('maps/page_with_map.html', name="No map to display")

    geojson_data = json2geojson(data)
    #print(geojson_data)

    maps_with_names = []
    for feature in geojson_data["features"]:
```

```
36              lat = feature["geometry"]["coordinates"][1]
37              lon = feature["geometry"]["coordinates"][0]
38              tags = feature["properties"].get("tags", {})
39              pop = int(tags.get("population", 0))
40
41              m = folium.Map(location=[lat, lon], zoom_start= get_zoom_
                from_population(pop), control_scale=True)
42
43              geojson_data_copy = geojson_data.copy()
44              geojson_data_copy["features"] = [feature]
45
46              folium.GeoJson(
47                  geojson_data_copy,
48                  name='City',
49                  marker=folium.Marker(icon=folium.Icon(color='red',
                    icon='info-sign'))
50              ).add_to(m)
51
52              folium.LayerControl().add_to(m)
53
54              maps_with_names.append([city + ", " + str(pop) + "
                inhabitants", m._repr_html_()])
55
56          return render_template('maps/page_with_several_maps.html',
                name="Maps of cities with the name of " + city + " and
                more than " + popmin + " inhabitants",
58              maps_with_names=maps_with_names)
59
60      except Exception as e:
61          flash("An error has been encountered:<br/>" + str(e),
            category="error")
62          return render_template('maps/page_with_map.html', name="No map
            to display")
```

- **Lines 3 and 4**: The `requests` and re libraries and the `osm2geojson()` function are imported.

- **Lines 6 to 8**: The `city` (name of the community to be searched for) and `popmin` (minimum population) parameters of the HTTP request are received and placed in variables.

  The `request` library, used here, is not imported in this snippet because it was already imported globally along with other libraries when the blueprint was created at the beginning of Part 2.

  The content of the `city` parameter is sanitized to prevent possible attacks, such as code injection.

- **Line 10**: This line starts a `try/except` block for error handling.
- **Lines 11 and 12**: The `popmin` parameter is checked – if it does not contain an integer value, 0 is used.
- **Lines 14 to 20**: The Overpass query is prepared.
- **Line 22**: The URL of the Overpass HTTP API is defined.
- **Line 23**: An HTTP request, using the Overpass query prepared earlier, is submitted to Overpass.
- **Line 25**: The result of the HTTP request is checked:
  - **Line 26**: If the request was successful, the response is parsed and placed in a JSON-structured object.
  - **Lines 27 to 29**: These lines deal with the case where the HTTP request has failed. Processing is aborted, and an error message is displayed on a web page.
- **Line 31**: The JSON-structured object is converted to a GeoJSON-structured object.
- **Line 32** (currently commented out): The GeoJSON content of the object returned by the `json2geojson()` function is printed in the terminal. This statement may be useful for debugging or fine-tuning the code: it enables the programmer to see the GeoJSON content obtained and analyze its structure, in order to identify the fields to be retrieved and their paths.
- **Line 34**: A list is created. It will hold the names and maps of all the communities present in the GeoJSON-structured object.

- **Lines 35 to 54**: A loop is executed over the features of the GeoJSON-structured object. For each feature:
    - **Lines 36 to 39**: The current feature's latitude, longitude, and population are retrieved.
    - **Line 41**: A map is created and centered on the feature's latitude and longitude coordinates. Its initial zoom level is calculated with the feature's population (using the `get_zoom_from_population()` function).
    - **Lines 43 and 44**: A copy of the GeoJSON-structured object containing only the current feature is created.
    - **Lines 46 to 50**: A `folium.GeoJson` object is created using the single-featured GeoJSON-structured object created previously, and it is added to the map. This data will add a red marker to the map at the center of the community.
    - **Line 52**: A layer control is added to enable the user to show or hide the marker identifying the community's position.
    - **Line 54**: The map, converted to HTML and JavaScript code, and an associated name ("<city>, <population> inhabitants") are added to the `map_with_names` list.
- **Lines 56 to 58**: A web page is rendered using the `maps/page_with_several_maps.html` template. Arguments are passed: the name "Maps of cities with the name <city> and more than <population> inhabitants" and the list of maps and names constructed by the previous statements.
- **Lines 60 to 62**: These lines form the `except` part of the `try/except` block. If an error is caught, an error message is prepared and displayed on a web page.

The preceding code is completed by the `isInt()` function, also placed in the `advanced_maps.py` blueprint. This function checks that the value received as an argument is an integer or can be converted to an integer.

```
def isInt(var):
    try:
        int(var)
```

```
        return True
    except ValueError:
        return False
```

The preceding code is also completed by the `get_zoom_from_population()` function, also placed in the `advanced_maps.py` blueprint. This function receives a number of inhabitants as an argument and returns an appropriate zoom level for a community with this population (beware that this logic has its limits: depending on the continent or country, communities may be more concentrated or more spread out).

```
def get_zoom_from_population(pop):
    if pop < 1000:
        return(16)
    elif pop < 10000:
        return(14)
    elif pop < 100000:
        return(13)
    elif pop < 1000000:
        return(12)
    elif pop < 5000000:
        return(11)
    else:
        return(10)
```

The preceding Python code is finally completed by the new HTML template `page_with_several_maps.html`, to be created in the `prototypes/templates/maps` directory. This template is responsible for displaying the maps for the communities returned by the Overpass query. Its contents are as follows:

```
{% extends 'base.html' %}

{% block title %}{{ name }}{% endblock %}

{% block header %}
<h2>{{ name }}</h2>
<style>
.accordion {
```

```
        background-color: #eee;
        color: #444;
        cursor: pointer;
        padding: 18px;
        width: 100%;
        border: none;
        text-align: left;
        outline: none;
        font-size: 15px;
        transition: 0.4s;
    }

    .active, .accordion:hover {
        background-color: #ccc;
    }

    .panel {
        padding: 0 18px;
        display: none;
        background-color: white;
        overflow: hidden;
    }
    </style>
    {% endblock %}

    {% block content %}
    {% if maps_with_names is defined %}
        {% for map_with_name in maps_with_names %}
            <button class="accordion">{{ map_with_name[0] }}</button>
            <div class="panel">
            <table frame="box"><tr>
                <td width="800px" rule="yes">{{ map_with_name[1] |
                safe }}</td>
            </tr></table>
            <p/>
```

```
43          </div>
44      {% endfor %}
45  {% endif %}
46  {% endblock %}
47
48  {% block script %}
49  <script>
50  let acc = document.getElementsByClassName("accordion");
51  let i;
52
53  for (i = 0; i < acc.length; i++) {
54      acc[i].addEventListener("click", function() {
55          this.classList.toggle("active");
56          let panel = this.nextElementSibling;
57          if (panel.style.display === "block") {
58              panel.style.display = "none";
59          } else {
60              panel.style.display = "block";
61          }
62      });
63  }
64  </script>
65  {% endblock %}
```

This page uses an "accordion" layout to display the maps (for more details, see `https://www.w3schools.com/howto/howto_js_accordion.asp`): each map is placed in a section initially reduced to a title, which map can be shown or hidden by clicking on the section title.

- **Lines 7 to 31**: These lines contain the style statements needed to use the accordion layout.
- **Line 35**: This Jinja statement checks whether a list of maps with names has been received from the Python blueprint.
- **Lines 36 to 44**: An iteration loop is executed over the list of maps with names (this is Jinja code executed on the server while an HTML page is generated from the template). For each item:

    - **Line 37**: A section title is created with the name extracted from the current item.
    - **Lines 38 to 43**: This is the section that can be shown or hidden on demand, thanks to the accordion layout.
    - **Line 40**: The map extracted from the current item is inserted in a table cell.
- **Lines 50 to 63**: These lines contain the JavaScript code needed to use the accordion layout.

Launching the server and opening the URL `http://127.0.0.1:5000/am/map_op4?city=Paris&popmin=5000` in the browser produces the following page:

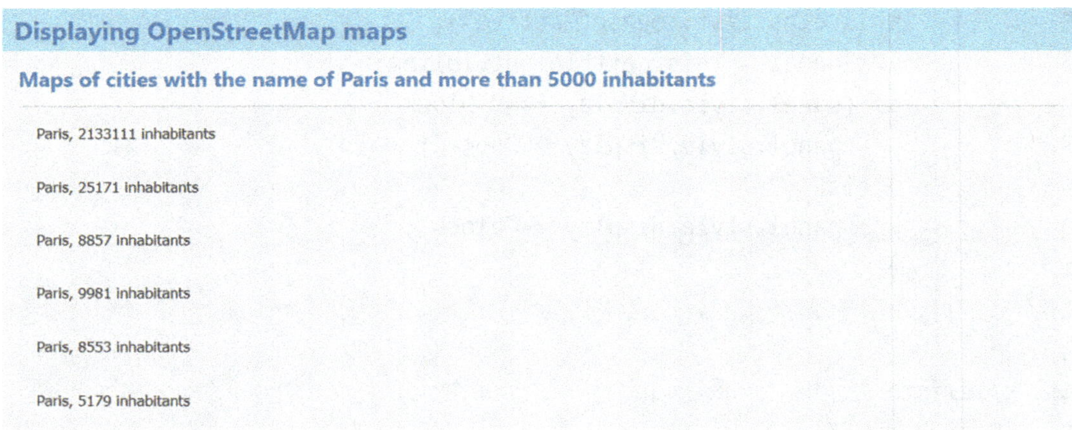

***Figure 6-12.** Cities named Paris with their populations*

Six communities named Paris and with more than 5,000 inhabitants have been found, but their maps are not currently displayed. Clicking on the first entry brings up a map of the French capital.

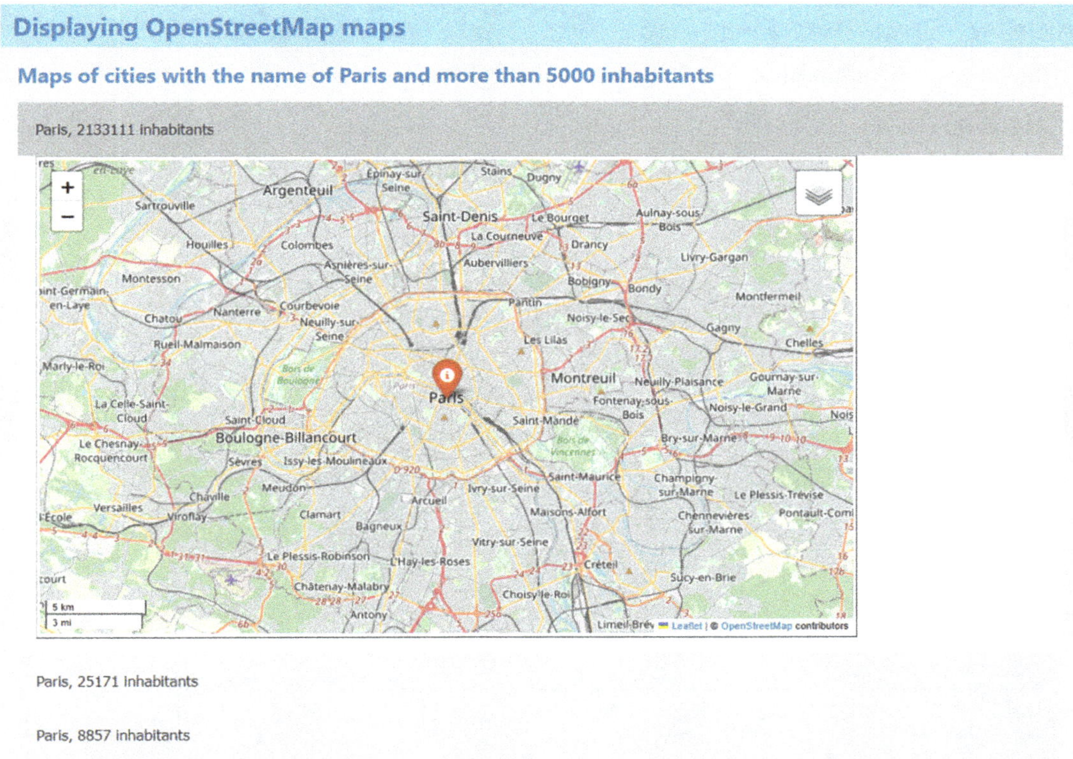

Figure 6-13. *Map created with Folium using OpenStreetMap tiles. Data © OpenStreetMap contributors, licensed under the ODbL. Overlaid data retrieved with Overpass*

Clicking on the second entry brings up a map of Wim Wenders's "Paris, Texas".

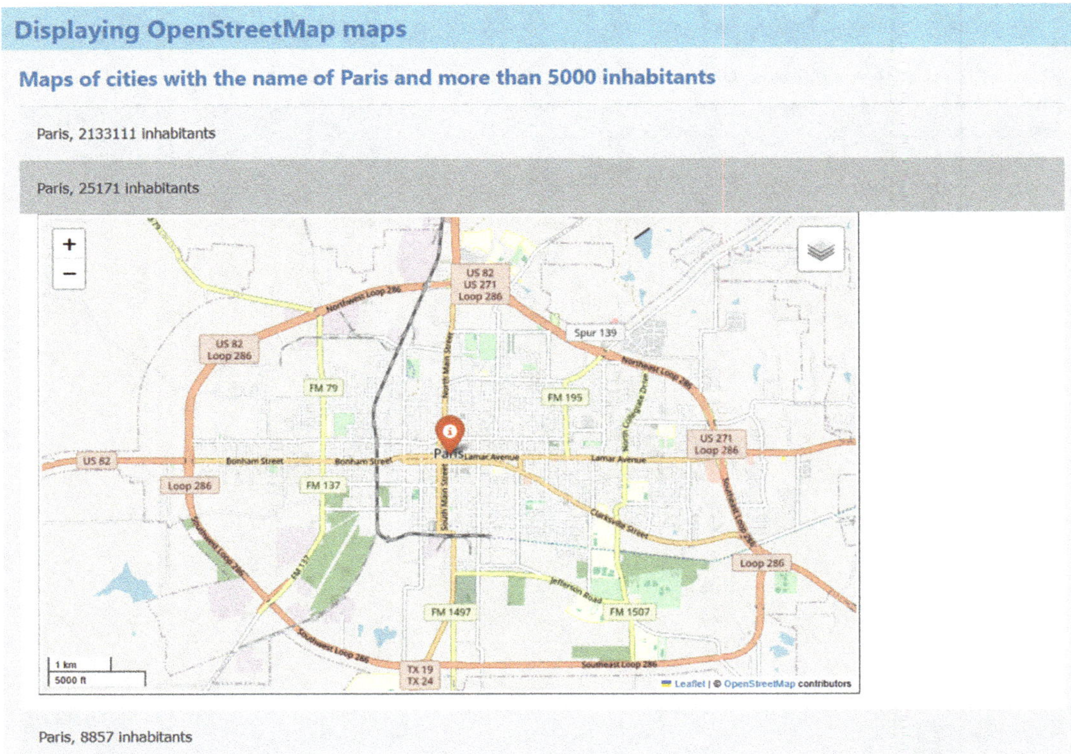

Figure 6-14. *Map created with Folium using OpenStreetMap tiles. Data © OpenStreetMap contributors, licensed under the ODbL. Overlaid data retrieved with Overpass*

Maps of the four other communities named Paris found (all in the USA) can also be displayed in the same way.

A final note: The above prototype finds many of the world's cities, but not all of them. For example, Beijing, Mexico City, and Rome are not found because the `name` tag of the Beijing node contains the value "北京市", that of Mexico City contains "Ciudad de México", and that of Rome contains "Roma". It is by these names in the local language and script that these cities should be searched. To work around this difficulty, the Overpass query can be modified as follows (in the `advanced_maps.py` blueprint), using the `name:en` tag:

```
[out: json][timeout:25];
(
    node[name="<city name>"][place~"^(city|town|village)$"]
```

```
        (if:(is_number(t["population"]) &&
        number(t["population"])>=5000));
    node["name:en"="<city_name>"][place~"^(city|town|village)$"
        (if:(is_number(t["population"])
        number(t["population"])>=5000));
);
out;
```

or using a regular expression:

```
[out: json][timeout:25];
node[~"^(name|name:en)$"~"^<city_name>$"][place~"^(city|town|village)$"
    (if:(is_number(t["population"])
    number(t["population"])>=5000));
out;
```

With this final modification (in **bold**) in our Python code:

```
...

        query = '''
            [out:json][timeout:25];
            node[~"^(name|name:en)$"~"^''' + city + '''$"][place~"^(ci
            ty|town|village)$"]
                (if:(is_number(t["population"]) &&
                number(t["population"])>=''' + popmin + '''));
            out;
        '''

...
```

communities are searched by their local name or by their English name (searching for communities by their English name only wouldn't work either, as many do not have one). Modified in this way, the query can find Beijing, Mexico City, and Rome.

The type of search we have just carried out – searching for communities with a given name, finding their coordinates, and displaying their maps – falls under a discipline called geocoding (which will be explained in the next chapter). Overpass is not particularly suited to this type of use. In the next chapter, we will use Nominatim, another OSM API designed for geocoding, to implement a similar scenario.

Alternative Solutions with Leaflet Plugins

In the previous sections, we demonstrated how to write Python and JavaScript code to query OpenStreetMap using an Overpass API request and how to use Leaflet or Folium to build a map that includes the data returned by Overpass. However, some Leaflet plugins also enable querying OSM with Overpass, sometimes directly from the map:

- **Leaflet Overpass Layer** (developed by Guillaume Amat): This plugin simplifies querying OSM via Overpass by introducing a new object named `L.OverpassLayer`. This object can be used to execute an Overpass request, then to add the retrieved data to a map (as markers or polylines). Both the Overpass query (based on the Overpass HTTP API) and the JSON to GeoJSON conversion are managed internally by the plugin.
- **leaflet.osmdatapicker** (developed by Amr Hamadeh): This plugin provides a new control that allows users to draw an area on the map and specify a selection criterion (e.g., `amenity: bar`). The plugin then queries OSM using Overpass and highlights the elements matching the selection criterion in the designated area on the map.

CHAPTER 7

Using Geocoded Data in Maps

Geocoding is the process of converting an address (e.g., "Via Veneto 202, Rome, Italy") into geographic coordinates (latitude and longitude) for use on a map or by a positioning system. The reverse process is called **reverse geocoding**: it obtains an address from geographic coordinates (latitude and longitude).

The aim of the prototypes in this chapter is to query OSM data via the **Nominatim API** to find either the coordinates of a place from its name or address (geocoding), or the name or address of a place from its coordinates (reverse geocoding), then use **Leaflet** or **Folium** to present the search result on a map.

Nominatim is a geocoding engine based on OpenStreetMap. Unlike Overpass, which is a general-purpose API, Nominatim is designed for geocoding (searching for a place's coordinates using its name or address) and reverse geocoding (searching for a place's address using its coordinates). Nominatim searches are not performed in the OSM database, but in a dedicated database that is optimized for geocoding and regularly synchronized with the OSM database.

Nominatim has a Web user interface, already seen in Chapter 3, and offers an API for programmatic (i.e., automatic from-application) searches. It is the latter that we are going to focus on.

The **Nominatim API** and its query syntax have already been described in Chapter 3. In the following prototypes, it will be used:

- Either by Python code on the server, via HTTP requests submitted using the `requests` Python library.
- Or from JavaScript code executed in the browser, via HTTP requests submitted using the `fetch()` JavaScript function.

P. Marie, *Modern Web Cartography*, https://doi.org/10.1007/979-8-8688-2335-0_7

As already explained in Chapter 3, there is a public endpoint for using the Nominatim API (`https://nominatim.openstreetmap.org/`), but it is intended for demonstration, education, and low-volume testing. For production use and high-throughput applications, it is recommended to self-host a Nominatim server – either by setting up the infrastructure within your organization or by using a third-party Nominatim service. For more details about Nominatim's usage policy, check the dedicated section in Chapter 3.

In the following examples, we will use `https://nominatim.openstreetmap.org/` endpoint.

Geocoding

Geocoding (or **direct geocoding**) consists of searching for the coordinates of a location using its name or address.

Finding Coordinates from an Address on the Server and Creating a Map on the Client with Leaflet

This prototype will search for communities (cities, towns, and villages) with a given name, passed as a request parameter, and will construct and display their maps. Community population will not be used as a selection criterion, unlike in the previous prototype (in the last chapter).

- Querying Nominatim for communities and finding information about them will be carried out in a Python function in the blueprint (server side).
- The maps will be created and displayed using Leaflet and client-side JavaScript code.

A suitable Nominatim query to search for all communities with the name "Rome" is as follows:

```
https://nominatim.openstreetmap.org/search?city=Rome&format=geojson&polyg
on_geojson=1&dedupe=1&extratags=1
```

- The `city=Rome` parameter requests a search for communities named Rome. This is not limited to large cities: towns and villages can also be included in the results.
- The `format=geojson` parameter requests that the response be returned in GeoJSON format.
- The `polygon_geojson=1` parameter requests that the coordinates of the vertices of a polygon delimiting the city's outline be included in the response (this will allow the city's territory to be visualized on the map).
- The `dedupe=1` parameter requests that, if cities are found more than once, they are only included once in the response.
- The `extratags=1` parameter requests that additional city properties be included in the response (notably their population, which we will display in a popup).

Such a request can easily be tested with a browser. The response is as follows (in GeoJSON format):


```
{
  "type": "FeatureCollection",
  "licence": "Data © OpenStreetMap contributors, ODbL 1.0. http://osm.org/
  copyright",
  "features": [
    {
      "type": "Feature",
      "properties": {
        "place_id": 73132244,
        "osm_type": "relation",
        "osm_id": 41485,
        "place_rank": 16,
        "category": "boundary",
        "type": "administrative",
        "importance": 0.8439277725492005,
        "addresstype": "city",
        "name": "Rome",
```

```
    "display_name": "Rome, Roma Capitale, Latium, Italie",
    "extratags": {
      "ele": "21",
      "rank": "0",
      "capital": "yes",
      "website": "https://www.comune.roma.it/",
      "gfoss_id": "5002",
      "wikidata": "Q220",
      "ref:ISTAT": "058091",
      "wikipedia": "it:Roma Capitale",
      "is_capital": "country",
      "population": "2848084",
      "ref:catasto": "H501",
      "linked_place": "city",
      "population:date": "2019",
      "capital_ISO3166-1": "yes",
      "old_official_name": "Comune di Roma",
      "source:population": "31.01.2013, http://www.demo.istat.it/
      bilmens2013gen/query.php?lingua",
      "wikimedia_commons": "Category:Rome"
    }
  },
  "bbox": [
    12.2344669,
    41.6556417,
    12.8557603,
    42.1410285
  ],
  "geometry": {
    "type": "MultiPolygon",
    "coordinates": [
      [
        [
          [
            12.2344669,
```

```
            41.7400461
          ],
          [
            12.2345933,
            41.7396553
          ],
          ...
```


This response includes a list of features: communities named "Rome". For each community, the returned data includes:

- A set of **properties**, mostly derived from the OSM `tags` field. These properties contain metadata such as the community's full name (including province, region, and country), its status as a capital, and its population.
- The coordinates of a **bounding box** that encloses the community's boundaries.
- A `geometry` section containing the precise coordinates that define the community's area boundaries.

Please remember that Nominatim queries do not always return all places matching the search criteria. When multiple matches are found, Nominatim prioritizes them and may return only those with higher importance.

To perform this type of search, the following code must be added to the `advanced_maps.py` blueprint:

```
@bp.route('/am/map_nom')
def map_nom():
    import requests, re, copy

    city = re.sub(r'[^a-zA-Z0-9\s\-]', '', request.args.get('city', 'Paris'))

    try:
        user_agent = {'User-Agent': 'CartoOSM/1.0 (patrick@carto-osm.fr)'}
        params = {
            "city": city,
```

```
            "format": "geojson",
            "polygon_geojson": 1,
            "dedupe": 1,
            "extratags": 1
        }
        response = requests.get(
            "https://nominatim.openstreetmap.org/search",
            params=params,
            headers = user_agent)

        if response.status_code == 200:
            data = response.json()
        else:
            flash("Error retrieving data with Nominatim API", category="error")
            return render_template('maps/page_with_map.html', name="No map to display")
        #print(response)

        geojson_data_with_names_and_locs = []
        for feature in data["features"]:
            if "addresstype" in feature["properties"] and \
            (feature["properties"]["addresstype"] == "city" or \
            feature["properties"]["addresstype"] == "town" or \
            feature["properties"]["addresstype"] == "village") and \
            "bbox" in feature and "display_name" in feature["properties"]:
                lat1 = feature["bbox"][1]
                lon1 = feature["bbox"][0]
                lat2 = feature["bbox"][3]
                lon2 = feature["bbox"][2]
                bbox = [[lat1, lon1], [lat2, lon2]]

                lat_med = (float(lat1) + float(lat2))/2
                lon_med = (float(lon1) + float(lon2))/2
                loc = [lat_med, lon_med]
```

```
44
45                  disp_name = feature["properties"]["display_name"]
46
47                  single_feature_data = copy.deepcopy(data)
48                  single_feature_data["features"] = [feature]
49
50                  geojson_data_with_names_and_locs.append([disp_name,
                    loc, bbox, single_feature_data])
51
52          return render_template('maps/page_with_several_maps_ll.html',
53              name="Maps of cities with the name " + city,
54              geojson_data_with_names_and_locs=geojson_data_with_names_
                and_locs)
55
56      except Exception as e:
57          flash("An error has been encountered:<br/>" + str(e),
            category="error")
58          return render_template('maps/page_with_map.html', name="No map
            to display")
```

- **Line 3**: The `requests`, `re`, and `copy` libraries are imported.
- **Line 5**: The `city` parameter of the HTTP request is received and sanitized against possible attacks (code injection).
- **Line 7**: This line starts a `try/except` block for error handling.
- **Lines 8 to 19**: These lines prepare the Nominatim request:
    - **Line 8**: An HTTP header is prepared to identify the client application "CartoOSM/1.0 (patrick@carto-osm.fr)" is used. It is required for the request to be accepted by the Nominatim server: without this header, the server blocks the request and returns a "403" error code.

      It is recommended to use an identifier specific to your application and preferably to include the version number and your email address as well. For example: `MyApp/1.0 (myemail@example.com)`.

- **Lines 9 to 15**: The parameters to be included in the request are prepared (they have already been explained above).
- **Lines 16 to 19**: A geocoding request is submitted to Nominatim – it searches for all communities matching the name provided as a parameter and specifies that the response should:
    - Be returned in GeoJSON format.
    - Include polygons delimiting the cities' territories.
    - Remove duplicates.
    - Contain additional properties (such as the cities' populations).

    The header prepared on line 8 is added to the request using the `headers` parameter.

  Here is an example of how to test this request with cURL:

  ```
  curl -s -G "https://nominatim.openstreetmap.org/search" ^
  --data-urlencode "city=Paris" ^
  --data-urlencode "format=geojson" ^
  --data-urlencode "polygon_geojson=1" ^
  --data-urlencode "dedupe=1" ^
  --data-urlencode "extratags=1" ^
  --user-agent "curl 8.13.0 (Windows)"
  ```

- **Line 21**: The result of the HTTP request is checked:
    - **Line 22**: If the request was successful, the response is parsed and placed in a GeoJSON-structured object.
    - **Lines 23 to 25**: These deal with the case where the request has failed. Processing is aborted, and an error message is displayed on a web page.
- **Line 26** (currently commented out): The response returned by Nominatim is printed in the terminal. This statement is useful for debugging and fine-tuning the code – it enables the programmer to see the GeoJSON content obtained and analyze its structure, in order to identify the fields of interest and their paths.

- **Line 28**: An empty list is created with the name `geojson_data_with_names_and_locs`. It will be used to store data to be passed to the HTML page and intended for Leaflet. Each item of the list will correspond to a community found and will hold:

  1. The full name of the community.
  2. Its location.
  3. The coordinates of an encompassing rectangle.
  4. GeoJSON data describing the contours of the community's territory.

- **Lines 29 to 50**: A loop is executed over the features of the GeoJSON-structured object (i.e., the communities found by Nominatim). For each of them:

  - **Lines 30 to 34**: Only communities with `addresstype`, `bbox`, and `display_name` attributes, and for which the `addresstype` attribute is "city", "town", or "village", are taken into account. The others are discarded.
  - **Lines 35 to 39**: The latitude and longitude of the bottom left and top right corners of a rectangle encompassing the current community ("bounding box") are retrieved and stored in a compound list named `bbox`.
  - **Lines 41 to 43**: The coordinates of the current community are calculated, using those of the rectangle encompassing it, and stored in a list (the coordinates of the community center are not returned by Nominatim, so they have to be calculated from those of the rectangle encompassing the community). We could have obtained these coordinates from Nominatim by using `"format=json"` (instead of `"format=geojson"`) in the request.
  - **Line 45**: The detailed name of the current community (also including the county, region, and country) is retrieved.
  - **Lines 47 and 48**: A copy of the JSON-structured object containing the data returned by Nominatim is created and modified to retain only the data related to the current community.

- **Line 50**: The data gathered for the current community is added to the `geojson_data_with_names_and_locs` list (created on line 28).

- **Lines 52 to 54**: A web page is rendered using the `maps/page_with_several_maps_ll.html` template. The name “Map of cities with the name <city >” and the `geojson_data_with_names_and_locs` list (including various data about the communities found) are passed as arguments.

- **Lines 56 to 58**: These lines form the `except` part of the `try/except` block. In the event of an error, an error message is prepared and displayed on a web page.

The template `page_with_several_maps_ll.html`, invoked by the previous Python function, must be created in the `prototypes/templates/maps` directory with the following contents:

```
<!doctype html>
<html lang="en">
<head>
    <meta charset="utf-8">
    <title>{{ name }} - Prototypes</title>
    <link rel="stylesheet" href="{{ url_for('static', filename='style.css') }}">
    <link rel="stylesheet" href="https://unpkg.com/leaflet@1.9.4/dist/leaflet.css" integrity="sha256-p4NxAoJBhIIN+hmNHrzRCf9tD/miZyoHS5obTRR9BMY=" crossorigin=""/>
</head>
<body>
<nav>
    <h1>Displaying OpenStreetMap maps</h1>
</nav>
<section class="content">
    <header>
    <h2>{{ name }}</h2>
    </header>
    {% for message in get_flashed_messages() %}
    <div class="flash">{{ message | safe }}</div>
```

```
    {% endfor %}
    {% if geojson_data_with_names_and_locs is defined %}
        {% for item in geojson_data_with_names_and_locs %}
            <h3>{{ item[0] }} [{{ item[1][0] }}, {{ item[1][1] }}]</h3>
            <div id="map{{ loop.index }}" style="width: 930px; height: 520px;"></div>
            <br/>
            <br/>
        {% endfor %}
    {% endif %}
</section>
<script src="https://unpkg.com/leaflet@1.9.4/dist/leaflet.js" integrity="sha256-20nQCchB9co0qIjJZRGuk2/Z9VM+kNiyxNV1lvTlZBo=" crossorigin=""></script>
<script type="text/javascript">
window.onerror = function(message, url, linenumber) {
    if (typeof map !== 'undefined' && map) {map.remove();}

    const elemDiv = document.createElement('div');
    elemDiv.innerHTML = '<p>' + message + ', ligne : ' + linenumber + '</p>';
    document.body.appendChild(elemDiv);
}
</script>

{% if geojson_data_with_names_and_locs is defined %}
    {% for item in geojson_data_with_names_and_locs %}
        <script id="geojson-data{{ loop.index }}" type="application/json">
        {{ item[3] | tojson }}
        </script>
    {% endfor %}
{% endif %}

```

```
<script>
{% if geojson_data_with_names_and_locs is defined %}
    {% for item in geojson_data_with_names_and_locs %}
        try {
            const geojsonData{{ loop.index }} = JSON.parse(document.getElementById("geojson-data{{ loop.index }}").textContent);

            map = L.map('map{{ loop.index }}').setView([{{ item[1][0] }}, {{ item[1][1] }}], 10);

            L.tileLayer('https://tile.openstreetmap.org/{z}/{x}/{y}.png', {
                maxZoom: 19,
                attribution: '&copy; <a href="http://www.openstreetmap.org/copyright">OpenStreetMap</a>'
            }).addTo(map);

            L.geoJSON(geojsonData{{ loop.index }}).addTo(map);

            if (geojsonData{{ loop.index }}.features && geojsonData{{ loop.index }}.features.length > 0) {
                const props = geojsonData{{ loop.index }}.features[0].properties;
                if (props && props.extratags && props.extratags.population) {
                    L.popup()
                    .setLatLng([{{ item[1][0] }}, {{ item[1][1] }}])
                    .setContent("Population: " + geojsonData{{ loop.index }}.features[0].properties.extratags.population + " inhabitants")
                    .openOn(map);
                }
            }

```

```
73              map.fitBounds({{ item[2] }});
74          } catch(e) {
75              if (typeof map !== 'undefined' && map) {map.remove();}
76
77              const div = document.getElementById('map{{ loop.index }}');
78              if (div) {div.innerHTML = `<p>${e}</p>`;}
79          }
80      {% endfor %}
81  {% endif %}
82  </script>
83  </body>
84  </html>
```

- **Lines 21 to 26**: A loop is executed over the data list (containing information about the communities found using Nominatim and passed by the calling Python function). Remember that this data list holds:

  1. The full name of the community.
  2. Its location.
  3. The coordinates of an encompassing rectangle.
  4. GeoJSON data describing the contours of the community's territory.

  This is Jinja code executed on the server while an HTML page is built from the template. For each community:

  - **Line 22**: A level 3 header (`<h3>`) is added with the community's detailed name and geographic coordinates.
  - **Line 23**: A division is added with a specific identifier (`map1`, `map2`, `map3`, etc.). It is intended to receive the city map built with Leaflet.

- **Lines 41 to 45**: Another loop is executed over the data list (containing information about the communities found using Nominatim and passed by the calling Python function). This is Jinja code executed on the server while an HTML page is built from the template. For each community:
    - **Lines 42 to 44**: A JSON script section with a specific identifier (`geojson-data1`, `geojson-data2`, `geojson-data3`, etc.) is added.
    - **Line 43**: The GeoJSON data returned by Nominatim for this community is stored in this script.
- **Lines 50 to 80**: Another loop is executed over the data list (containing information about the communities found using Nominatim and passed by the calling Python function). For each community:
    - **Line 52**: The GeoJSON data is retrieved in a script section.
    - **Line 54**: A map is created and centered on the coordinates of the community. The identifier of the division in which the map is to be placed is specified.
    - **Lines 56 to 59**: A tile layer is associated with the map.
    - **Line 61**: An `L.geoJSON` object is created using the GeoJSON data retrieved on line 52. Then it is added to the map. This will result in Leaflet drawing the contours of the community's territory.
    - **Lines 63 to 71**: These lines check whether the population is included in the community's attributes. If it is, a popup is placed in the center of the community with the population information.
    - **Line 73**: The zoom level is adjusted on the rectangle encompassing the community.

Launching the server and opening the URL `http://127.0.0.1:5000/am/map_nom?city=Rome` in the browser produces a list of eight communities named Rome, with their coordinates, maps, and populations.

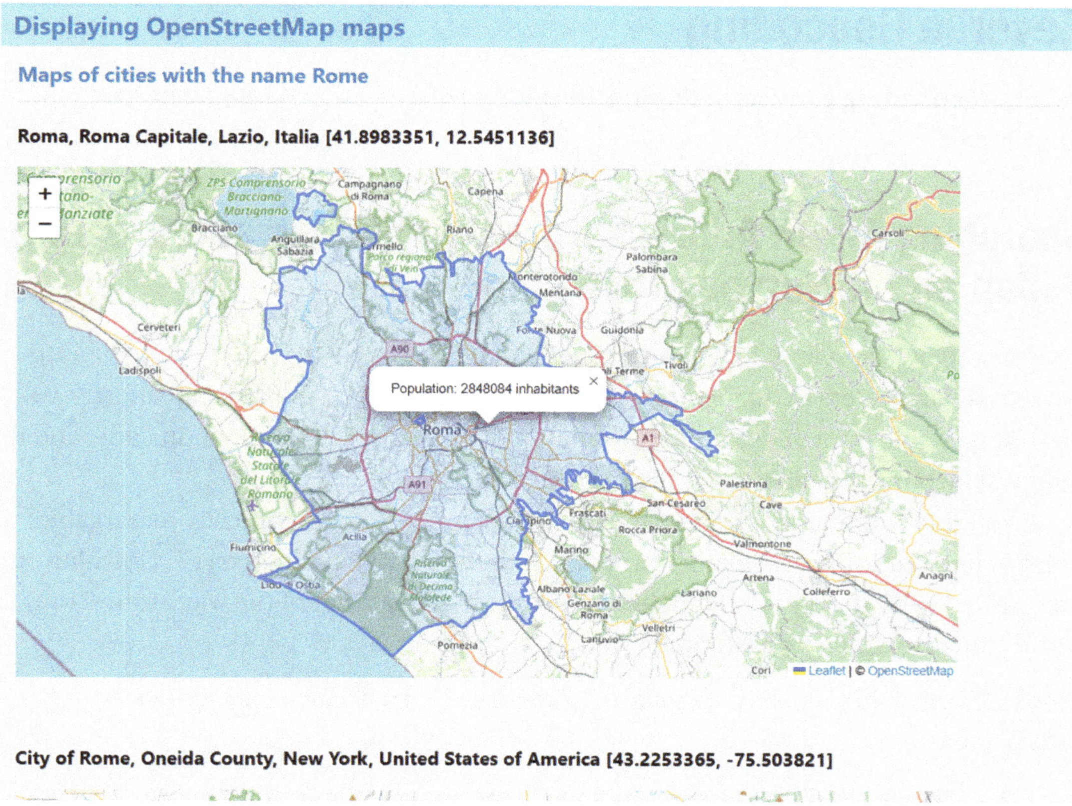

Figure 7-1. *Map created with Leaflet using OpenStreetMap tiles. Data © OpenStreetMap contributors, licensed under the ODbL. Overlaid data retrieved with Nominatim*

Using the browser's scroll bar, you can see the seven other communities named Rome.

The problem of communities not found because of their local names, which was encountered with Overpass in the previous prototype, does not occur here: Nominatim searches for the community's name provided in the request in all languages (it compares it with the values of all `name:<language>` tags in OSM). As a result, cities like Beijing, Mexico City, and Rome can be found with their English name.

Reverse Geocoding

Reverse geocoding consists in finding the address of a location based on its geographic coordinates.

Searching for an Address from Coordinates and Creating a Map on the Server with Folium

This prototype will search for the address of a point with a given latitude and longitude, passed as parameters to the request, and will display it on a map of that location. It will use a server-side implementation: Nominatim will be queried from Python code, and the map will be built with Folium within the blueprint.

A suitable Nominatim query to find the address of the geographically remarkable point where the 45th parallel north and the 90th meridian west intersect – a location halfway between the equator and the North Pole and a quarter of the way around the world from the Greenwich Meridian – is:

```
https://nominatim.openstreetmap.org/reverse?lat=45&lon=-90&zoom=18&f
ormat=json
```

- The `lat` and `lon` parameters specify the geographic coordinates of the point whose address is searched for.
- The `zoom` parameter specifies the desired level of detail in the returned address. The value "18" requests the highest possible level of address detail.
- The `format=json` parameter requests that the response be returned in JSON format.

Such a request can easily be tested with a browser. The response is as follows (in JON format):


```
{
  "place_id": 346732582,
  "licence": "Data © OpenStreetMap contributors, ODbL 1.0. http://osm.org/
  copyright",
  "osm_type": "node",
  "osm_id": 11100430798,
```

```
  "lat": "45.0000000",
  "lon": "-90.0000000",
  "class": "historic",
  "type": "memorial",
  "place_rank": 30,
  "importance": 5.000152590214416e-05,
  "addresstype": "historic",
  "name": "45x90",
  "display_name": "45x90, Lovers Lane, Town of Rietbrock, Marathon County,
  Wisconsin, 54411, États-Unis d'Amérique",
  "address": {
    "historic": "45x90",
    "road": "Lovers Lane",
    "municipality": "Town of Rietbrock",
    "county": "Marathon County",
    "state": "Wisconsin",
    "ISO3166-2-lvl4": "US-WI",
    "postcode": "54411",
    "country": "United States",
    "country_code": "us"
  },
  "boundingbox": [
    "44.9999500",
    "45.0000500",
    "-90.0000500",
    "-89.9999500"
  ]
}
```


This response includes the full address of the place in the `display_name` field, which includes the street, city, county, state, ZIP code, and country. The same information is also provided in separate fields such as `road`, `municipality`, `county`, `state`, `postcode`, and `country`.

To perform this type of search, the following code must be added to the advanced_maps.py blueprint:

```
@bp.route('/am/map_nom_rev1')
def map_nom_rev1():
    import requests

    lat = request.args.get('lat', None)
    lon = request.args.get('lon', None)

    try:
        if (lat != None and lon != None and isFloat(lat) and isFloat(lon)):
            user_agent = {'User-Agent': 'CartoOSM/1.0 (patrick@carto-osm.fr)'}
            params = {
                "lat": float(lat),
                "lon": float(lon),
                "zoom": 16,
                "format": "json"
            }
            response = requests.get(
                "https://nominatim.openstreetmap.org/reverse",
                params=params,
                headers = user_agent)

            if response.status_code == 200:
                data = response.json()
                address = data.get("display_name", "In the middle of nowhere")
            else:
                flash("Error retrieving data with Nominatim API", category="error")
                return render_template('maps/page_with_map.html', name="No map to display")

```

```
29              m = folium.Map(location=[float(lat), float(lon)], zoom_start=8)
30
31              folium.Marker(
32                  location=[float(lat), float(lon)],
33                  popup=folium.Popup(address + "<br/>[" + lat + ", " + lon + "]", show=True)
34              ).add_to(m)
35
36              return render_template('maps/page_with_map.html',
37                  name="Address of the point with latitude " + lat + " and longitude " + lon,
38                  map_html=m._repr_html_())
39
40          else:
41              m = folium.Map(location=[45, -90], zoom_start=4)
42
43              return render_template('maps/page_with_map.html',
44                  name="No coordinates provided",
45                  map_html=m._repr_html_())
46
47      except Exception as e:
48          flash("An error has been encountered:<br/>" + str(e), category="error")
49          return render_template('maps/page_with_map.html', name="No map to display")
```

- **Line 3**: The `requests` library is imported.
- **Lines 5 and 6**: The `lat` and `lon` parameters of the HTTP request are received.
- **Line 8**: This line starts a `try/except` block for error handling.
- **Line 9**: This line checks that a latitude and longitude have been received in the HTTP request and contain decimal values. The next statements, querying Nominatim, will only be executed if valid coordinates have been received.

- **Lines 10 to 20**: These lines prepare and submit the Nominatim request:
    - **Line 10**: An HTTP header is prepared to identify the client-application. It is required for the request to be accepted by the Nominatim server (without this header, the server blocks the request and returns a "403" error code).

      To identify the client-application, it is recommended to use an identifier specific to your application and preferably to include the version number and your email address as well. For example: `MyApplication/1.0 (myemail@example.com)`.
    - **Lines 11 to 16**: The parameters to be passed to the request are set.
    - **Lines 17 to 20**: A reverse geocoding request is submitted to Nominatim with the latitude and longitude as parameters. A detailed address is also requested (with zoom level 16), and the response is expected in JSON format.

  Here is an example of how to test this request with cURL:

  ```
  curl -s -G "https://nominatim.openstreetmap.org/reverse" ^
  --data-urlencode "lat=0" ^
  --data-urlencode "lon=30" ^
  --data-urlencode "zoom=16" ^
  --data-urlencode "format=json" ^
  --user-agent "curl 8.13.0 (Windows)"
  ```
- **Line 22**: The result of the request is checked:
    - **Line 23**: If the request was successful, the response is parsed and placed in a JSON-structured object.
    - **Line 24**: If a detailed address has been found, it is extracted from Nominatim's response. Otherwise (e.g., if the search coordinates are in the middle of the sea), the default address "In the middle of nowhere" is used.
    - **Lines 25 to 27**: These lines deal with the case where the HTTP request has failed. Processing is aborted, and an error message is displayed on a web page.

- **Line 29**: A map is created and centered on the point searched for, with a medium zoom level (8).
- **Lines 31 to 34**: A marker is added to the map, positioned on the point searched for, with a popup showing the point's detailed address and geographic coordinates.
- **Lines 36 to 38**: A web page is rendered using the `page_with_map.html` template. The name "Address of latitude point <lat> and longitude point <lon>" and the map, converted to HTML and JavaScript code, are passed as arguments.
- **Lines 40 to 45**: These lines are executed when the initial HTTP request has not provided valid coordinates. In this case, a map of North America is created and displayed on a web page.
- **Lines 47 to 49**: These lines form the `except` part of the `try/except` block. In the event of an error, an error message is prepared and displayed on a web page.

The preceding code is completed by the `isFloat()` function, also placed in the `advanced_maps.py` blueprint. This function checks that a variable contains a decimal number or a value that can be converted to a decimal.

```
def isFloat(var):
    try:
        float(var)
        return True
    except ValueError:
        return False
```

Launching the server and opening the URL `http://127.0.0.1:5000/am/map_nom_rev1?lat=45&lon=-90` in the browser produces the following map, with the searched point indicated by a marker and its address presented in a popup:

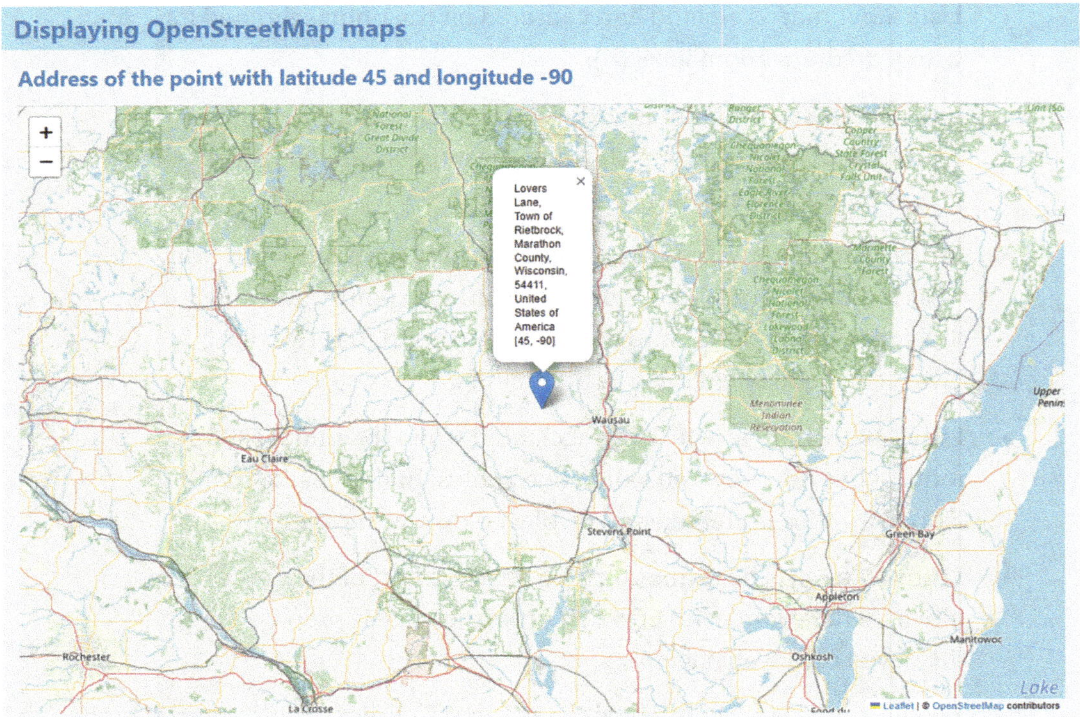

Figure 7-2. *Map created with Folium using OpenStreetMap tiles. Data © OpenStreetMap contributors, licensed under the ODbL. Overlaid data retrieved with Nominatim. Geographic data retrieved with Nominatim*

This geographically remarkable point is located in Rietbrock, Wisconsin (USA), where it is marked by a monument.

Figure 7-3. *Photo: Royalbroil, via Wikimedia Commons. Licensed under CC BY-SA 3.0*

Searching for an Address by Clicking on a Map with Folium

The shortcoming of the previous prototype is that it requires knowing the latitude and longitude of a location in order to search for its address. Often, we would prefer to simply point and click on a location on the map to find out its address. Our new challenge will be to enhance the previous prototype so that we can also perform reverse geocoding from a point on the map selected with the mouse.

A first approach is to use Folium's `LatLngPopup()` function, which adds to the map the ability, when clicked on any point, to display its latitude and longitude in a popup. This information can then be used to manually complete the HTTP request with the `lat` and `lon` parameters, before sending it again.

To implement this enhancement, duplicate the preceding blueprint code (route /am/map_nom_rev1) and associate it with a new route: /am/map_nom_rev2. In this duplicated code, add the following statement:

```
folium.features.LatLngPopup().add_to(m)
```

just after the two statements that create a map, on lines 29 and 41 (the statement must be inserted twice: after each of these two map creation statements):

```
1   @bp.route('/am/map_nom_rev2')
2   def map_nom_rev2():
    ...
29              m = folium.Map(location=[float(lat), float(lon)], zoom_
                start=8)
30
31              folium.features.LatLngPopup().add_to(m)
32
33              folium.Marker(
    ...
43              m = folium.Map(location=[45, -90], zoom_start=4)
44
45              folium.features.LatLngPopup().add_to(m)
46
47              return render_template('maps/page_with_map.html',
48                  name="No coordinates provided",
49                  map_html=m._repr_html_())
    ...
```

When this change is done, restart the server and open the URL http://127.0.0.1:5000/am/map_nom_rev2. Clicking anywhere on the map brings up the coordinates of the designated location.

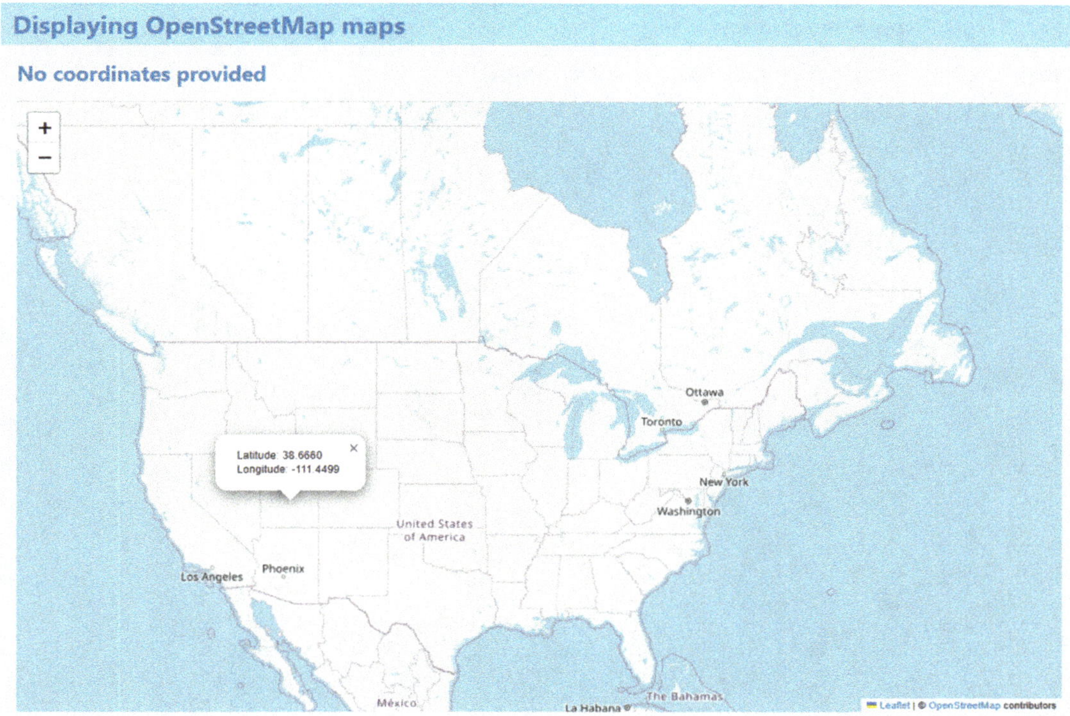

Figure 7-4. *Map created with Folium using OpenStreetMap tiles. Data © OpenStreetMap contributors, licensed under the ODbL*

Then, if you open the same URL again, completed with the coordinates found, `http://127.0.0.1:5000/am/map_nom_rev2?lat=38.6660&lon=-111.4499`, you get the address you are looking for, and the map zooms in on that location.

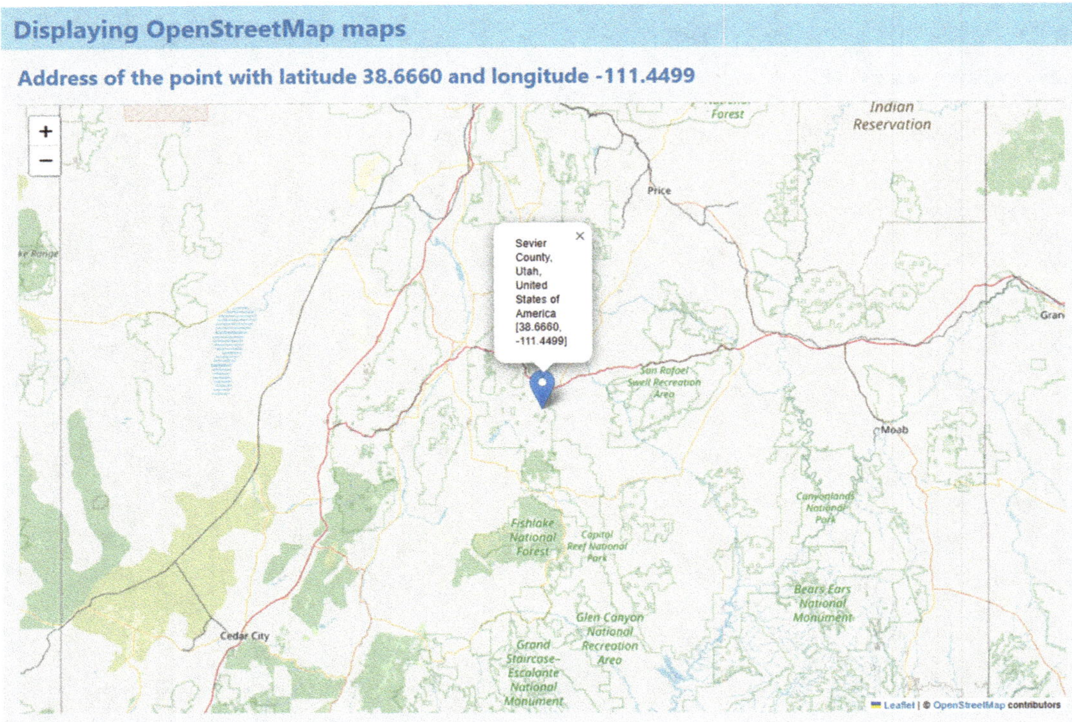

Figure 7-5. *Map created with Folium using OpenStreetMap tiles. Data © OpenStreetMap contributors, licensed under the ODbL. Overlaid data retrieved with Nominatim*

However, this approach is not satisfactory, as it requires a two-step process and, above all, copying the coordinates displayed in a popup at the end of the first step and pasting them into the URL before resending the request for the second step. Folium, quite simply, does not go much further. Let's try to push back these limits.... Our next approach will be to inject some JavaScript code into the Folium map to automate the preceding process (retrieving the coordinates of the clicked point and resubmitting the request with these coordinates). This JavaScript code will:

1. Intercept mouse clicks on the map.
2. Extract the coordinates of the point on which the user has clicked.
3. Invoke the same URL again, with the latitude and longitude found as parameters.

To implement this solution, the following code (in **bold**) must be added to the /am/map_nom_rev2 route in the advanced_maps.py blueprint:

```
@bp.route('/am/map_nom_rev2')
def map_nom_rev2():
    import requests

    lat = request.args.get('lat', None)
    lon = request.args.get('lon', None)
    click_event_script = """
        <script>
        document.addEventListener("DOMContentLoaded", function() {
            const map = Object.values(window).find(obj => obj instanceof L.Map);

            if (map) {
                map.on('click', function(e) {
                    const lat = e.latlng.lat.toFixed(6);
                    const lng = e.latlng.lng.toFixed(6);

                    window.top.location.href = '/am/map_nom_rev2?lat=' + lat + '&lon=' + lng;
                });
            }
        });
        </script>
    """

    try:
        if (lat != None and lon != None and isFloat(lat) and isFloat(lon)):
            user_agent = {'User-Agent': 'CartoOSM/1.0 (patrick@carto-osm.fr)'}
            params = {
                "lat": float(lat),
                "lon": float(lon),
                "zoom": 16,
                "format": "json"
```

```
            }
            response = requests.get(
                "https://nominatim.openstreetmap.org/reverse",
                params=params,
                headers = user_agent)

            if response.status_code == 200:
                data = response.json()
                address = data.get("display_name", "In the middle of nowhere")
            else:
                flash("Error retrieving data with Nominatim API", category="error")
                return render_template('maps/page_with_map.html', name="No map to display")

            m = folium.Map(location=[float(lat), float(lon)], zoom_start=8)

            #folium.features.LatLngPopup().add_to(m)

            folium.Marker(
                location=[float(lat), float(lon)],
                popup=folium.Popup(address + "<br/>[" + lat + ", " + lon + "]" , show=True)
            ).add_to(m)

            m.get_root().html.add_child(folium.Element(click_event_script))

            return render_template('maps/page_with_map.html',
                name=f"Address at [{lat}, {lon}] - Click on the map",
                map_html=m._repr_html_())

        else:
            m = folium.Map(location=[45, -90], zoom_start=4)
```

```
62
63                #folium.features.LatLngPopup().add_to(m)
64
65                m.get_root().html.add_child(folium.Element(click_event_
                  script))
66
67                return render_template('maps/page_with_map.html',
68                    name="No coordinates provided - Click on the map",
69                    map_html=m._repr_html_())
70
71        except Exception as e:
72            flash("An error has been encountered:<br/>" + str(e),
              category="error")
73            return render_template('maps/page_with_map.html', name="No map
              to display")
```

- **Lines 7 to 22**: The JavaScript code used to handle click events is stored in a variable named `click_event_script`.
    - **Line 9**: An event listener is added to the HTML document and associated with a function defined below. This ensures that the following statements are executed only after the content of the web page has fully loaded. Otherwise, the map element might not be found.
    - **Line 10**: The map built by Folium is searched.
    - **Line 12**: This line checks whether the map has been found.
    - **Lines 13 to 18**: If the map has been found, an event listener is defined: the associated function will be executed when the user clicks on the map:
        - **Lines 14 and 15**: The latitude and longitude of the point clicked by the user are retrieved.
        - **Line 17**: The URL of the route associated with this Python function is called, with the latitude and longitude retrieved included in the request as parameters.

- **Lines 47 and 63**: The previously used `folium.features.LatLngPopup()` function is commented out because it is no longer necessary.
- **Line 54**: The map's HTML structure is accessed and completed with the preceding JavaScript code.
- **Line 65**: When valid coordinates are not included in the request, the map's HTML structure is also accessed and completed with the JavaScript code.

After restarting the server, opening the URL `http://127.0.0.1:5000/am/map_nom_rev2` in a browser, shifting the map to the south, and clicking on a point on the Brazilian coast, we get the following result:

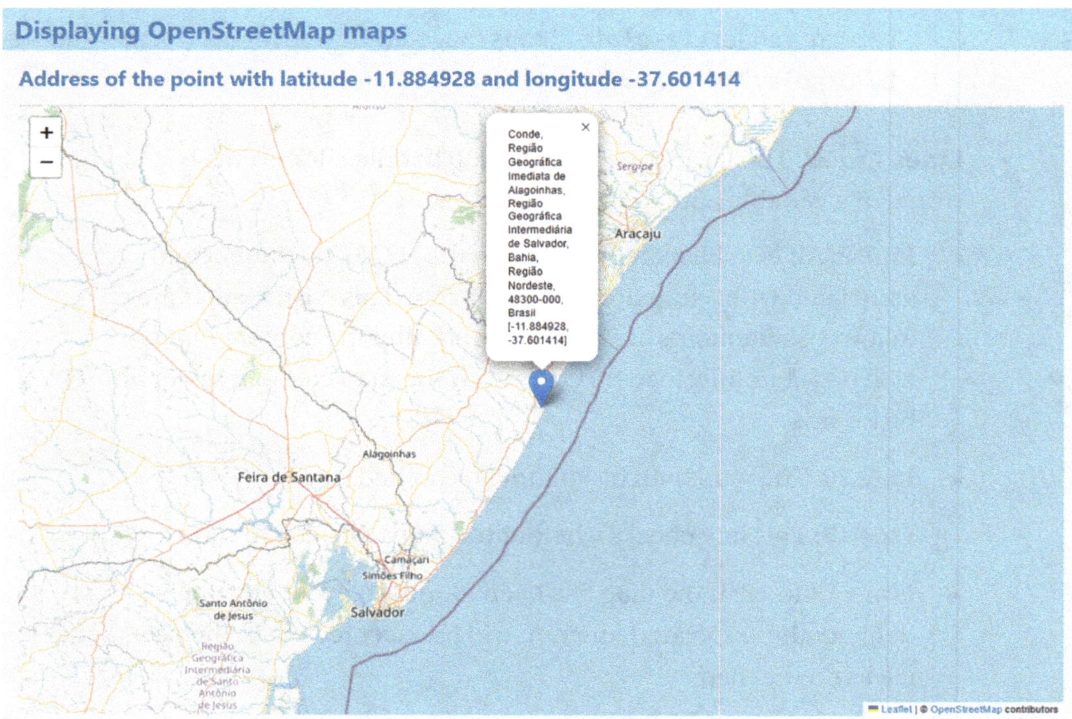

***Figure 7-6.** Map created with Folium using OpenStreetMap tiles. Data © OpenStreetMap contributors, licensed under the ODbL. Overlaid data retrieved with Nominatim*

Another possible solution would be to avoid returning to the Python blueprint on the server when the user clicks on the map, and, instead, perform the reverse geocoding search directly from JavaScript using the `fetch()` function to send a request to the Nominatim server.

This is the very limit of what can reasonably be achieved with Folium. An implementation of this functionality with Leaflet.js and JavaScript code running in the browser would be much more suitable, as it would avoid having to go back to the server and rebuild the entire map. This approach would also be easier to implement and more efficient in terms of response time. We will put it into practice in the next section.

Searching for an Address by Clicking on a Map with Leaflet

In this new prototype, when the user provides coordinates in the request or clicks on a point on the map, Nominatim will be queried for the detailed address of that location, which will then be displayed on the map in a popup. These operations will be carried out using JavaScript and Leaflet in the browser, while the Python route and view function will simply receive the initial HTTP request and render an HTML page via a template, passing it the necessary arguments.

The Nominatim request to be used is the same as in the previous examples. For example, to search for the detailed address of a location with latitude 45 and longitude 0 and receive the response in JSON format:

```
https://nominatim.openstreetmap.org/reverse?lat=45&lon=0&zoom=16&format=json
```

To implement the prototype, the following code must be added to the `advanced_maps.py` blueprint:

```
@bp.route('/am/map_nom_rev3')
def map_nom_rev3():

    lat = request.args.get('lat', None)
    lon = request.args.get('lon', None)

    try:
```

```
8           if (lat != None and lon != None and isFloat(lat) and
            isFloat(lon)):
9               return render_template('maps/page_with_address_rev.html',
10                  name="Click on the map",
11                  loc=[float(lat), float(lon)],
12                  zoom=8)
13          else:
14              return render_template('maps/page_with_address_rev.html',
15                  name="Click on the map",
16                  loc=["", ""],
17                  zoom=4)
18
19      except Exception as e:
20          flash("An error has been encountered:<br/>" + str(e),
            category="error")
21          return render_template('maps/page_with_map.html', name="No map
            to display")
```

- **Lines 4 and 5**: The `lat` and `lon` arguments of the HTTP request are received.
- **Lines 8 to 12**: If a latitude and a longitude have been supplied in the request and are valid, a web page is built using the template `page_with_address_rev.html`. The arguments passed are a name to be displayed, coordinates (those received with the request), and a zoom level (8).
- **Lines 13 to 17**: If a latitude and a longitude have not been supplied, or are not valid, no geographic coordinates are passed to the template (a map of North America will be displayed with a zoom level of 4).
- **Lines 19 to 21**: These lines handle any errors that may occur.

The template `page_with_address_rev.html`, invoked by the previous function, must be created in the `prototypes/templates/maps` directory with the following contents:

```
{% extends 'base_ll.html' %}

{% block title %}{{ name }}{% endblock %}
```

```

{% block header %}
<h2>{{ name }}</h2>
{% endblock %}

{% block script %}
<script type="text/javascript">
let map;
let currentMarker;

async function getAddress(lat, lon) {
    try {
        const response = await fetch(
            `https://nominatim.openstreetmap.org/reverse?lat=${lat}&lon=${lon}&zoom=16&format=json`
        );

        if (response.ok) {
            const data = await response.json();

            return data.hasOwnProperty('display_name') ? data.display_name : "In the middle of nowhere";
        } else {
            return "Communication error with Nominatim!";
        }
    } catch (error) {
        return "Communication error with Nominatim!";
    }
}

async function initializeMap() {
    const loc = {{ loc | tojson }};

    if ("{{ loc[0] }}" !== "" && "{{ loc[1] }}" !== "") {
        const initLat = Number(loc[0]);
```

```
        const initLon = Number(loc[1]);

        map = L.map('map').setView([initLat, initLon], {{ zoom }});

        L.tileLayer('https://tile.openstreetmap.org/{z}/{x}/{y}.png', {
            maxZoom: 19,
            attribution: '&copy; <a href="http://www.openstreetmap.org/copyright">OpenStreetMap</a>'
        }).addTo(map);

        const address = await getAddress(initLat, initLon);

        currentMarker = L.marker(loc).addTo(map)
        .bindPopup(`${address}<br/>[${initLat.toFixed(5)}, ${initLon.toFixed(5)}]`)
        .openPopup();
    } else {
        map = L.map('map').setView([45, -90], {{ zoom }});

        L.tileLayer('https://tile.openstreetmap.org/{z}/{x}/{y}.png', {
            maxZoom: 19,
            attribution: '&copy; <a href="http://www.openstreetmap.org/copyright">OpenStreetMap</a>'
        }).addTo(map);
    }

    async function onMapClick(e) {
        const latlng = e.latlng;
        const address = await getAddress(latlng.lat, latlng.lng);

        let zoom = map.getZoom();
        if (zoom < 8) {
            zoom = 8;
        }
```

```
68          map.setView(latlng, zoom);
69
70          if (currentMarker) {
71              map.removeLayer(currentMarker);
72          }
73
74          currentMarker = L.marker(latlng).addTo(map)
75          .bindPopup(`${address}<br/>[${latlng.lat.toFixed(5)},
            ${latlng.lng.toFixed(5)}]`)
76          .openPopup();
77      }
78
79      map.on('click', onMapClick);
80  }
81
82  initializeMap();
83  </script>
84  {% endblock %}
```

- **Lines 11 and 12**: The global variables `map` and `currentMarker` are declared.
- **Lines 14 to 30**: The `getAddress()` function is defined. It queries Nominatim to obtain the address of a location from its coordinates. This function is asynchronous, as it uses the asynchronous `fetch()` function.
    - **Lines 16 to 18**: Nominatim is queried using the `fetch()` function, passing the coordinates of the location as parameters of the HTTP request.

      No identification header is needed since the browser will add one when executing the resulting HTML file.

      Depending on the environment used, invoking Nominatim from JavaScript in the browser may cause a CORS (cross-origin resource sharing) error. This is unlikely with the `https://nominatim.openstreetmap.org/` public endpoint, but may occur with other endpoints. In such cases, a common solution is to use a server-side proxy that adds a contact User-Agent (typically of the form `MyApp/1.0 (me@example.com)`) and appropriate headers. Appendix C explains how to do so. Another possible solution is to route the request through a public CORS proxy service (such as **corsproxy.io**).

- **Lines 20 to 23**: If the request was successful, the Nominatim's response is parsed and placed in a JSON-structured object.

  If the Nominatim's response includes a `display_name` property, its value is returned as the address of the location. Otherwise, "In the middle of nowhere" is returned as the address (this occurs typically when the address of a point located in the sea is searched for).

- **Lines 24 and 25**: If the request failed, an error message is returned.

  As the asynchronous `fetch()` function is called with the `await` statement, a `try/catch` block can be used to catch and process errors.

- **Lines 32 to 79**: The `initializeMap()` function is defined. It is responsible for creating the map.
  - **Line 33**: The geographic coordinates passed by the Python function are stored in a variable. The Jinja filter `tojson` is used to ensure that the latitude and longitude are inserted by the Jinja engine rather than being left blank (which would cause a syntax error).
  - **Line 35**: If these coordinates are not empty strings:
    - **Lines 36 and 37**: The latitude and longitude are retrieved and verified to be numeric.
    - **Line 39**: A map is created and centered on the coordinates and with the zoom level received from the calling Python function.
    - **Lines 41 to 44**: A tile layer is associated with the map.
    - **Lines 46**: Nominatim is queried (via the `getAddress()` function) to retrieve the address corresponding to these coordinates.

- **Lines 48 to 50**: Then a marker is positioned on this point with its address (returned by Nominatim) and coordinates displayed in a popup.
- **Line 51**: If coordinates have been passed as empty strings:
    - **Line 52**: A map is created and centered on the point [45, -90] and with the zoom level received from the calling Python function.
    - **Lines 54 to 57**: A tile layer is associated with the map.
- **Lines 60 to 77**: The `onMapClick()` function is defined. It will be executed when the user clicks on the map.
    - **Line 61**: The coordinates of the point clicked by the user are retrieved.
    - **Line 62**: Nominatim is queried (via the `getAddress()` function) to retrieve the address corresponding to these coordinates.
    - **Lines 64 to 67**: The map's current zoom level is checked. If it is less than 8, it is increased to level 8.
    - **Line 68**: The map is centered on the coordinates where the user clicked, and the zoom level determined by the preceding statements is applied.
    - **Lines 70 to 72**: If there is a marker on the map, it is deleted.
    - **Lines 74 to 76**: A new marker is added at the point clicked by the user, and a popup containing the address, supplied by Nominatim and returned by the `getAddress()` function, and the coordinates of the point, is associated with the marker.
- **Line 79**: An event listener is defined – when the user clicks on the map, the `onMapClick()` function will be executed.
- **Line 82**: The `initializeMap()` function is executed. It will be executed just once, when the page is opened (unless the user requests a page reload):

After restarting the server, opening the URL `http://127.0.0.1:5000/am/map_nom_rev3` in a browser, shifting the map to the east, and clicking on a point on the Adriatic coast, we get the following resul:

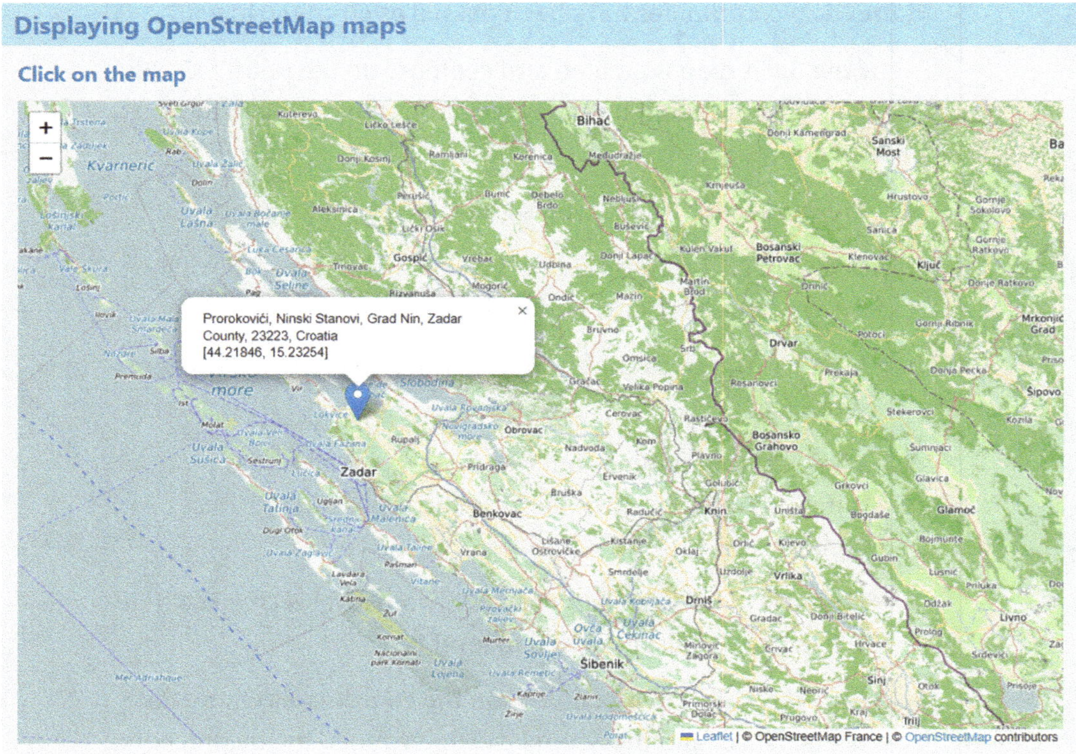

***Figure 7-7.** Map created with Leaflet using OpenStreetMap tiles. Data © OpenStreetMap contributors, licensed under the ODbL. Overlaid data retrieved with Nominatim*

You can continue to click on the map to obtain the addresses of other points.

If the map fails to display, it may be caused by a CORS error (as explained above). Verify this possibility, and, if necessary, implement a server-side proxy as described in Appendix C or use a public CORS proxy service (such as **corsproxy.io**).

We have just implemented the same functionality with Leaflet that we previously created with Folium. However, the implementation code with Leaflet is much simpler, and the addresses of the points clicked on are displayed faster (in particular, when the user clicks on the map, it is not recreated, but only moved and zoomed - if necessary). This confirms that, when you want to offer interaction capabilities beyond what is natively possible with maps built with Folium, you should consider using Leaflet.

Alternative Solutions with Leaflet and Folium Plugins

In the previous sections, we demonstrated how to write Python and JavaScript code to use the Nominatim API together with Leaflet or Folium for geocoding searches. However, some Leaflet and Folium plugins can also be used for geocoding.

Numerous Leaflet geocoding plugins are available - some use Nominatim, while others rely on different geocoding engines such as the ArcGIS Online Geocoder, MapTiler Geocoding API, Bing, Geoapify, GeoNames, OpenCage Geosearch, and the Pelias Geocoder API. Among the plugins that use Nominatim are:

- **L.Highlight** (developed by Maciej Kowalski): This plugin simplifies the process of issuing Nominatim geocoding requests by introducing a new object named L.Layer.Highlight. This object can perform a geocoding search using Nominatim (searching for an address) and automatically highlight the resulting location on the map.
- **Leaflet Control OSM Geocoder** (developed by Karsten Hinz): This plugin adds a new control to the map that allows users to enter an address for search. It then performs a search using Nominatim and recenters the map on the location found.
- **Leaflet GeoSearch** (developed by Stephan Meijer): Another plugin that adds a new control to the map in the form of an input field where users can enter an address. The plugin then searches for the address, recenters the map on that location, and highlights it with a marker.

A Folium plugin that also supports geocoding with Nominatim is the following:

- **Geocoder**: This plugin adds a search box to the map to search for geographic features like cities, countries, etc. Users can search by names or addresses. The places found are highlighted with a marker and a popup.

CHAPTER 8

Creating Advanced Choropleth Maps

Choropleth maps are thematic maps in which each geographic area is shaded with a color representing an economic or statistical value specific to that area.

Preparing Data on the Server and Creating a Choropleth Map on the Client with Leaflet

This prototype will build a map of Canada, where the provinces are shaded in lighter or darker colors according to their unemployment rates. The province name and unemployment rate will also be displayed in a tooltip when the mouse hovers over it.

A mixed processing scenario will be used:

1. On the server, Python code will prepare GeoJSON data describing the contours of the provinces and CSV data with the unemployment rates by province. An HTML page will then be built using a template, with the GeoJSON and CSV data injected into it, and the page will be sent to the browser.

2. In the browser, the HTML page will be received and executed, and a choropleth map will be built with Leaflet using the GeoJSON and CSV data included in the page.

Contours of Canadian provinces in GeoJSON format are available at `https://data.opendatasoft.com/api/explore/v2.1/catalog/datasets/georef-canada-province@public/exports/geojson`. This file was stored and can also be found at `https://github.com/Apress/Modern-Web-Cartography` (file: `carto_osm/prototypes/static/Canada_provincial_limits.geojson`).

P. Marie, *Modern Web Cartography*, https://doi.org/10.1007/979-8-8688-2335-0_8

Retrieve it and download it into the prototypes/static directory of the workspace.

Unemployment rates by province are available on Statistics Canada's Labour Force Survey website. This data was stored and can be found at https://github.com/Apress/Modern-Web-Cartography (file: carto_osm/prototypes/static/Unemployment_Rates_by_Canadian_Province.csv).

Retrieve it and download it into the prototypes/static directory of your workspace.

The contents of the Unemployment_Rates_by_Canadian_Province.csv file are as follows:

```
Province,2021,2022,2023,2024
Alberta,7.3,6.5,5.8,6.5
British Columbia,5.3,5.1,4.6,5.5
Manitoba,5.4,5.3,4.7,5.4
Newfoundland and Labrador,11.6,12.9,10.3,10.1
New Brunswick,7.9,7.7,5.8,7.9
Nova Scotia,7.9,6.5,5.8,6.5
Ontario,6.0,5.3,5.2,6.7
Prince Edward Island,7.8,8.1,6.7,7.4
Quebec,4.4,4.1,4.2,4.9
Saskatchewan,7.1,5.0,4.6,5.4
Yukon,4.2,4.5,3.6,4.3
```

The code to be added to the advanced_maps.py blueprint to build this map is shown below:

```
@bp.route('/am/map_cho1')
def map_cho1():
    import json

    try:
        with open("prototypes/static/Canada_provincial_limits.
        geojson", 'r', encoding='utf-8') as f:
            province_geo = json.load(f)

        with open("prototypes/static/Unemployment_Rates_by_Canadian_
        Province.csv", 'r', encoding='utf-8') as f:
            province_data = f.read()
```

```
11
12            return render_template('maps/page_with_choropleth_map_
              canada.html',
13                name="Choropleth map: unemployment rates by province in
                  Canada",
14                province_geo=province_geo,
15                province_data=province_data)
16
17        except Exception as e:
18            flash("An error has been encountered:<br/>" + str(e),
              category="error")
19            return render_template('maps/page_with_map.html', name="No map
              to display")
```

- **Lines 6 and 7**: The local GeoJSON file containing the province contours is read and parsed.
- **Lines 9 and 10**: The local CSV file containing unemployment rates by province is read.
- **Lines 12 to 15**: A web page is rendered using the `page_with_choropleth_map_canada.html` template. A name, province outline data, and unemployment data are passed as arguments.
- **Lines 17 to 19**: These lines handle any errors that may occur by displaying an error message on a web page.

The template `page_with_choropleth_map_canada.html`, invoked by the previous function, must be created in the `prototypes/templates/maps` directory with the following contents:

```
{% extends 'base_ll.html' %}

{% block title %}{{ name }}{% endblock %}

{% block header %}
<h2>{{ name }}</h2>
{% endblock %}

```

```
{% block script %}
<script id="geojson-data" type="application/json">
{{ province_geo | tojson }}
</script>

<script id="csv-data" type="application/csv">
{{ province_data | safe }}
</script>

<script src="//d3js.org/d3.v7.min.js" charset="utf-8"></script>

<script type="text/javascript">
let map;
try {
    const geojsonData = JSON.parse(document.getElementById("geojson-data").textContent);

    const rawCsvData = document.getElementById("csv-data").textContent
    .replace(/\r/g, '')
    .trim();
    const csvData = d3.csvParse(rawCsvData);

    const csvDataLookup = {};
    csvData.forEach(row => {
        csvDataLookup[row["Province"]] = +row["2024"];
    });

    const allValues = Object.values(csvDataLookup);
    const valueExtent = d3.extent(allValues);
    const getColor = d3.scaleSequential()
        .domain(valueExtent)
        .interpolator(d3.interpolateYlOrRd);

    function style(feature) {
        const provinceName = feature.properties.prov_name_en;
```

```
43          const value = csvDataLookup[provinceName];
44          return {
45              fillColor: value != null ? getColor(value) : '#999999',
46              weight: 1,
47              opacity: 1,
48              color: 'white',
49              fillOpacity: 0.7
50          };
51      }
52
53      map = L.map('map').setView([58.5, -96.7], 4);
54
55      L.tileLayer('https://tile.openstreetmap.org/{z}/{x}/{y}.png', {
56          maxZoom: 19,
57          attribution: '&copy; <a href="http://www.openstreetmap.org/
            copyright">OpenStreetMap</a>'
58      }).addTo(map);
59
60      L.geoJSON(geojsonData, { style: style }).addTo(map)
61  } catch(e) {
62      if (typeof map !== 'undefined' && map) {map.remove();}
63
64      const div = document.getElementById("map")
65      if (div) {div.remove();}
66
67      const elemDiv = document.createElement('div');
68      elemDiv.innerHTML = `<p>${e}</p>`;
69      document.body.appendChild(elemDiv);
70  }
71  </script>
72  {% endblock %}
```

- **Lines 10 to 12**: These lines contain the JSON script section in which the province outline data (in GeoJSON format) will be placed by the Jinja engine.

- **Lines 14 to 16**: These lines contain the CSV script section in which the unemployment data (in CSV format) will be placed by the Jinja engine.
- **Line 18**: This line is a link to the site where the code of the `d3.csvParse()` function (used to parse the CSV data on line 28) can be found.
- **Line 23**: The GeoJSON data with the province contours is retrieved, parsed (with a JSON parser), and placed in a GeoJSON-structured object.
- **Lines 25 to 28**: The CSV data with unemployment rates by province is retrieved.
    - First, the carriage return characters (which would interfere with the `d3.csvParse()` function) are eliminated, as well as leading and trailing blanks (if any).
    - Then the data is parsed (with a CSV parser) and placed in an array.
- **Lines 30 to 33**: An object named `csvDataLookup` is created with a key/value structure, containing, for each province, a key with the province's name and a value with its unemployment rate in 2024 (unemployment data for 2021, 2022, and 2023 are also present, but will not be used).
- **Line 35**: This line extracts all the unemployment rate values of the `csvDataLookup` object.
- **Line 36**: This line retrieves the lower and higher unemployment rates.
- **Lines 37 to 39**: These lines define an adaptive color scale from yellow to red for the range of unemployment rates.
- **Lines 41 to 51**: These lines define the `style()` function, which associates layout properties with a feature of the GeoJSON data (i.e., a province): fill color, outline color and thickness, opacity level, etc.
    - The `csvDataLookup` object is first used to find the unemployment rate from the province name.

- Then the `getColor()` function is called to determine the fill color to use for the unemployment rate using the adaptive color scale defined earlier. If no unemployment rate is found for a province, gray is used.

- **Line 53**: A map is created and positioned on the coordinates of a point at the center of Canada.
- **Lines 55 to 58**: A tile layer is associated with the map.
- **Line 60**: An `L.geoJSON` object is created with the GeoJSON-structured object (containing the province contours), and the `style()` function is used with the `style` option to specify how each province should be represented. This function will be applied to each feature (i.e., province) present in the GeoJSON data. The `L.geoJSON` object is then added to the map.

Launching the server and opening the URL `http://127.0.0.1:5000/am/map_cho1` in the browser produces the following map:

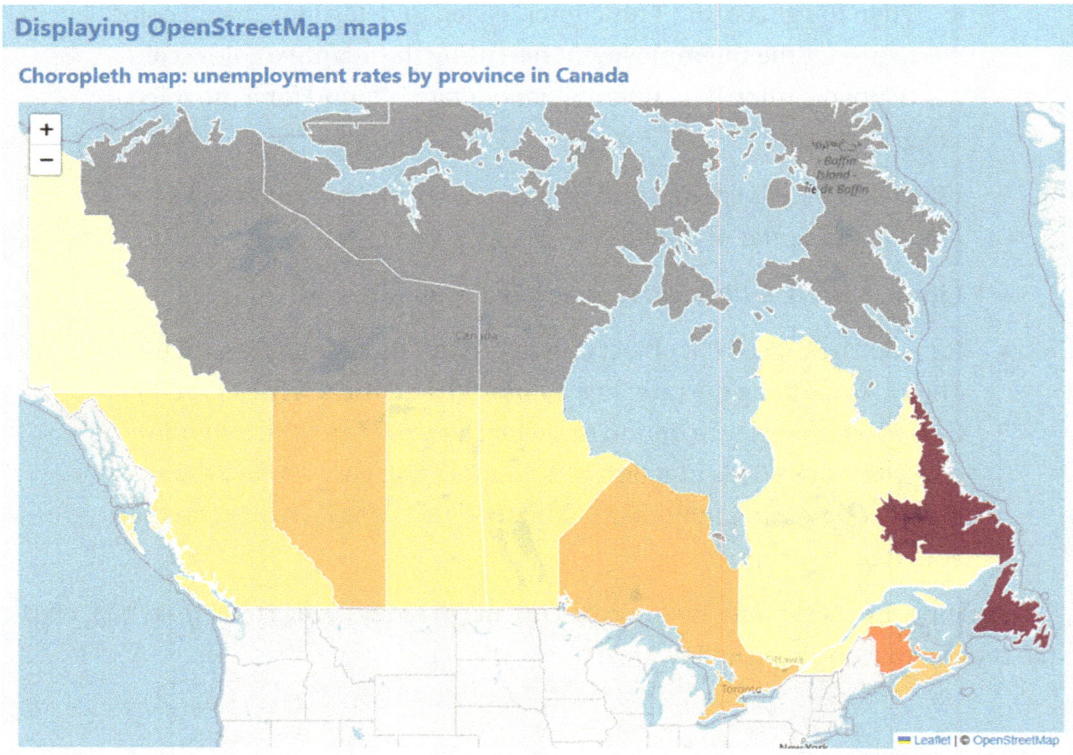

Figure 8-1. *Map created with Leaflet using OpenStreetMap tiles. Data © OpenStreetMap contributors, licensed under the ODbL. Province boundaries from Opendatasoft, "Provinces and Territories - Canada" dataset, derived from Statistics Canada, licensed under the Open Government Licence - Canada. Unemployment rates from Statistics Canada, Labour Force Survey, licensed under the Statistics Canada Open Licence*

Low unemployment rates are indicated by light colors and high unemployment rates by dark colors.

We are now going to make this map more interactive by adding code so that, when the mouse cursor hovers over a province, its outline is highlighted and its name and unemployment rate displayed in a tooltip.

In the previous code for the template `page_with_choropleth_map_canada.html`, let's insert or modify the following statements (changes are shown in **bold**):

```
    ...
        }).addTo(map);
```

```
        function highlightFeature(e) {
            const layer = e.target;

            layer.setStyle({
                weight: 5,
                color: '#666',
                dashArray: '',
                fillOpacity: 0.7
            });

            if (!L.Browser.ie && !L.Browser.opera) {
                layer.bringToFront();
            }
        }

        let geojson;

        function resetHighlight(e) {
            geojson.resetStyle(e.target);
        }

        function onEachFeature(feature, layer) {
            layer.bindTooltip(`${feature.properties.prov_name_en}:
            ${csvDataLookup[feature.properties.prov_name_en] ??
            'Unknown'} %`, {
                sticky: true
            });
            layer.on({
                mouseover: function (e) {
                    highlightFeature(e);
                },
                mouseout: function(e) {
                    resetHighlight(e);
                }
```

```
92              });
93          }
94
95          geojson = L.geoJSON(geojsonData, {
96              style: style,
97              onEachFeature: onEachFeature
98          }).addTo(map)
99
100     } catch(e) {
...
```

- **Lines 60 to 73**: The `highlightFeature()` function is defined. It changes the color and thickness of a province's outline and brings it to the foreground (relative to neighboring provinces). The province to be changed is found in the `e` (for event) variable received as an argument.
- **Line 75**: A `geojson` variable is declared. It will be used to set an `L.geoJSON` object and add it to the map.
- **Lines 77 to 79**: The `resetHighlight()` function is defined. It cancels changes to the layout of a province made by the `highlightFeature()` function. The province to be changed is found in the `e` (for event) variable received as an argument.
- **Lines 81 to 93**: The `onEachFeature()` function is defined. It:
    - Associates with a feature (province) a tooltip containing its name and unemployment rate.
    - Defines two events that will trigger actions:
        - When the mouse cursor hovers over a province, the `highlightFeature()` function changes its layout style to highlight it.
        - When the mouse cursor leaves a province, the `resetHighlight()` function returns its layout to normal.

- **Lines 95 to 98**: An `L.geoJSON` object is created with:
    - The GeoJSON-structured object (containing province contours).
    - The `style` option set with the `style()` function, to specify the layout properties of the provinces (this function will be executed for each of the provinces).
    - The `onEachFeature` option set with the `onEachFeature()` function, to assign additional functionalities to the provinces (this function will be executed for each of the provinces).

This object is set in the `geojson` variable (initialized on line 75 and used on line 78) and added to the map.

Launching the server and opening the URL `http://127.0.0.1:5000/am/map_cho1` in the browser produces the same map as above, but, when the mouse cursor hovers over a province, its name and unemployment rate are displayed, and its outline changes color.

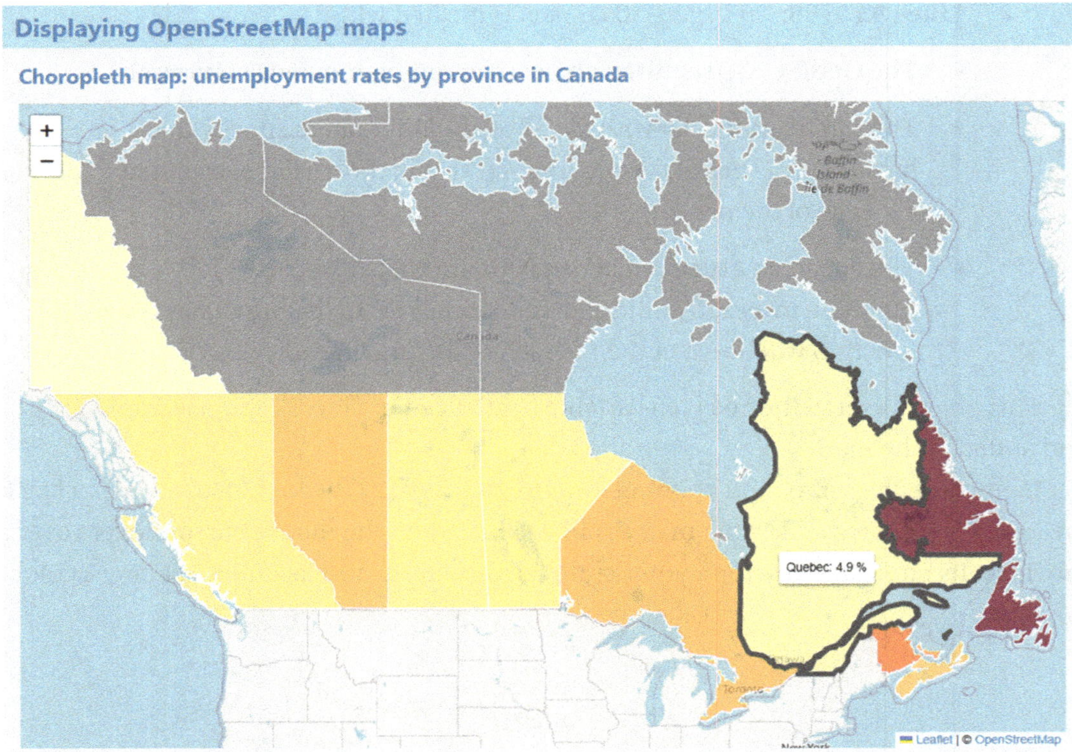

Figure 8-2. *Map created with Leaflet using OpenStreetMap tiles. Data © OpenStreetMap contributors, licensed under the ODbL. Province boundaries from Opendatasoft, "Provinces and Territories – Canada" dataset, derived from Statistics Canada, licensed under the Open Government Licence – Canada. Unemployment rates from Statistics Canada, Labour Force Survey, licensed under the Statistics Canada Open Licence*

Let's also add a legend to the map. To do this, insert the following statements (in **bold**):

```
...
      }).addTo(map)

      const legend = L.control({position: 'bottomleft'});
      legend.onAdd = function(map) {
          const div = L.DomUtil.create('div', 'info legend');
          div.innerHTML += `<strong>Unemployment rate (%):</strong><br>`;
```

```
104
105             const grades = d3.range(valueExtent[0], valueExtent[1],
                (valueExtent[1]-valueExtent[0])/5);
106             grades.push(valueExtent[1]);
107
108             for (let i = 0; i < grades.length - 1; i++) {
109                 const from = grades[i].toFixed(1);
110                 const to = grades[i+1].toFixed(1);
111                 div.innerHTML += `<i style="background:${getColor((grades
                    [i] + grades[i+1])/2)}"></i> ${from} &ndash; ${to}<br>`;
112             }
113
114             return div;
115         };
116
117         legend.addTo(map);
118
119     } catch(e) {
    ...
```

- **Line 100**: A legend is created in the bottom left-hand corner of the map.
- **Lines 101 to 115**: The contents of the legend are defined (but will be created only when the legend is added to the map):
    - **Line 102**: A division that will contain the legend is created.
    - **Line 103**: A title is added to the legend.
    - **Lines 105 and 106**: These lines prepare an array of unemployment rate intervals to be represented in the legend.
    - **Lines 108 to 112**: A loop is executed over the unemployment rate intervals, adding a line for each one with:
        - A small colored rectangle (<i> element) using the color corresponding to the midpoint of that interval from the color scale.
        - A text label showing the minimum and maximum unemployment rates for that interval.

- **Line 117**: The legend is added to the map (which triggers the creation of its contents).

Let's add a final touch to our map: we will modify the code so that, when the user clicks on a province, the map smoothly zooms in on that province.

To implement this, in the previous code of the `page_with_choropleth_map_canada.html` template, insert or modify statements as shown in **bold** below:

```
...
53      map = L.map('map', {zoomSnap: 0.1}).setView([58.5, -96.7], 4);
...
81      function onEachFeature(feature, layer) {
82          layer.bindTooltip(`${feature.properties.prov_name_en}:
            ${csvDataLookup[feature.properties.prov_name_en] ||
            'Unknown'} %`, {
83              sticky: true
84          });
85          layer.on({
86              mouseover: function (e) {
87                  highlightFeature(e);
88              },
89              mouseout: function(e) {
90                  resetHighlight(e);
91              },
92              click: function(e) {
93                  map.flyToBounds(e.target.getBounds());
94              }
95          });
96      }
...
```

- **Line 53**: The map can be zoomed in by steps of 0.1, for finer adjustment of the zoom level when clicking on a province.

- **Lines 92 to 94**: An event handler defines the action to be performed when the user clicks on a province: smoothly zoom in to adjust the map to the boundaries of that province.

Displaying the map again and clicking on a province produces the following view:

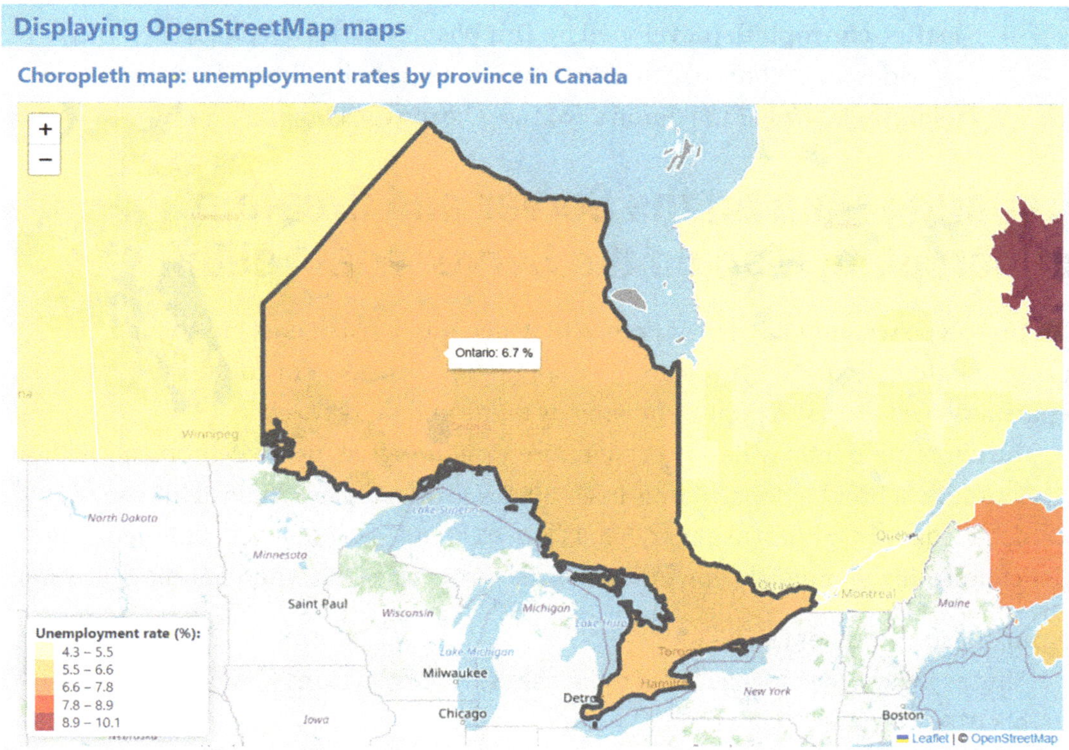

Figure 8-3. *Map created with Leaflet using OpenStreetMap tiles. Data © OpenStreetMap contributors, licensed under the ODbL. Province boundaries from Opendatasoft, "Provinces and Territories - Canada" dataset, derived from Statistics Canada, licensed under the Open Government Licence - Canada. Unemployment rates from Statistics Canada, Labour Force Survey, licensed under the Statistics Canada Open Licence*

The legend has been added at the bottom left, and the map has been zoomed in on the province of Ontario.

Alternative Solution with a Leaflet Plugin

In the previous prototype, we demonstrated how to write JavaScript code to build a choropleth map with Leaflet, which is slightly more complex than doing so with Folium. However, there is a Leaflet plugin that can make the process easier:

- **leaflet-choropleth** (developed by Tim Wisniewski): This plugin provides an `L.choropleth` object, which helps build choropleth maps using logic similar to Folium's `folium.Choropleth` object.

Preparing Data on the Server and Creating a Choropleth Map on the Server with Folium

This prototype will build a map of France in which the departments are shaded in darker or lighter colors according to their population densities. In addition, clicking on a department will display a popup showing its name and density.

The choropleth map will be created entirely on the server using Python and Folium.

Contours of French departments in GeoJSON format are available at `https://www.data.gouv.fr/fr/datasets/r/90b9341a-e1f7-4d75-a73c-bbc010c7feeb`. The application will dynamically download this data from this URL when building the map:

```
regions_geo = requests.get(
    "https://www.data.gouv.fr/fr/datasets/r/90b9341a-e1f7-4d75-a73c-
    bbc010c7feeb"
).json()
```

Population density data by department are available on the INSEE website: `https://www.insee.fr/fr/statistiques/5544529?sommaire=5435421#tableau-figure3`.

This data was downloaded, reworked with a text editor to reformat it for ease of use. The resulting file was stored and can be found at `https://github.com/Apress/Modern-Web-Cartography` (file: `carto_osm/prototypes/static/Densities per department.csv`).

Retrieve it and download it into the `prototypes/static` directory of your workspace.

The contents of the `Densities per department.csv` file are as follows:

```
Department,Density
Ain,115
Aisne,72
```

```
Allier,46
Alpes-de-Haute-Provence,24
Alpes-Maritimes,253
Ardèche,59

...

Vaucluse,157
Vendée,101
Vienne,63
Vosges,62
Yonne,45
Yvelines,631
```

The code to be added to the advanced_maps.py blueprint to build the map is as follows:

```
@bp.route('/am/map_cho2')
def map_cho2():
    import requests, pandas

    try:
        response = requests.get(
            "https://www.data.gouv.fr/fr/datasets/r/90b9341a-
            e1f7-4d75-a73c-bbc010c7feeb"
        )

        if response.status_code == 200:
            departments_geo = response.json()
        else:
            flash("Error downloading department contours from the
            Internet", category="error")
            return render_template('maps/page_with_map.html', name="No
            map to display")

        departments_data = pandas.read_csv("prototypes/static/
        Densities per department.csv")
```

```
17
18            m = folium.Map(location=[46.7696, 2.4331], zoom_start=6)
19
20            folium.Choropleth(
21                geo_data= departments_geo,
22                data=departments_data,
23                columns=["Department", "Density"],
24                key_on="feature.properties.nom",
25                threshold_scale=[0, 20, 46, 79.7, 123.6, 180.5, 254.5,
                  350.6, 475.4, 637.6, 848.2, 1121.8, 1477.2, 1939.9,
                  2538.6, 3317.6, 4329.6, 5644.2, 7351.8, 9570, 12451.4,
                  16194.3, 21056.5],
26                fill_color="BuPu",
27                name="Densities",
28                legend_name="Densities (inhabitants per km2)"
29            ).add_to(m)
30
31            folium.LayerControl().add_to(m)
32
33            return render_template('maps/page_with_map.html',
34                name="Choropleth map: densities per department in France",
35                map_html=m._repr_html_())
36
37        except Exception as e:
38            flash("An error has been encountered:<br/>" + str(e),
              category="error")
39            return render_template('maps/page_with_map.html', name="No map
              to display")
```

- **Line 3**: The `requests` and pandas libraries are imported.
- **Line 5**: A `try/except` block is started for error handling.
- **Lines 6 to 8**: The department contours in GeoJSON format are downloaded from an Internet site via an HTTP request.

In case this data is no longer available on the Internet, it has also been downloaded and can be found at `https://github.com/Apress/Modern-Web-Cartography` (file: `carto_osm/prototypes/static/French_department_contours.json`).

- **Lines 10 to 14**: The result of the request is checked.
    - If the request was successful, the response is parsed and placed in a GeoJSON-structured object.
    - Otherwise, a web page is rendered with an error message.
- **Line 16**: The local CSV file containing densities by department is loaded into a Pandas DataFrame object (with a multidimensional table structure).
- **Line 18**: A map is created and centered on a position specified by latitude and longitude coordinates (the center of France). An initial zoom level (6) is assigned.
- **Lines 20 to 29**: A `folium.Choropleth` object is created with the following parameters:
    - The `geo_data` parameter with the JSON-structured object containing department contours in GeoJSON format.
    - The `data` parameter with the Pandas DataFrame object containing densities by department.
    - The `columns` parameter indicating which two columns of the Pandas DataFrame object are to be used to construct the map:
        - The first one to find the department's name and correlate it with a department name in the JSON-structured object (containing department contours).
        - The second one to find the department's population density to be represented on the map via a color shade.
    - The `key_on` parameter specifies the key containing the department names in the JSON-structured object, to be correlated with the department names appearing in the first

of the columns of the Pandas DataFrame object (`columns` parameter). The field name here is `nom` (meaning "name" in French) because the GeoJSON file is French and partially uses French names.

- The `threshold_scale` parameter indicates the value thresholds to be used to change color shades when filling in departments on the map. Because population densities vary greatly between departments, a logarithmic scale is used, since a linear scale would not provide sufficient detail and a satisfactory rendering.
- The `fill_color` parameter indicates the color palette from which shades will be used to fill in the departments. The `BuPu` palette (i.e., blue-purple) has been chosen.
- The `name` parameter defines the name of the choropleth layer as it will appear in the layer control.
- The `legend_name` parameter defines the title of the legend.

- **Line 29**: The `folium.Choropleth` object is added to the map.
- **Line 31**: A layer control is added to enable the user to show or hide the density choropleth.
- **Lines 33 to 35**: A web page is rendered using the `page_with_map.html` template. The name "Choropleth map: densities per department in France" and the map, converted to HTML and JavaScript code, are passed as arguments.
- **Lines 37 to 39**: These lines handle any errors that may occur.

Launching the server and opening the URL `http://127.0.0.1:5000/am/map_cho2` in the browser produces the following map:

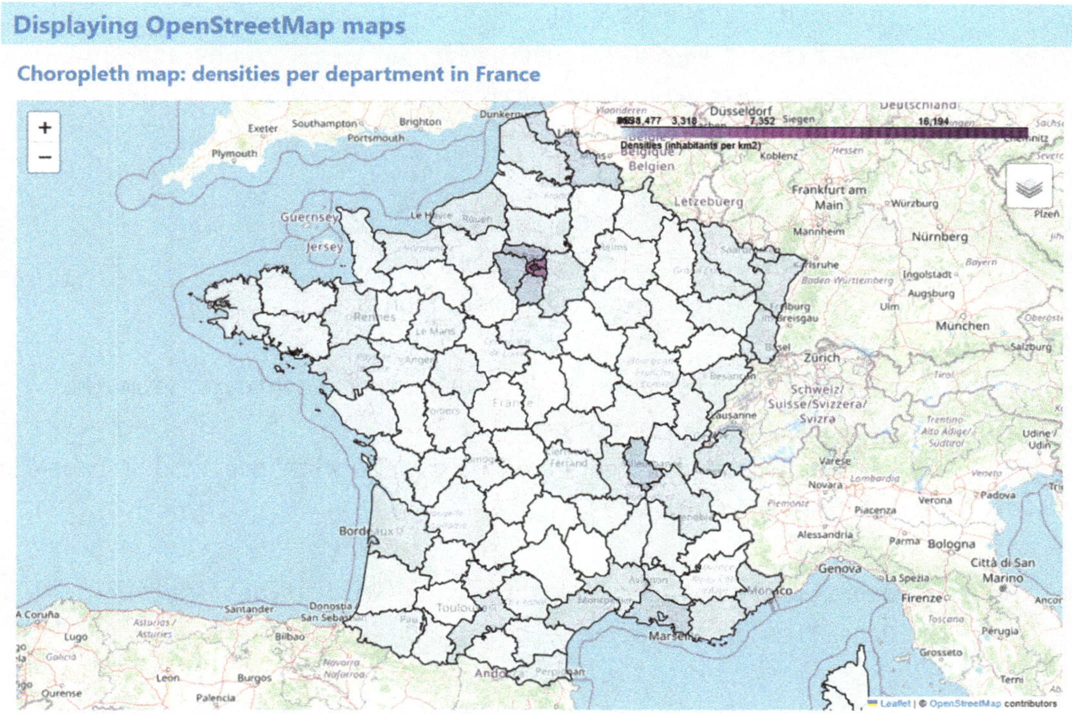

Figure 8-4. *Map created with Folium using OpenStreetMap tiles. Data © OpenStreetMap contributors, licensed under the ODbL. Department boundaries from data.gouv.fr, "Départements de France," available under the Open Licence 2.0 (Etalab). Department densities from INSEE, "Population par département – France, portrait social," licensed under the Open Licence 2.0 (Etalab)*

As in the previous prototype (choropleth map created with Leaflet), we will enhance this map with the ability to display department names and densities. To do this, we will use a `folium.GeoJsonPopup` object, which will display information when a department is clicked.

The `map_cho2()` function, modified in this way (added or modified lines are shown in **bold**), becomes:

```
@bp.route('/am/map_cho2')
def map_cho2():
    import requests, pandas

    try:
```

```
    response = requests.get(
        "https://www.data.gouv.fr/fr/datasets/r/90b9341a-e1f7-4d75-a73c-bbc010c7feeb"
    )

    departments_geo = None
    if response.status_code == 200:
        departments_geo = response.json()
    else:
        flash("Error downloading department contours from the Internet", category="error")
        return render_template('maps/page_with_map.html', name="No map to display")

    departments_data = pandas.read_csv("prototypes/static/Densities per department.csv")

    m = folium.Map(location=[46.769567, 2.433057], zoom_start=6)

    crpl = folium.Choropleth(
        geo_data=departments_geo,
        data=departments_data,
        columns=["Department", "Density"],
        key_on="feature.properties.nom",
        threshold_scale=[0, 20, 46, 79.7, 123.6, 180.5, 254.5, 350.6, 475.4, 637.6, 848.2, 1121.8, 1477.2, 1939.9, 2538.6, 3317.6, 4329.6, 5644.2, 7351.8, 9570, 12451.4, 16194.3, 21056.5],
        fill_color="BuPu",
        name="Densities",
        legend_name="Densities (inhabitants per km2)",
        highlight=True
    ).add_to(m)

```

```
33          indexed_department_data = departments_data.set_
            index('Department')
34
35          for dep in crpl.geojson.data['features']:
36              dep_name = dep['properties'].get('nom', 'Unknown')
37              if dep_name in indexed_department_data.index:
38                  dep['properties']['density'] = str(indexed_department_
                    data.loc[dep_name, 'Density'])
39              else:
40                  dep['properties']['density'] = "Unavailable"
41
42          folium.GeoJsonPopup(fields=['nom', 'density'],
            aliases=['Name:', 'Density:'], labels=True, localize=True,
            safe=False).add_to(crpl.geojson)
43
44          folium.LayerControl().add_to(m)
45
46          return render_template('maps/page_with_map.html',
47              name="Choropleth map: densities per department in France",
48              map_html=m._repr_html_())
49
50      except Exception as e:
51          flash("An error has been encountered:<br/>" + str(e),
            category="error")
52          return render_template('maps/page_with_map.html', name="No map
            to display")
```

- **Line 30**: The `highlight=True` parameter is added so that the department over which the mouse cursor passes is highlighted by a thicker outline.
- **Line 33**: The Pandas DataFrame object, containing the densities with the department names, is indexed to make it easier to find the density of a department from its name.

- **Lines 35 to 40**: A loop is executed over the features in the GeoJSON data part of the choropleth object. For each feature (i.e., department):
    - The department's name is retrieved.
    - An additional property, named `density`, is added, containing the department's density (retrieved using the indexed Pandas DataFrame object).
- **Line 42**: A `folium.GeoJsonPopup` object is created to display the name and density of a department in a popup (when the department is clicked). Then it is added to the GeoJSON part of the choropleth object.

Launching the server and opening the browser at the URL `http://127.0.0.1:5000/am/map_cho2` produces the same map as above, but:

- When the mouse cursor hovers over a department, its outline becomes thicker, and it is brought to the foreground relative to neighboring departments.
- Clicking on a department displays its name and population density.

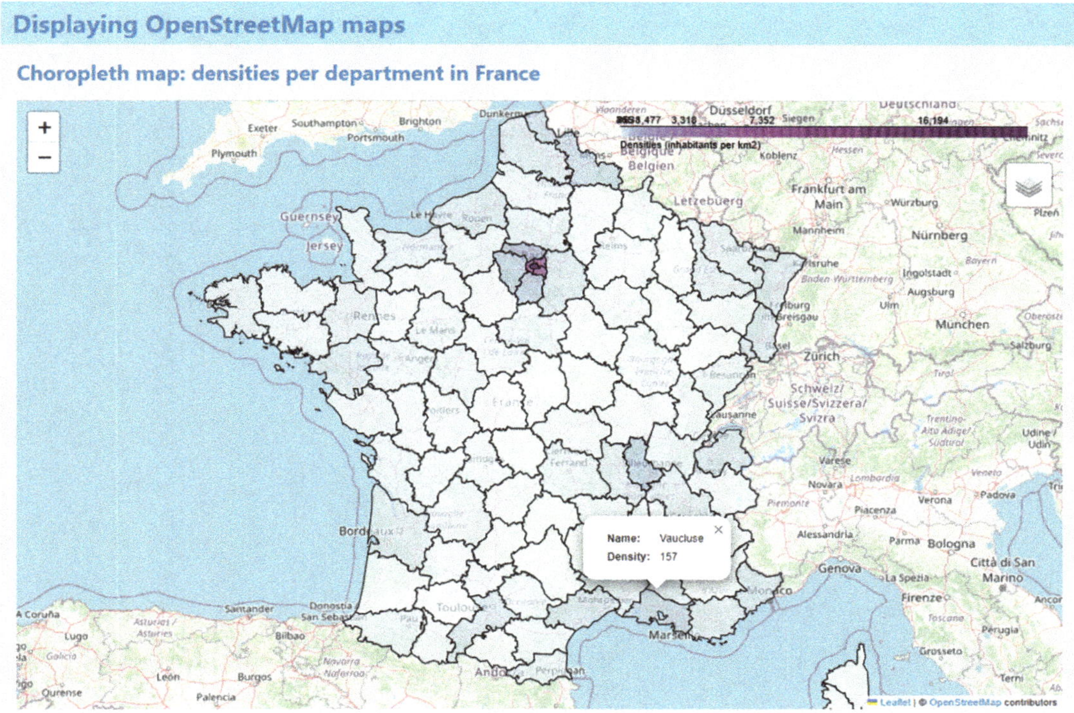

Figure 8-5. *Map created with Folium using OpenStreetMap tiles. Data © OpenStreetMap contributors, licensed under the ODbL. Department boundaries from data.gouv.fr, "Départements de France," available under the Open Licence 2.0 (Etalab). Department densities from INSEE, "Population par département – France, portrait social," licensed under the Open Licence 2.0 (Etalab)*

CHAPTER 9

Route Import, Calculation, and Visualization

A frequent functionality of web (and cell phone) mapping tools is **route planning**, i.e., finding routes between two locations (possibly with intermediate waypoints). OpenStreetMap's user interface offers this service.

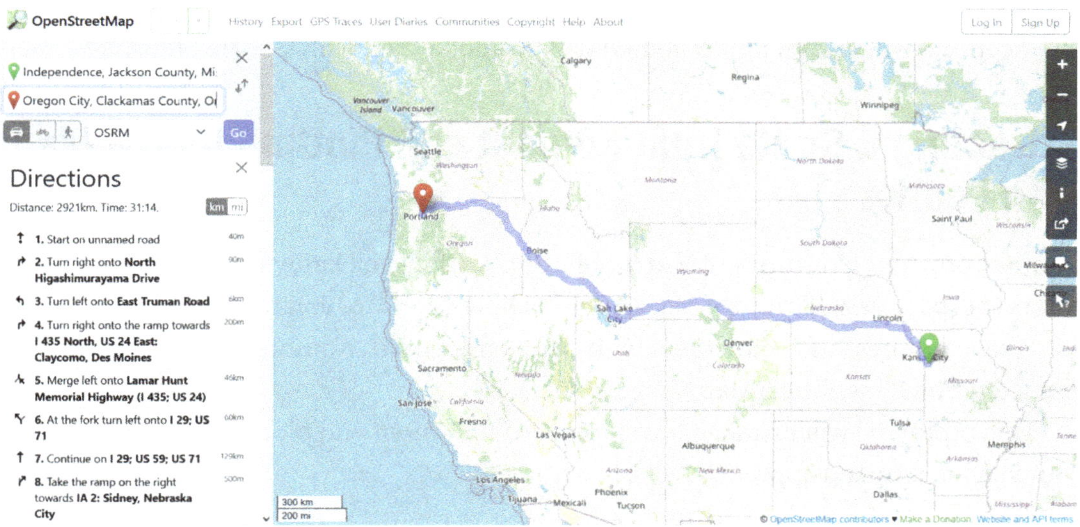

Figure 9-1. Screenshot of openstreetmap.org using map data © OpenStreetMap contributors, under ODbL

However, route calculations are not carried out by the OSM site itself, but are entrusted to external route planning engines (OSRM, GraphHopper, Valhalla, Mapbox Direction API, or BRouter, depending on requirements).

Many other sites also offer this feature: Google Maps, Microsoft Bing Maps, Mappy, Komoot, etc.

P. Marie, *Modern Web Cartography*, https://doi.org/10.1007/979-8-8688-2335-0_9

Route planning is not a functionality offered by the Overpass API, and it would be inappropriate to try to implement a routing function on the basis of OSM data obtained dynamically via Overpass requests. However, some routers do work with OSM data (OSRM, GraphHopper, OpenRouteService, Komoot, etc.), but, to do so, they periodically download data from OSM and store it in their own databases, organized, optimized, and completed for route planning calculations. The requests they receive are processed using these dedicated databases.

Below, we will first **export** routes found using web-based (or mobile) route planning tools and then **plot** them on a map built with Leaflet or Folium. Next, we will **call a route planning service** via an API to request a route and then **plot** it on a map built with Leaflet.

Exporting a Route and Plotting It on a Map

This scenario will first be implemented with Komoot, then with Google Maps.

Extracting a Route from Komoot and Plotting It on a Map with Leaflet

Komoot provides a web-based and mobile application well known among cyclists and hikers (`https://www.komoot.com`). This tool allows users to plan travel routes, navigate using saved routes, and share them with other enthusiasts. Komoot is based on data extracted from OpenStreetMap.

This prototype will extract a travel route from Komoot and plot it on a map created with Leaflet. A mixed approach will be used:

1. The route will be extracted from Komoot in GPX format and then converted to GeoJSON format by Python code executed on the server. This code will then render an HTML page via a template, passing the route's GeoJSON data to the page.

2. The browser will receive the HTML page, access the embedded GeoJSON data, and use JavaScript and Leaflet to build a map and plot the route.

Komoot enables you to export routes in GPX format (a data exchange format between GPS systems). After selecting a route in Komoot, the **Download GPX file** action can be used to export it in GPX format. However, to export a route in GPX format, you need to **unlock** the region (or regions) in which it is located. Komoot allows you to unlock a single region free of charge; to unlock others, you need a paid subscription. After unlocking the Paris, France, region, the GPX file for a route in this area can be downloaded.

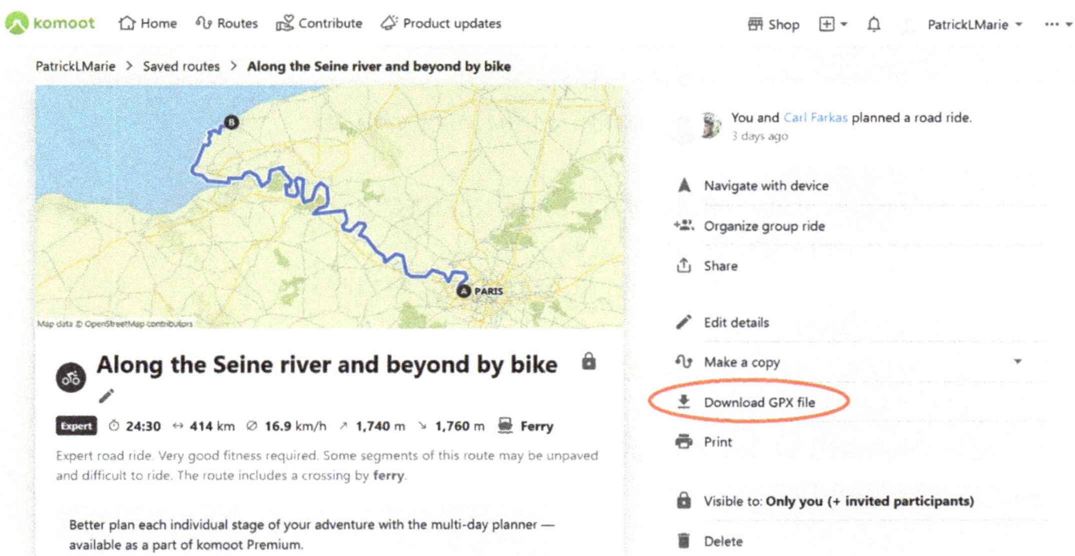

Figure 9-2. *Screenshot of Komoot. Maplibre | © Komoot | Map data © OpenStreetMap-Mitwirkende*

A GPX file is thus obtained (whose contents are shown below in truncated form):

```
<?xml version='1.0' encoding='UTF-8'?>
<gpx version="1.1" creator="https://www.komoot.de"
xmlns="http://www.topografix.com/GPX/1/1"
xmlns:xsi="http://www.w3.org/2001/XMLSchema-instance"
xsi:schemaLocation="http://www.topografix.com/GPX/1/1
http://www.topografix.com/GPX/1/1/gpx.xsd">
    <metadata>
        <name>Along the Seine river and beyond by bike</name>
        <Author>
            <link href="https://www.komoot.de">
```

```
                <text>komoot</text>
                <type>text/html</type>
            </link>
        </author>
    </metadata>
    <trk>
        <name>Along the Seine river and beyond by bike</name>
        <trkseg>
            <trkpt lat="48.885456" lon="2.156155">
                <ele>30.207216</ele>
                <time>2025-06-09T16:11:04.490Z</time>
            </trkpt>
            <trkpt lat="48.885499" lon="2.156237">
                <ele>30.207216</ele>
                <time>2025-06-09T16:11:05.577Z</time>
            </trkpt>
            <trkpt lat="48.885615" lon="2.156347">
                <ele>30.207216</ele>
                <time>2025-06-09T16:11:07.656Z</time>
            </trkpt>
            ...
        </trkseg>
    </trk>
</gpx>
```

General information is followed by a route (trk) described in the form of a series of waypoints, each with its latitude and longitude (and, incidentally, its elevation and projected transit time). These are the latitudes and longitudes we will use to plot the route on a map.

This GPX file can also be checked and visualized graphically (by plotting the route on a map) with online tools such as GPS Visualizer (https://www.gpsvisualizer.com/).

This file was stored and can be found at https://github.com/Apress/Modern-Web-Cartography (file: carto_osm/prototypes/static/Along_the_Seine_river_and_beyond_by_bike.gpx).

Retrieve it and download it into the prototypes/static directory of your workspace.

The GPX format is not directly usable by Leaflet and Folium, but there is a Python library named `geopandas` that reads GPX files and saves their data in a GeoDataFrame object. This GeoDataFrame object can also be converted to GeoJSON format that can be used by Leaflet and Folium. So, in summary, here is the sequence we are following:

Komoot → GPX → GeoDataFrame → GeoJSON → Leaflet

In JavaScript, there is also a library named `togeojson` that performs the same conversion.

The following code must be added to the `advanced_maps.py` blueprint to prepare the data for Leaflet, which will create a map and plot the route on it:

```
@bp.route('/am/map_rout1')
def map_rout1():
    import geopandas, json

    try:
        track = geopandas.read_file("prototypes/static/Along_the_Seine_river_and_beyond_by_bike.gpx", layer="tracks")
        if track.crs is None:
            track = track.set_crs("EPSG:4326")
        track_geo = track[['geometry']]
        track_geoJSON = json.loads(track_geo.to_json())

        return render_template('maps/page_with_route1.html',
            name=" Map with route extracted from Komoot: along the Seine river and beyond by bike",
            loc=[49.2167, 1.1667],
            zoom=10,
            track_geoJSON=track_geoJSON)

    except Exception as e:
        flash("An error has been encountered:<br/>" + str(e), category="error")
        return render_template('maps/page_with_map.html', name="No map to display")
```

- **Line 3**: The geopandas library is imported. It will be used to read a GPX file and load it into memory and then generate its data in GeoJSON format. This library must first be installed in the workspace (via the `pip install geopandas` command, executed with the terminal from the `carto_osm` directory).

  Installing GeoPandas is straightforward in recent Python environments - the preceding `pip` command is generally sufficient. However, installation may be more difficult in older environments because GeoPandas has several dependencies, including C/C++ libraries such as GDAL, PROJ, and GEOS. While modern installation tools usually handle these dependencies automatically when you install GeoPandas, older environments may not, and you might need to install these libraries separately. Using **Conda** rather than pip can often simplify the process.

- **Line 6**: The GPX file data is read and loaded into memory as a GeoDataFrame object, using the `geopandas` library. The `layer="tracks"` parameter specifies which layer of the GPX data we want to read. Here, we are interested in the "tracks" layer, which contains the geometry of the route.

- **Lines 7 and 8**: The CRS (Coordinate Reference System) used by the GPX data is checked. If none is found, EPSG:4326 is set (EPSG:4326 is the CRS used by GPS, while OpenStreetMap uses EPSG:3857).

- **Line 9**: The GeoDataFrame object is truncated to keep only the useful part (the geometric data containing the coordinates of the route's waypoints).

- **Line 10**: The GeoDataFrame object is converted to GeoJSON.

- **Lines 12 to 16**: A web page is rendered using the `maps/page_with_route1.html` template. A name, the coordinates on which to position the map, an initial zoom level (10), and the route in GeoJSON format are passed as arguments.

- **Lines 18 to 20**: These lines handle any errors that may occur by displaying an error message on a web page.

The template `page_with_route1.html`, invoked by the previous function, must be created in the `prototypes/templates/maps` directory with the following contents:

```
{% extends 'base_ll.html' %}

{% block title %}{{ name }}{% endblock %}

{% block header %}
<h2>{{ name }}</h2>
{% endblock %}

{% block script %}
<script id="geojson-data" type="application/json">
{{ track_geoJSON | tojson }}
</script>

<script type="text/javascript">
let map;
try {
    const geojsonData = JSON.parse(document.getElementById("geojson-
    data").textContent);

    map = L.map('map').setView({{ loc }}, {{ zoom }});

    L.tileLayer('https://tile.openstreetmap.org/{z}/{x}/{y}.png', {
        maxZoom: 19,
        attribution: '&copy; <a href="http://www.openstreetmap.org/
        copyright">OpenStreetMap</a>'
    }).addTo(map);

    const routeLayer = L.geoJSON(geojsonData, {
        style: {
            color: 'darkgreen',
            weight: 4
        }
    }).addTo(map);
```

```
        let start = null;
        let end = null;
        if (geojsonData.features.length > 0 && geojsonData.features[0].geometry.type === "LineString") {
            const coords = geojsonData.features[0].geometry.coordinates;
            start = coords[0];
            end = coords[coords.length - 1];
        } else if (geojsonData.features.length > 0 && geojsonData.features[0].geometry.type === "MultiLineString") {
            const coords = geojsonData.features[0].geometry.coordinates;
            const startArray = coords[0];
            start = startArray[0];
            const endArray = coords[coords.length - 1];
            end = endArray[endArray.length - 1];
        }

        if (start != null && end != null) {
            const greenIcon = L.icon({
                iconUrl: 'https://raw.githubusercontent.com/pointhi/leaflet-color-markers/master/img/marker-icon-green.png',
                shadowUrl: 'https://unpkg.com/leaflet@1.9.4/dist/images/marker-shadow.png',
                iconSize: [25, 41],
                iconAnchor: [12, 41],
                popupAnchor: [1, -34],
                shadowSize: [41, 41]
            });

            const redIcon = L.icon({
                iconUrl: 'https://raw.githubusercontent.com/pointhi/leaflet-color-markers/master/img/marker-icon-red.png',
                shadowUrl: 'https://unpkg.com/leaflet@1.9.4/dist/images/marker-shadow.png',
                iconSize: [25, 41],
```

```
61                iconAnchor: [12, 41],
62                popupAnchor: [1, -34],
63                shadowSize: [41, 41]
64            });
65
66            L.marker([start[1], start[0]], {icon: greenIcon}).addTo(map).
              bindPopup("Enjoy you ride!");
67            L.marker([end[1], end[0]], {icon: redIcon}).addTo(map).
              bindPopup("This is the end");
68        }
69
70        map.fitBounds(routeLayer.getBounds());
71
72    } catch(e) {
73        if (typeof map !== 'undefined' && map) {map.remove();}
74
75        const div = document.getElementById("map")
76        if (div) {div.remove();}
77
78        const elemDiv = document.createElement('div');
79        elemDiv.innerHTML = `<p>${e}</p>`;
80        document.body.appendChild(elemDiv);
81    }
82    </script>
83    {% endblock %}
```

- **Lines 10 to 12**: The JSON script section in which the route's GeoJSON data will be set by the Jinja engine.
- **Line 17**: The route's GeoJSON data is retrieved (from the script section on lines 10 to 12), parsed, and set into a JSON-structured object.
- **Line 19**: A map is created, positioned, and zoomed according to the arguments supplied by the calling Python function.
- **Lines 21 to 24**: A tile layer is associated with the map.

- **Lines 26 to 31**: An `L.geoJSON` object is created with the JSON-structured object, and a color and thickness to plot the route are specified. The object is then added to the map, plotting the route as a polyline.
- **Lines 33 to 45**: The coordinates of the start and end points of the route are retrieved. Depending on whether the route contains a `LineString` or a `MultiLineString`, different statements are executed.
- **Lines 47 to 68**: If valid coordinates have been found, markers are added at the start and end points of the route:
    - **Lines 48 to 64**: A green icon (for the start marker) and a red icon (for the end marker) are defined.
    - **Lines 66 and 67**: Two markers are added on the map at the start and end positions of the route, using the coordinates previously obtained and the two icons just defined. Popups are associated with these markers.
- **Line 70**: The map zoom level is adjusted to the route just plotted.
- **Lines 72 to 81**: These lines handle any errors that may occur.

Launching the server and opening the URL `http://127.0.0.1:5000/am/map_rout1` with the browser produces the following:

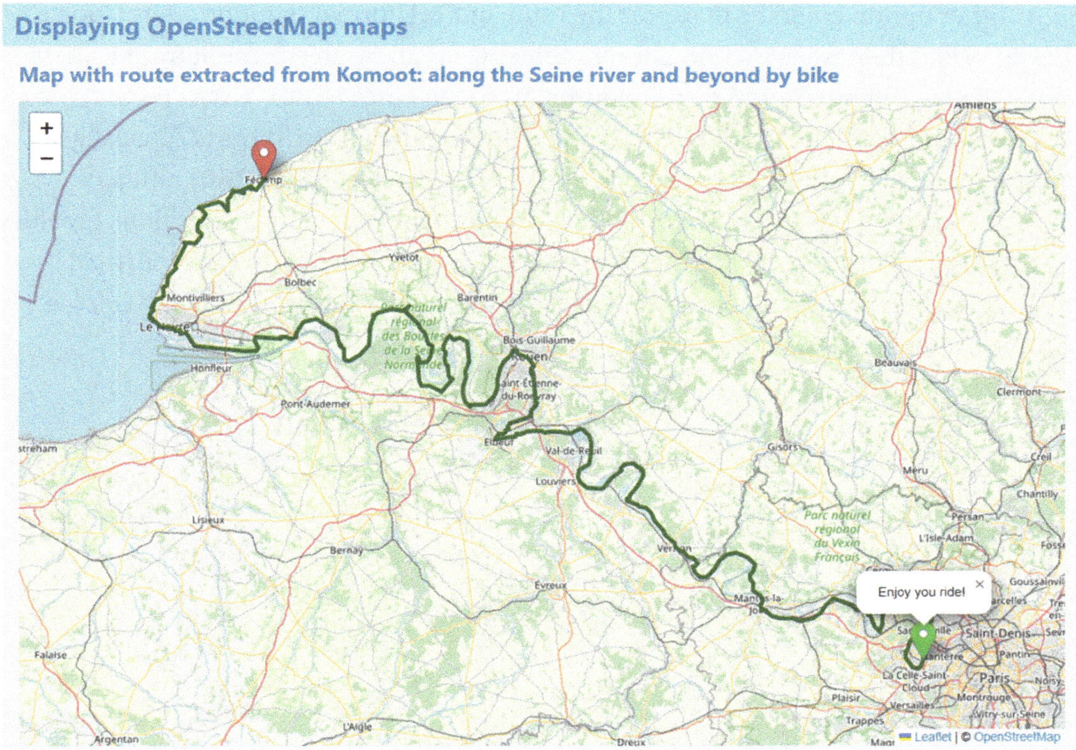

Figure 9-3. *Map created with Leaflet using OpenStreetMap tiles. Data © OpenStreetMap contributors, licensed under the ODbL. Route found by Komoot. Maplibre | © Komoot | Map data © OpenStreetMap-Mitwirkende*

Extracting a Route from Google Maps and Plotting It on a Map with Folium

Google Maps is a website universally known for its maps and immersive panoramic view capabilities ("Google Street View" functionality). It can also be used to search for routes between two locations or, more generally, to plan routes with several waypoints.

This new prototype will extract a route created with Google Maps, convert it to GeoJSON format, and plot it on a map built with Folium. All these operations will be performed on the server using Python code.

Google Maps does not have a native functionality for exporting routes in GPX format, but additional tools make this possible. For example, **GMaps2GPX**, a free Google Chrome extension, can be used in this role (actually, its base version is free, but advanced versions are also available and must be purchased).

Another option could be to export the route in KML format (supported by Google My Maps) and then use a Python library such as GeoPandas to convert it from KML to GeoJSON (we will not prototype this approach).

The GMaps2GPX installation code can be downloaded from `https://gmaps2gpx.com/` (on the main page of this site, click the **Add to Chrome (IT'S FREE)** button).

After installing this extension in Google Chrome (following the instructions provided at `https://gmaps2gpx.com/how-to-install-gmaps2gpx-chrome-extension.html`), the function becomes available as an icon in the top right-hand corner of the Google Chrome screen (circled in red below).

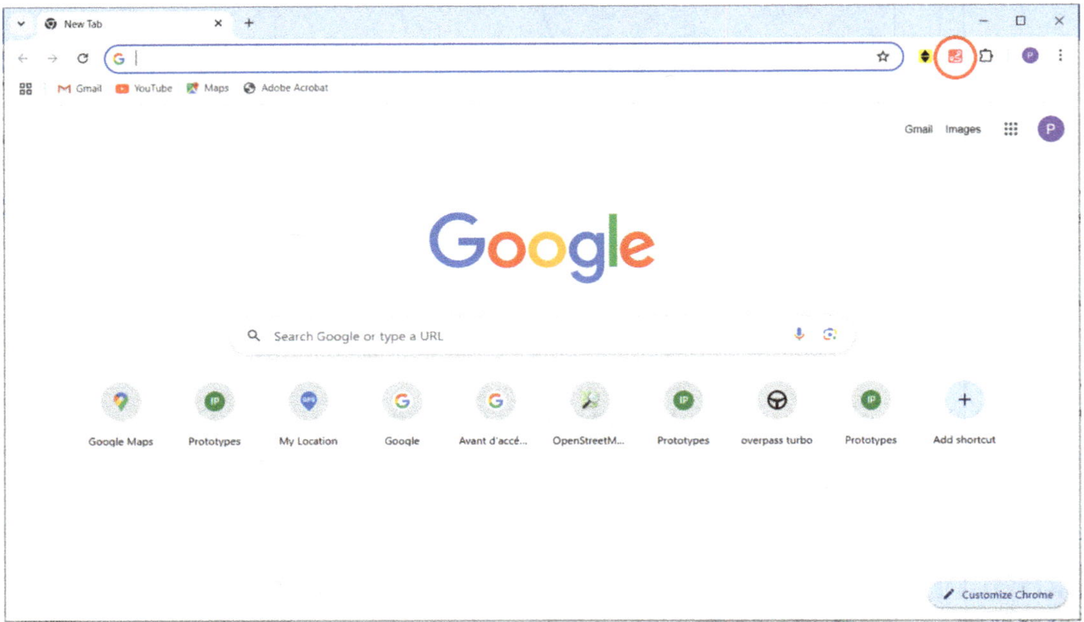

Figure 9-4. *Screenshot of Google Chrome with the GMaps2GPX icon. Google Chrome is a trademark of Google LLC. Google and Google Chrome are trademarks of Google LLC, and this book is not endorsed by or affiliated with Google in any way*

Clicking this icon opens Google Maps, where a route can be searched for (as shown below).

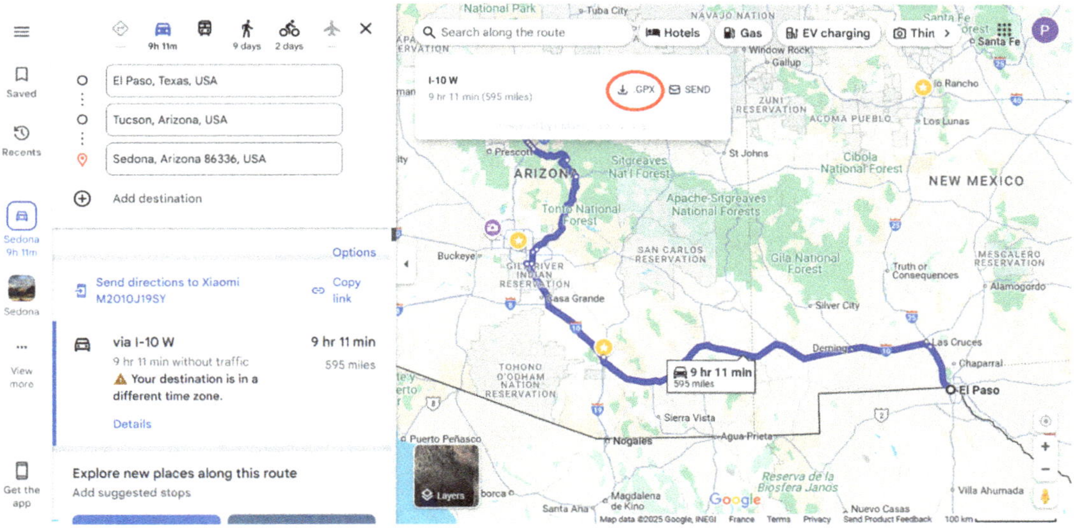

Figure 9-5. *Screenshot of Google Maps with the GMaps2GPX panel. Google Maps is a trademark of Google LLC. Google and Google Maps are trademarks of Google LLC, and this book is not endorsed by or affiliated with Google in any way*

Once a route has been found, clicking the **Download (represented as an icon) .GPX** link (circled in red in the screen above) downloads the route as a GPX file.

The resulting file can be displayed with a text editor (contents shown below in truncated form):

```
<?xml version="1.0" encoding="UTF-8" standalone="no" ?>
    <gpx xmlns="http://www.topografix.com/GPX/1/1"
    xmlns:gpxx="http://www.garmin.com/xmlschemas/GpxExtensions/v3"
    xmlns:gpxtpx="http://www.garmin.com/xmlschemas/TrackPointExtension/v1"
    creator="getwebooster.com" version="1.1"
    xmlns:xsi="http://www.w3.org/2001/XMLSchema-instance"
    xsi:schemaLocation="http://www.topografix.com/GPX/1/1 http://www.
    topografix.com/GPX/1/1/gpx.xsd http://www.garmin.com/xmlschemas/
    GpxExtensions/v3 http://www.garmin.com/xmlschemas/GpxExtensionsv3.xsd
    http://www.garmin.com/xmlschemas/TrackPointExtension/v1 http://www.
    garmin.com/xmlschemas/TrackPointExtensionv1.xsd">
        <metadata>
            <link href="https://gmaps2gpx.extensionsbox.com/">
                <text>GMaps2GPX</text>
```

```
        </link>
        <time>2025-05-15T11:31:29.782Z</time>
    </metadata>

    <wpt lat="31.761877799999997" lon="-106.48502169999999">
        <name><![CDATA[El Paso, Texas, États-Unis]]></name>
        <desc><![CDATA[El Paso, Texas, USA]]></desc>
    </wpt>

    <wpt lat="32.2539787" lon="-110.97417689999999">
        <name><![CDATA[Tucson, Arizona, USA]]></name>
        <desc><![CDATA[Tucson, Arizona, USA]]></desc>
    </wpt>

    <wpt lat="34.862614" lon="-111.8137258">
        <name><![CDATA[Sedona, Arizona, États-Unis]]></name>
        <desc><![CDATA[Sedona, Arizona 86336, USA]]></desc>
    </wpt>

    <trk>
        <name><![CDATA[El Paso, Texas, États-Unis to Sedona, Arizona,
        États-Unis]]></name>
            <trkseg>

                <trkpt lat="31.761489" lon="-106.4848548">
                </trkpt>

                <trkpt lat="31.7617079" lon="-106.4857149">
                </trkpt>

                <trkpt lat="31.7619204" lon="-106.4859062">
                </trkpt>

                <trkpt lat="31.7619661" lon="-106.4859429">
                </trkpt>

                ...
```

It can also be checked and visualized graphically (with the route plotted on a map) with online tools such as GPS Visualizer (`https://www.gpsvisualizer.com/`).

This file was stored and can be found at `https://github.com/Apress/Modern-Web-Cartography` (file: `carto_osm/prototypes/static/From_El_Paso_to_Sedona.gpx`).

Retrieve it and download it into the `prototypes/static` directory of your workspace.

The following code must be added to the `advanced_maps.py` blueprint to convert the route data, build a map, and plot the route on it:

```
@bp.route('/am/map_rout2')
def map_rout2():
    import geopandas as gpd

    try:
        track = gpd.read_file("prototypes/static/From_El_Paso_to_Sedona.gpx", layer="tracks")
        if track.crs is None:
            track = track.set_crs("EPSG:4326")
        track_geo = track[['geometry']]
        track_geoJSON = track_geo.to_json()

        m = folium.Map(location=[33.542498, -111.870257], zoom_start=8)

        folium.GeoJson(
            data=track_geoJSON,
            style_function=lambda x: {'color': 'darkblue', 'weight': 5}
        ).add_to(m)

        track_geoJSONDict = track_geo.__geo_interface__

        geom_type = track_geoJSONDict["features"][0]["geometry"]["type"]
        coords = track_geoJSONDict["features"][0]["geometry"]["coordinates"]
        if geom_type == 'LineString':
            start_coords = [coords[0][1], coords[0][0]]  # [lat, lon]
            end_coords = [coords[-1][1], coords[-1][0]]
        elif geom_type == 'MultiLineString':
```

```
27              start_coords = [coords[0][0][1], coords[0][0][0]]
28              end_coords = [coords[-1][-1][1], coords[-1][-1][0]]
29          else:
30              start_coords = None
31              end_coords = None
32
33          if (start_coords != None or end_coords != None):
34              folium.Marker(start_coords, popup="Departure: El Paso",
                icon=folium.Icon(color='green')).add_to(m)
35              folium.Marker(end_coords, popup="Arrival: Sedona",
                icon=folium.Icon(color='red')).add_to(m)
36
37          bounds = track_geo.total_bounds
38          m.fit_bounds([[bounds[1], bounds[0]], [bounds[3], bounds[2]]],
39              padding=(30, 30))
40
41          return render_template('maps/page_with_map.html',
42              name = "Map with route extracted from Google Maps: from El
                Paso to Sedona",
43              map_html=m._repr_html_())
44
45      except Exception as e:
46          flash("An error has been encountered:<br/>" + str(e),
            category="error")
47          return render_template('maps/page_with_map.html', name = "No
            map to display")
```

- **Line 3**: The geopandas library is imported.
- **Line 6**: The data from the GPX file (containing the route extracted from Google Maps) is read and put into a GeoDataFrame object.
- **Lines 7 and 8**: The CRS (Coordinate Reference System) used by the GPX data is checked. If none is found, EPSG:4326 is set (EPSG:4326 is the CRS used by GPS, while OpenStreetMap uses EPSG:3857).

- **Line 9**: This object is truncated to retain only the data of interest (the geographic positions).
- **Line 10**: The truncated GeoDataFrame object is converted to GeoJSON.
- **Line 12**: A map is created and centered on a position with suitable coordinates and an initial zoom level (8).
- **Lines 14 to 17**: A `folium.GeoJson` object is created with the route's GeoJSON data. The `style_function` parameter is used to apply style properties to all features in the GeoJSON data (i.e., polylines).

  Then this `folium.GeoJson` object is added to the map, thus plotting a polyline.
- **Lines 19 to 31**: The GeoDataFrame object is converted into a GeoJSON-like dictionary. Its geometry is examined to determine whether it contains a `LineString` or a `MultiLineString`. Then the coordinates of the first and last points are extracted accordingly (the index -1 refers to the last element in the list).
- **Lines 33 to 35**: If valid coordinates have been found, markers are added at these coordinates to indicate the start and end of the route.
- **Lines 37 to 39**: The map view is adjusted to fit the geographical bounds of the route just plotted. The padding attribute ensures a 30-pixel margin around all sides of the fitted area, helping to keep both markers (start and end points) visible. Without this padding, one or both markers might be partially cut off.
- **Lines 41 to 43**: A web page is rendered using the `page_with_map.html` template. A name and the map, converted to HTML and JavaScript code, are passed as arguments.
- **Lines 45 to 47**: These lines handle any errors that may occur.

Launching the server and opening the address `http://127.0.0.1:5000/am/map_rout2` in the browser produces the following map:

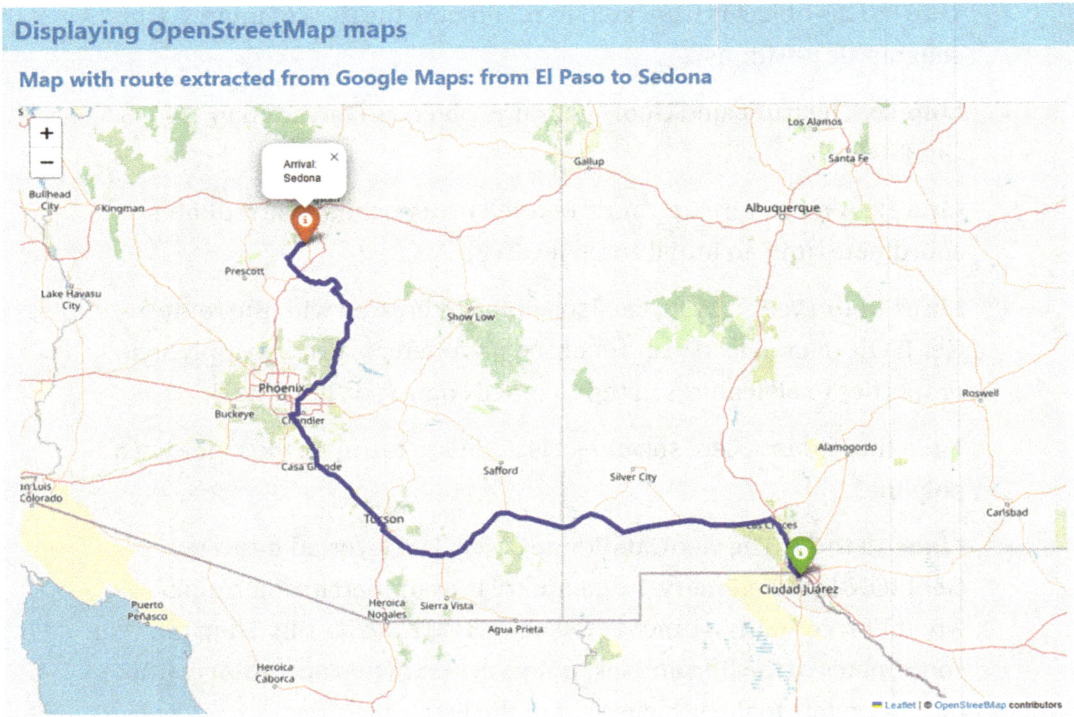

Figure 9-6. *Map created with Folium using OpenStreetMap tiles. Data © OpenStreetMap contributors, licensed under the ODbL. Route found by Google Maps. Google Maps is a trademark of Google LLC. Google and Google Maps are trademarks of Google LLC, and this book is not endorsed by or affiliated with Google in any way*

Alternative Solutions with Leaflet Plugins

In the previous sections, we demonstrated how to write Python and JavaScript code to convert GPX data and plot it on a map using Leaflet or Folium. Alternatively, some Leaflet plugins can also be used for that purpose:

- **Leaflet.FileLayer** (developed by Mathieu Leplatre): This plugin provides a new control on the map that allows users to load a GPX file and plot its data on the map. The map automatically repositions to fit the plotted polyline.
- **Leaflet GPX** (developed by Maxime Petazzoni): This plugin provides a new object called `L.GPX` that can be used to load GPX data from a file and add it to the map, thereby plotting a polyline.

Invoking a Route Planning Service

It is also possible to call a **route planning service** with an API to search for a route directly from our application code (either from Python code running on the server or from JavaScript code running in the browser). The route found can then be plotted on a map built with Leaflet or Folium.

Several online route planning services via public APIs are available:

- In the open source domain: OSRM, Valhalla, GraphHopper, BRouter (for bicycle routes), etc. Rather than using these services via a public web API, it is also possible to download the service code and install it on a server.
- In the field of proprietary software and services: Google Maps Directions API, Microsoft Bing Maps Routes API, Mapbox Directions API, HERE Routing API, Tomtom Routing API, Esri ArcGIS Routing, etc. These services apply usage-based pricing, typically charging per request beyond a free quota.

The routing calculations performed by these tools use specialized algorithms such as Dijkstra's algorithm, A*, Contraction Hierarchies, and Bidirectional Search (explaining these algorithms is beyond the scope of this book).

Most of these tools propose more than basic route planning:

- Routing can be customized by selecting a transport type (car, public transportation, bicycle, pedestrian, etc.) and by setting options such as avoiding toll roads, unpaved roads, and ferry crossings or using bike paths only.

  Detailed step-by-step navigation instructions can also be obtained in addition to the routes.
- These services sometimes offer additional functionalities, such as solvers for the Traveling Salesman Problem (which finds the shortest possible route to visit a set of points and return to the starting location) and distance matrix generators. A distance matrix is a square matrix (or, in an optimized form, a triangular matrix, to avoid storing duplicate data) containing the distances between every pair of points in a set.

Searching for a Route with the OSRM Router on the Server and Plotting It on a Map with Leaflet

This prototype will search for a route using the OSRM route planning service, queried from Python code on the server. The resulting route will then be plotted on a map with Leaflet in the browser.

Other possible approaches are:

- Search for a route with OSRM and plot it on a map from Python code on the server.
- Search for a route with OSRM and plot it on a map from JavaScript code executed in the browser.

Open Source Routing Machine (**OSRM**) is an open source project that provides a web-based user interface for route searching (`https://map.project-osrm.org/`).

***Figure 9-7.** Screenshot from OSRM (Open Source Routing Machine), © Project OSRM contributors. Map data © OpenStreetMap contributors (ODbL)*

Note that this web user interface is provided as a demo and is not intended for production use.

It also offers an HTTP API (route searches are submitted as HTTP requests) for programmatic route searching. This API is documented at `https://project-osrm.org/docs/v5.24.0/api`. We will use it for our prototype.

Finally, it should be noted that:

- OSRM works with geographic data extracted from the OpenStreetMap database.
- For route calculation, OSRM uses the bidirectional Dijkstra algorithm and the Contraction Hierarchies (better suited for large distance matrices).
- OSRM's public HTTP API, which we will use below, is offered for demonstration and testing purposes (the usage policy states: "One request per second max. No scraping, no heavy usage").
- For production use, it is recommended to install the OSRM code on a private server or use this service provided by a third party.

To search for a pedestrian route from Saint-Jean-Pied-de-Port (France) to Santiago de Compostela (Spain) via OSRM's public API, the following query is suitable:

```
https://router.project-osrm.org/route/v1/foot/-1.238317,43.162498;
-8.540143,42.881940?geometries=geojson
```

- The first part of the URL path, `/route`, indicates that a route is being searched.
- The second part of the path, `/v1`, specifies the OSRM version to be used (at the time of writing, only v1 is available).
- The third part of the path, `/foot`, specifies that a pedestrian route is requested.
- The last part of the path provides the geographical coordinates of the starting and ending points of the route (Saint-Jean-Pied-de-Port in France and Santiago de Compostela in Spain).
- The `geometries=geojson` parameter requests that the response include the route's geometric data in GeoJSON format.
- As the `alternatives` parameter is not used, it defaults to `false`, and one single route will be returned.

This request is accepted by OSRM; however, it returns a driving route rather than a pedestrian one, because the "foot" profile is not supported on the public OSRM demo server. We will nevertheless use it in our example and recalculate the duration to match a pedestrian pace.

Here is the response returned by OSRM (contents shown below in truncated form):


```
{
  "code": "Ok",
  "routes": [
    {
      "legs": [
        {
          "steps": [],
          "weight": 30652.3,
          "summary": "",
          "duration": 30652.3,
          "distance": 760394.4
        }
      ],
      "weight_name": "routability",
      "geometry": {
        "coordinates": [
          [
            -1.238327,
            43.162502
          ],
          [
            -1.342896,
            43.033302
          ],
          [
            -1.330949,
            42.992523
          ],

          ...

          [
            -8.540159,
            42.88194
          ]
```

```
        ],
        "type": "LineString"
      },
      "weight": 30652.3,
      "duration": 30652.3,
      "distance": 760394.4
    }
  ],
  "waypoints": [
    {
      "hint": "y26jk3oRK5QNAAAAAwAAABoAAAAAAAAAEHV1QWEZPOCwfuRBAAAAAAOAAAA
      DAAAAGgAAAAAAAAC8FgEAyRrt_4abkgLTGu3_gpuSAgIA_wEAAAAA",
      "location": [
        -1.238327,
        43.162502
      ],
      "name": "Place Floquet",
      "distance": 0.9266044703
    },
    {
      "hint": "PTZ7gvtQrYUPAAAAAQAAAFkAAAAIAAAAgBogQXyiqj4KbThCj7q
      oQA8AAAABAAAAWQAAAAgAAAC8FgEAAbB9_5RTjgIRsH3_lFOOAgYALwAAAAAA",
      "location": [
        -8.540159,
        42.88194
      ],
      "name": "Rúa de Aller Ulloa",
      "distance": 1.307297059
    }
  ]
}
```


The OSRM response is in JSON format, but not GeoJSON. However, it does include the geometric details of the route in GeoJSON format, in the `geometry` section. To use it with Leaflet or Folium, we need to extract the parts we are interested in (`geometry` and `distance`) and build a valid GeoJSON object with this information. Our resulting GeoJSON object will look like this:

```
{
    "type": "FeatureCollection",
    "features": [
        {
            "type": "Feature",
            "properties": {
                "distance_miles": "472.48",
                "duration_hours": "152.1",
            },
            "geometry": {
                "coordinates": [
                    [
                        -1.238327,
                        43.162502
                    ],
                    [
                        -1.342896,
                        43.033302
                    ],
                    [
                        -1.330949,
                        42.992523
                    ],

                    ...

                    [
                        -8.540159,
                        42.88194
                    ]
                ],
                "type": "LineString"
            }
        }
    ]
}
```

Note that tools such as GeoJSONLint (`https://geojsonlint.com/`), which verifies and graphically displays GeoJSON data, can be useful to check and fine-tune the structure of the GeoJSON object (so that it can be displayed on a map using Leaflet or Folium).

Finally, the code to be added to the `advanced_maps.py` blueprint to build this map is the following:

```
@bp.route('/am/map_rout3')
def map_rout3():
    import requests, json

    try:
        response = requests.get(
            "https://router.project-osrm.org/route/v1/foot/-1.238317,43.162498;-8.540143,42.881940?geometries=geojson"
        )

        if response.status_code == 200:
            response_json = response.json()
        else:
            flash("Error querying OSRM", category="error")
            return render_template('maps/page_with_map.html', name="No map to display")

        if not response_json.get("routes"):
            flash("No route found.", category="error")
            return render_template('maps/page_with_map.html', name="No map to display")

        route_geo = response_json['routes'][0]
        distance_miles = round(route_geo['distance'] / 1609.344, 2)
        duration_hours = round((distance_miles / 3.106856), 1)
        geom = route_geo['geometry']

```

```
        geojson_data = json.loads('{"type": "FeatureCollection",
        "features": [{"type": "Feature", "properties": {}}]}')
        geojson_data['features'][0]['properties'] = {
            'distance_miles': distance_miles,
            'duration_hours': duration_hours
        }
        geojson_data['features'][0]['geometry'] = geom

        gtype = geom['type']
        coords = geom['coordinates']

        if not coords:
            flash("No geometry coordinates in route", category="error")
            return render_template('maps/page_with_map.html', name="No
            map todisplay")
        if gtype == 'LineString':
            start_coords = coords[0][::-1]
            end_coords = coords[-1][::-1]
        elif gtype == 'MultiLineString':
            start_coords = coords[0][0][::-1]
            end_coords = coords[-1][-1][::-1]
        else:
            flash("No suitable geometry coordinates in route",
            category="error")
            return render_template('maps/page_with_map.html', name="No
            map to display")

        return render_template('maps/page_with_route3.html',
            name="Map with route searched with OSRM: from Saint-Jean-
            Pied-de-Port to Santiago-de-Compostela",
            loc=[42.874, -4.598],
            zoom=7,
            track_geo=geojson_data,
            start=start_coords,
            end=end_coords
```

```
55          )
56
57      except Exception as e:
58          flash("An error has been encountered:<br/>" + str(e), category="error")
59          return render_template('maps/page_with_map.html', name="No map to display")
```

- **Line 3**: The `requests` and `json` libraries are imported.
- **Lines 6, 7, and 8**: A request is submitted to the OSRM server for a pedestrian route between Saint-Jean-Pied-de-Port (France) and Santiago de Compostela (Spain). A detailed explanation of the parameters has already been provided above.
- **Lines 10 to 14**: The result of the request is checked:
    - If the request was successful, the OSRM response is parsed and placed in a JSON-structured object.
    - Otherwise, an error message is displayed on a web page.
- **Lines 16 to 18**: These lines check that a route has been found (there are points which cannot be connected by any route). If none has been found, an error message is displayed on a web page.
- **Lines 20 to 23**: The distance and the geometry of the route are extracted from the JSON-structured object containing the OSRM response data. The duration is recalculated from the distance, because the duration returned by OSRM reflects driving time, not walking time. We estimate the walking time in hours, assuming a pace of 3.106856 mph (5 km/h).
- **Lines 25 to 30**: A valid GeoJSON object is built using the data extracted from the OSRM response or calculated: distance, duration, and geometry of the route.
- **Lines 32 to 46**: The coordinates of the first and last points of the JSON object are extracted.

  `coords[-1]` means the last item of the list (the first one starting from the end).

`[::-1]` means that the two items (longitude and latitude) will be flipped.

- **Lines 48 to 55**: A web page is rendered using the `maps/page_with_route3.html` template. The following arguments are passed: a name, the coordinates on which to position the map (those of a point in northern Spain), the initial zoom level (7), the route data in GeoJSON format, and the coordinates of the start and end points.
- **Lines 57 to 59**: These lines handle any errors that may occur by displaying an error message on a web page.

The template `page_with_route3.html` must be created in the `prototypes/templates/maps` directory with the following contents:

```
{% extends 'base_ll.html' %}

{% block title %}{{ name }}{% endblock %}

{% block header %}
<h2>{{ name }}</h2>
{% endblock %}

{% block script %}
<script id="geojson-data" type="application/json">
{{ track_geo |  tojson }}
</script>

<script type="text/javascript">
let map;
try {
    const geojsonData = JSON.parse(document.getElementById("geojson-data").textContent);

    map = L.map('map').setView({{ loc }}, {{ zoom }});

    L.tileLayer('https://tile.openstreetmap.org/{z}/{x}/{y}.png', {
        maxZoom: 19,
```

```
        attribution: '&copy; <a href="http://www.openstreetmap.org/copyright">OpenStreetMap</a>'
    }).addTo(map);

    const routeLayer = L.geoJSON(geojsonData, {
        style: {
            color: 'purple',
            weight: 4
        }
    }).addTo(map);

    const center = routeLayer.getBounds().getCenter();

    L.popup()
    .setLatLng(center)
    .setContent(`Distance: ${geojsonData.features[0].properties.distance_miles} miles<br>Duration: ${geojsonData.features[0].properties.duration_hours} hours`)
    .openOn(map);

    const greenIcon = L.icon({
        iconUrl: 'https://raw.githubusercontent.com/pointhi/leaflet-color-markers/master/img/marker-icon-green.png',
        shadowUrl: 'https://unpkg.com/leaflet@1.9.4/dist/images/marker-shadow.png',
        iconSize: [25, 41],
        iconAnchor: [12, 41],
        popupAnchor: [1, -34],
        shadowSize: [41, 41]
    });

    const redIcon = L.icon({
        iconUrl: 'https://raw.githubusercontent.com/pointhi/leaflet-color-markers/master/img/marker-icon-red.png',
```

```
51            shadowUrl: 'https://unpkg.com/leaflet@1.9.4/dist/images/
              marker-shadow.png',
52            iconSize: [25, 41],
53            iconAnchor: [12, 41],
54            popupAnchor: [1, -34],
55            shadowSize: [41, 41]
56        });
57
58        L.marker({{ start|tojson }}, {icon: greenIcon}).addTo(map).
          bindPopup("Start: Saint-Jean-Pied-de-Port");
59        L.marker({{ end|tojson }}, {icon: redIcon}).addTo(map).
          bindPopup("End: Santiago de Compostela");
60
61        map.fitBounds(routeLayer.getBounds());
62
63    } catch(e) {
64        if (typeof map !== 'undefined' && map) {map.remove();}
65
66        const div = document.getElementById("map")
67        if (div) {div.remove();}
68
69        const elemDiv = document.createElement('div');
70        elemDiv.innerHTML = `<p>${e}</p>`;
71        document.body.appendChild(elemDiv);
72    }
73    </script>
74    {% endblock %}
```

- **Lines 10 to 12**: A JSON script section in which the route's GeoJSON data will be inserted by the Jinja engine.
- **Line 17**: The GeoJSON data contained in the script section (lines 10 to 12) is retrieved, parsed, and placed in a JSON-structured object.
- **Line 19**: A map is created, centered on a position received from the blueprint and with an initial zoom level also received from the blueprint.

- **Lines 21 to 24**: A tile layer is associated with the map.
- **Lines 26 to 31**: An `L.geoJSON` object is created with the JSON-structured object and style settings; then it is added to the map.
- **Line 33**: The coordinates of the center of the route are calculated.
- **Lines 35 to 38**: A popup is added at the center of the route with the distance and duration of the route.
- **Lines 40 to 56**: Two icons are defined, one green and the other red. They will be used by the markers created next.
- **Lines 58 and 59**: Two markers are added to the map – a green one at the start and a red one at the end of the route – to visually indicate the route's origin and destination.
- **Line 61**: The zoom level is adjusted to the route plotted.
- **Lines 63 to 72**: These lines handle any errors that may occur.

After launching the server, opening the URL `http://127.0.0.1:5000/am/map_rout3` in the browser and clicking on the route, we get the following map:

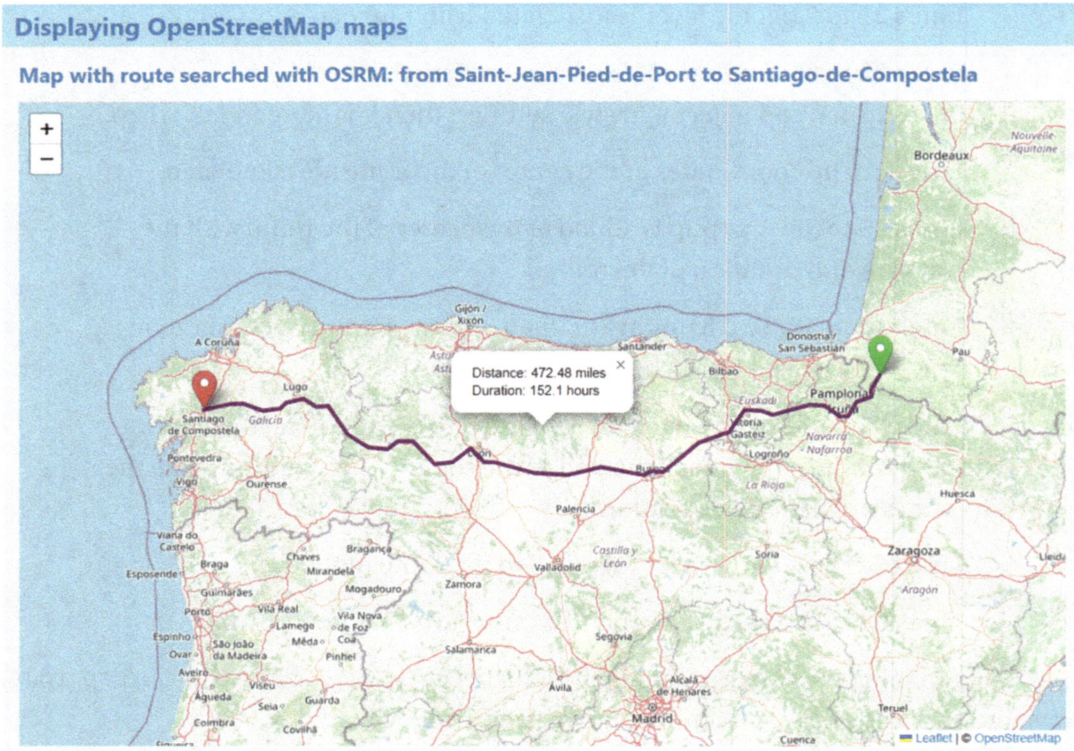

__Figure 9-8.__ Map created with Leaflet using OpenStreetMap tiles. Data © OpenStreetMap contributors, licensed under the ODbL. Route generated by OSRM (Open Source Routing Machine), © Project OSRM contributors. Map data © OpenStreetMap contributors (ODbL).

It is also possible to retrieve **navigation instructions** from OSRM when searching for a route. We will now add this functionality to our prototype. While OSRM does not provide this feature for pedestrian routes, it does support it for driving routes. Therefore, we will switch to searching for a driving route for demonstration purposes.

A suitable OSRM query that returns a driving route from Saint-Jean-Pied-de-Port (France) to Santiago de Compostela (Spain) with navigation instructions is:

```
https://router.project-osrm.org/route/v1/driving/-1.238317,43.162498;
-8.540143,42.881940?steps=true&geometries=geojson
```

- The `driving` part of the URL path specifies that we are searching for a driving itinerary.
- The `steps=true` parameter instructs OSRM to return route steps with turn-by-turn navigation instructions.

The relevant part of the response containing navigation instructions looks like this:


```
{
    "code": "Ok",
    "routes": [
        {
            "geometry": {
                "coordinates": [ ... ],
                "type": "LineString"
            },
            "legs": [
                {
                    "steps": [
                        {
                            "geometry": {
                                "coordinates": [ ... ],
                                "type": "LineString"
                            },
                            "maneuver": {
                                "bearing_after": 211,
                                "bearing_before": 0,
                                "location": [
                                    -1.238327,
                                    43.162502
                                ],
                                "type": "depart"
                            },
                            "mode": "driving",
                            "ref": "D 933",
                            "driving_side": "right",
                            "name": "",
                            "intersections": [ ... ],
                            "weight": 6.7,
                            "duration": 6.7,
                            "distance": 73
                        },
```

```
{
    "geometry": {
        "coordinates": [ ... ],
        "type": "LineString"
    },
    "maneuver": {
        "bearing_after": 278,
        "bearing_before": 223,
        "location": [
            -1.238832,
            43.161961
        ],
        "modifier": "right",
        "type": "continue"
    },
    "mode": "driving",
    "ref": "D 933",
    "driving_side": "right",
    "name": "Place Floquet",
    "intersections": [ ... ],
    "weight": 12.5,
    "duration": 12.5,
    "distance": 137.5
},
...
```

The itinerary consists of one or more legs (several legs are received when waypoints are specified in the request), each containing multiple steps that represent basic movements. Each step includes navigation details such as:

- The type of maneuver (`maneuver.type`).
- A modifier indicating the direction (`maneuver.modifier`).
- The road reference (`ref`).
- The road name (`name`).
- The distance of the step (`distance`).

We will now modify our prototype to display a basic list of navigation instructions below the map, implementing it as a new route (`/am/map_rout4`) in the blueprint. The code for `/am/map_rout4` is identical to that of `/am/map_rout3`, with a few modifications or additions (highlighted in **bold** below):

```
@bp.route('/am/map_rout4')
def map_rout4():
    import requests, json

    try:
        response = requests.get(
            "https://router.project-osrm.org/route/v1/driving/-1.238317,43.162498;-8.540143,42.881940?steps=true&geometries=geojson"
       )

        if response.status_code == 200:
            response_json = response.json()
        else:
            flash("Error querying OSRM", category="error")
            return render_template('maps/page_with_map.html', name="No map to display")

        if not response_json.get("routes"):
            flash("No route found.", category="error")
            return render_template('maps/page_with_map.html', name="No map to display")

        route_geo = response_json['routes'][0]
        distance_miles = round(route_geo['distance'] / 1609.344, 2)
        duration_hours = round(route_geo['duration'] / 3600, 1)
        geom = route_geo['geometry']

        geojson_data = json.loads('{"type": "FeatureCollection", "features": [{"type": "Feature", "properties": {}}]}')
        geojson_data['features'][0]['properties'] = {
```

```
            'distance_miles': distance_miles,
            'duration_hours': duration_hours
        }
        geojson_data['features'][0]['geometry'] = geom

        gtype = geom['type']
        coords = geom['coordinates']

        if not coords:
            flash("No geometry coordinates in route",
            category="error")
            return render_template('maps/page_with_map.html',
            name="No map to display")
        if gtype == 'LineString':
            start_coords = coords[0][::-1]
            end_coords = coords[-1][::-1]
        elif gtype == 'MultiLineString':
            start_coords = coords[0][0][::-1]
            end_coords = coords[-1][-1][::-1]
        else:
            flash("No suitable geometry coordinates in route",
            category="error")
            return render_template('maps/page_with_map.html',
            name="No map to display")

        instructions = []
        i = 0
        for leg in route_geo['legs']:
            for step in leg['steps']:
                i = i + 1
                maneuver_type = step['maneuver'].get('type', '')
                modifier = step['maneuver'].get('modifier', '')
                road_ref = step.get('ref', '')
                road_name = step.get('name', '')
                dist = int(step.get('distance', 0)//1)

```

```
            if (maneuver_type == 'depart'):
                instruction_text = 'Start from departure point'
            elif (maneuver_type == 'arrive'):
                instruction_text = 'Arrive at destination'
            elif (maneuver_type == 'roundabout'):
                instruction_text = 'Enter roundabout'
            elif (maneuver_type == 'exit roundabout'):
                instruction_text = 'Exit roundabout'
            elif (modifier == 'left'):
                instruction_text = 'Turn left'
            elif (modifier == 'right'):
                instruction_text = 'Turn right'
            elif (modifier == 'slight left'):
                instruction_text = 'Make a slight left'
            elif (modifier == 'slight right'):
                instruction_text = 'Make a slight right'
            else:
                instruction_text = 'Continue'

            if (road_ref != ''):
                instruction_text = f"{instruction_text} on {road_ref}"
                if (road_name != ''):
                    instruction_text = f"{instruction_text} ({road_name})"
            else:
                if (road_name != ''):
                    instruction_text = f"{instruction_text} on {road_name}"

            instructions.append([i, instruction_text, dist])

    return render_template('maps/page_with_route4.html',
        name="Map with route searched with OSRM: from Saint-Jean-Pied-de-Port to Santiago-de-Compostela",
```

```
90                loc=[42.874, -4.598],
91                zoom=7,
92                track_geo=geojson_data,
93                start=start_coords,
94                end=end_coords,
95                instructions=instructions
96            )
97
98        except Exception as e:
99                flash("An error has been encountered:<br/>" + str(e),
                  category="error")
100           return render_template('maps/page_with_map.html', name="No
              map to display")
```

- **Line 7**: The OSRM request now includes `driving` and `steps=true` to retrieve navigation steps.
- **Line 22**: The calculation of the trip duration is modified to use the value returned in the response, which is suitable for a travel by car.
- **Lines 50 to 86**: Nested loops are used to iterate over all steps in all legs:
    - **Lines 53 to 57**: Navigation information is extracted: maneuver type, modifier, road reference, road name, and distance.
    - **Lines 59 to 84**: A readable instruction string is built from this data.
    - **Line 86**: The resulting readable instruction string is added to a list with a step number and a distance.
- **Lines 88 to 96**: While rendering an HTML page, an additional argument is passed – the list of navigation instructions.

The `page_with_rout3.html` template must be duplicated with the name `page_with_rout4.html`, and this new template must be modified as shown below (changes are highlighted in **bold**):

```
{% extends 'base_ll.html' %}

{% block title %}{{ name }}{% endblock %}

{% block header %}
<style>
table, th, td {
   border: 1px solid black;
   border-collapse: collapse;
}
</style>
<h2>{{ name }}</h2>
{% endblock %}

{% block content %}
<br/>
<table style="width:100%">
<tr>
<th>Steps</th>
<th>Navigation instructions</th>
<th>Distances<br/>(m)</th>
</tr>
{% for instruction in instructions %}
<tr>
<td style='text-align: center'>{{ instruction[0] }}</td>
<td>{{ instruction[1] | safe }}</td>
<td style='text-align: right'>{{ instruction[2] | safe }}</td>
</tr>
{% endfor %}
</table>
{% endblock %}

{% block script %}
...
```

- **Lines 6 to 11**: A `<style>` block defines basic table formatting - rules will surround the table and separate rows and cells.
- **Lines 17 to 30**: A table is created to display the step numbers, navigation instructions, and distances by looping over the list passed from Python.

After relaunching the server, opening the URL `http://127.0.0.1:5000/am/map_rout4` in a browser, the map is displayed along with a table of navigation instructions (only the first 23 steps out of 108 are shown here under).

Displaying OpenStreetMap maps

Map with route searched with OSRM: from Saint-Jean-Pied-de-Port to Santiago-de-Compostela

Steps	Navigation instructions	Distances (m)
1	Start from departure point on D 933 (Place Floquet)	61
2	Turn right on D 933 (Place Floquet)	145
3	Make a slight left on D 933 (Place Juan de Huarte)	58
4	Continue on D 933 (Route d'Estella)	132
5	Continue on D 933 (Route d'Arnéguy)	7482
6	Enter roundabout on D 933	9
7	Exit roundabout on D 933	32
8	Continue on N-135 (Iruña - Frantzia (Luzaide) errepidea / Carretera Pamplona - Francia (Valcarlos))	19320
9	Make a slight right on N-135 (Iruñeko errepidea)	136
10	Continue on N-135 (Iruña - Frantzia (Luzaide) errepidea / Carretera Pamplona - Francia (Valcarlos))	1843
11	Continue on N-135 (San Nicolas karrika)	1195
12	Continue on N-135 (Iruña - Frantzia (Luzaide) errepidea / Carretera Pamplona - Francia (Valcarlos))	3340
13	Continue on N-135 (San Bartolome karrika)	642
14	Continue on N-135 (Iruña - Frantzia (Luzaide) errepidea / Carretera Pamplona - Francia (Valcarlos))	10013
15	Continue on N-135 (Erroko zeharkarrika)	698
16	Continue on N-135 (Iruña - Frantzia (Luzaide) errepidea / Carretera Pamplona - Francia (Valcarlos))	8333
17	Continue on N-135 (Orreagako etorbidea)	1060
18	Continue on N-135 (Iruña - Frantzia (Luzaide) errepidea / Carretera Pamplona - Francia (Valcarlos))	7423
19	Continue on N-135 (Calle Landa / Landa kalea)	235
20	Continue on N-135 (Iruña - Frantzia (Luzaide) errepidea / Carretera Pamplona - Francia (Valcarlos))	3189
21	Continue on N-135 (Carretera Pamplona - Francia (Valcarlos) / Iruña - Frantzia (Luzaide) errepidea)	1348
22	Enter roundabout on N-135	58
23	Exit roundabout on N-135	241

Figure 9-9. *Map created with Leaflet using OpenStreetMap tiles. Data © OpenStreetMap contributors, licensed under the ODbL. Route and navigation instructions generated by OSRM (Open Source Routing Machine), © Project OSRM contributors. Map data © OpenStreetMap contributors (ODbL)*

Finally, we will add the capability to export the route found by OSRM as a GPX file from our application. This will allow the route to be shared with any GPS system that supports this format, as well as navigation and route planning applications (such as Komoot). Let's insert or modify the following statements (in **bold**) in the Python `map_rout4()` function:

```
@bp.route('/am/map_rout4')
def map_rout4():
import requests, json, gpxpy, gpxpy.gpx

...

                    instructions.append([i, instruction_text, dist])

        if gtype == 'LineString':
            gpx = gpxpy.gpx.GPX()
            gpx_track = gpxpy.gpx.GPXTrack()
            gpx.tracks.append(gpx_track)
            gpx_segment = gpxpy.gpx.GPXTrackSegment()
            gpx_track.segments.append(gpx_segment)

            for lon, lat in coords:
                gpx_segment.points.append(gpxpy.gpx.
                GPXTrackPoint(latitude=lat, longitude=lon))

            gpx_data = gpx.to_xml()
        elif gtype == 'MultiLineString':
            gpx = gpxpy.gpx.GPX()
            gpx_track = gpxpy.gpx.GPXTrack()
            gpx.tracks.append(gpx_track)

            for part in coords:
                gpx_segment = gpxpy.gpx.GPXTrackSegment()
                gpx_track.segments.append(gpx_segment)

                for lon, lat in part:
```

```
109                     gpx_segment.points.append(gpxpy.gpx.
                        GPXTrackPoint(latitude=lat, longitude=lon))
110
111             gpx_data = gpx.to_xml()
112
113         return render_template('maps/page_with_route4.html',
114             name="Map with route searched with OSRM: from Saint-Jean-
                Pied-de-Port to Santiago-de-Compostela",
115             loc=[42.873698, -4.598225],
116             zoom=7,
117             track_geo=json.dumps(route_geo),
118             start=start_coords,
119             end=end_coords,
120             instructions=instructions,
121             gpx_data=gpx_data
122         )
...
```

- **Line 3**: The gpxpy and gpxpy.gpx libraries, needed to convert the data to GPX format, are imported.
- **Line 88**: If the geometry of the route returned by OSRM is of type LineString:
    - **Line 89**: An empty GPX object is created.
    - **Lines 90 and 91**: An empty GPX track object is created and added to the GPX object.
    - **Lines 92 and 93**: An empty GPX segment object is created and added to the GPX track object.
    - **Lines 95 and 96**: A loop is executed over the coordinates of the points in the LineString. For each point, its latitude and longitude are extracted, and a corresponding track point is added to the GPX segment.
    - **Line 98**: The GPX object is converted to XML and set in the gpx_data variable.

- **Line 99**: If the geometry of the route returned by OSRM is of type `MultiLineString`:
    - **Line 100**: An empty GPX object is created.
    - **Lines 101 and 102**: An empty GPX track object is created and added to the GPX object.
    - **Lines 104 to 109**: A loop is executed over the parts of the `MultiLineString`. For each of these parts:
        - **Lines 105 and 106**: An empty GPX segment object is created and added to the GPX track object.
        - **Lines 108 and 109**: A loop is executed over the coordinates of the points in the `MultiLineString` part. For each point, its latitude and longitude are extracted, and a corresponding track point is added to the GPX segment.
    - **Line 111**: The GPX object is converted to XML and set in the `gpx_data` variable.
- **Lines 113 to 122**: An additional argument, `gpx_data`, is added to inject the GPX data into the web page to be built.

The template `page_with_rout4.html` must also be modified (changes are shown in **bold** below):

```
...

{% block content %}
<br/>
{% if gpx_data %}
<a id="download-link" download="route_export.gpx">Export GPX</a>
<p/>
{% endif %}
<table style="width:100%">

...

</script>

```

```
42  {% if gpx_data %}
43  <script id="gpx-data" type="application/xml">
44  {{ gpx_data | safe }}
45  </script>
46  {% endif %}
47
48  <script type="text/javascript">
107 </script>
108
109 {% if gpx_data %}
110 <script>
111     const blob = new Blob([document.getElementById("gpx-data").
        textContent], { type: "application/xml" });
112     const url = URL.createObjectURL(blob);
113
114     const link = document.getElementById("download-link");
115     link.href = url;
116 </script>
117 {% endif %}
118 {% endblock %}
```

- **Lines 17 to 20**: If GPX data is passed by the Python function, a link is added to the page. This link will allow users to export the GPX data.
- **Lines 42 to 46**: If GPX data is passed by the Python function, it is inserted into an XML script section. The `safe` filter is used to keep the XML content unchanged.
- **Lines 109 to 117**: If GPX data is passed by the Python function, this JavaScript code completes the link of line 18:
    - **Line 111**: A binary large object (Blob) is created from the XML code of the route's GPX data (retrieved from the XML script section of lines 43 to 45).
    - **Line 112**: A URL is created with the Blob.
    - **Lines 114 and 115**: The link of line 18 is retrieved and the preceding URL is set as its target (`href`).

To execute this new code, you must first install the `gpxpy` library by running `pip install gpxpy` in the VS Code terminal from within the `carto_osm` directory. Then, start the server and open the URL `http://127.0.0.1:5000/am/map_rout4` in a browser. Between the map and the navigation instruction table, a new link labeled "Export GPX" appears.

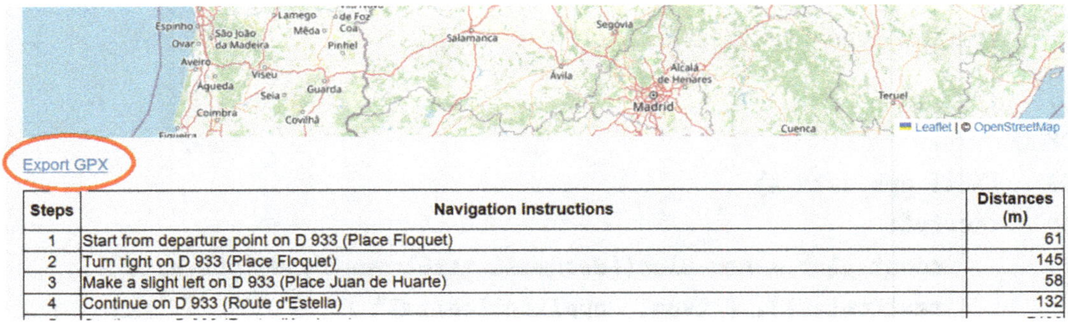

Steps	Navigation instructions	Distances (m)
1	Start from departure point on D 933 (Place Floquet)	61
2	Turn right on D 933 (Place Floquet)	145
3	Make a slight left on D 933 (Place Juan de Huarte)	58
4	Continue on D 933 (Route d'Estella)	132

Figure 9-10. *Map created with Leaflet using OpenStreetMap tiles. Data © OpenStreetMap contributors, licensed under the ODbL. Route and navigation instructions generated by OSRM (Open Source Routing Machine), © Project OSRM contributors. Map data © OpenStreetMap contributors (ODbL)*

Clicking this link downloads the route's GPX data as a file named "route_export.gpx" into your workstation's default download directory. This file can then be imported into GPS systems or navigation applications, enabling the route found by OSRM to be shared.

Alternative Solutions with Leaflet Plugins

In the preceding section, we demonstrated how to write Python and JavaScript code to use OSRM together with Leaflet for route planning. Several Leaflet plugins are available for route planning using external engines. Some use OSRM, while others rely on different routing engines such as the Mapbox Directions API, AMAP, and GraphHopper. Among the plugins that use OSRM are:

- **Leaflet Routing Machine** (developed by Per Liedman): This plugin provides a new control for searching a route between two locations (you can also define waypoints). It then calculates the route and displays it on the map. A list of navigation instructions is also shown in the control. OSRM, GraphHopper, and the Mapbox Directions API can be used as the routing engine.

- **Leaflet RouteToAddress** (developed by Astrid Günther): This plugin provides a new control for searching a route from a custom address to a fixed destination. It uses Nominatim to locate places by address and is well-suited for use cases such as "Find your way to us" descriptions on websites. By default, it uses OSRM for route planning but also supports the Mapbox Directions API.

CHAPTER 10

User Geolocation

Geolocation is the process of determining the precise geographical position of a person, a device (such as a cell phone or computer), a vehicle, or any other connected object (or object equipped with a geolocation device). Generally, this position is expressed in coordinates (latitude and longitude) and enables the person or object to be positioned on a map.

The prototypes in this chapter will search (programmatically) the geographical position of the user of our applications and display it on a map (actually, it is the position of the laptop through which he or she accesses the application via a browser that will be searched). Several approaches are possible:

- Obtain the geographic coordinates of the user's current location, **based on the IP address** with which he is connected to the Internet. This generally gives inaccurate results, with an error of up to several kilometers:
    - If this operation is performed on the **server side**, we can at best hope to obtain the server's position, but not that of the user connecting to the server via a browser and a network. However, it is often the coordinates of another server on the network that are obtained.
    - Even if this query is carried out on the **client side**, via JavaScript code executed by the browser, we cannot be sure of obtaining the correct position. The result is often the coordinates of the server that has assigned its IP address to the workstation, rather than those of the workstation itself.
- Alternatively, geolocation coordinates can be obtained client-side (where the browser is running) through JavaScript code embedded in HTML pages and **leveraging the workstation's hardware** (such as GPS, Wi-Fi, or Bluetooth). Geolocation performed using this method generally yields more accurate results.

We will now implement these different approaches and see the results they produce.

P. Marie, *Modern Web Cartography*, https://doi.org/10.1007/979-8-8688-2335-0_10

Geolocation Based on IP Address

This type of geolocation will be tested first on the server, then on the client.

Using Geolocation by IP Address on the Server and Creating a Map with Folium

This first approach consists of obtaining the geographical coordinates of the current location using Python code that performs a geolocation **based on the IP address on the server**. The position found will then be represented on a map built with Folium on the server.

The following Python statements can be used to search for geographic coordinates from the IP address:

```
1   position = geocoder.ip('me')
2   lat, lon = position.latlng
```

- The `geocoder` library must first be installed in the workspace (`pip install geocoder` run with the VS Code terminal from the `carto_osm` directory).
- **Line 1**: The `geocoder.ip()` function is used to query the location of the machine executing the code, which, in our case, is the server.
- **Line 2**: The latitude and longitude are extracted from the position found.

The code added to the `advanced_maps.py` blueprint to build this map is as follows:

```
@bp.route('/am/map_gl1')
def map_gl1():
    import geocoder

    try:
        position = geocoder.ip('me')

        if not position.ok or not position.latlng:
            flash("Unable to retrieve your position from IP address",
            category="error")
```

```
10              return render_template('maps/page_with_map.html', name="No map to display")
11
12          lat, lon = position.latlng
13
14          m = folium.Map(location=[lat, lon], zoom_start=14)
15
16          folium.Marker(
17              location=[lat, lon],
18              tooltip="I",
19              popup=f"I'm here!<br>[{lat:.5f}, {lon:.5f}]",
20              icon=folium.Icon(icon="user"),
21          ).add_to(m)
22
23          return render_template('maps/page_with_map.html',
24              name="Map with geolocation via IP address (server side)",
25              map_html=m._repr_html_())
26
27      except Exception as e:
28          flash("An error has been encountered:<br/>" + str(e), category="error")
29          return render_template('maps/page_with_map.html', name="No map to display")
```

- **Line 3**: The geocoder library is imported.
- **Line 6**: Our position (actually, the server's position) is looked up from our IP address (actually, the server's).
- **Lines 8 to 10**: These statements check whether a position has been found. If this is not the case, an error message is displayed on a web page.
- **Line 12**: The latitude and longitude found are extracted.
- **Line 14**: A map is created and centered on the geographic coordinates found in the previous step. The initial zoom level is set to 14.

- **Lines 16 to 21**: A marker is created and positioned at the latitude and longitude found, with a popup and tooltip. Then it is added to the map.
- **Lines 23 to 25**: A web page is rendered using the `page_with_map.html` template. A name and the map, converted to HTML and JavaScript code, are passed as arguments.

Launching the server and opening the URL `http://127.0.0.1:5000/am/map_gl1` with the browser produces the following map.

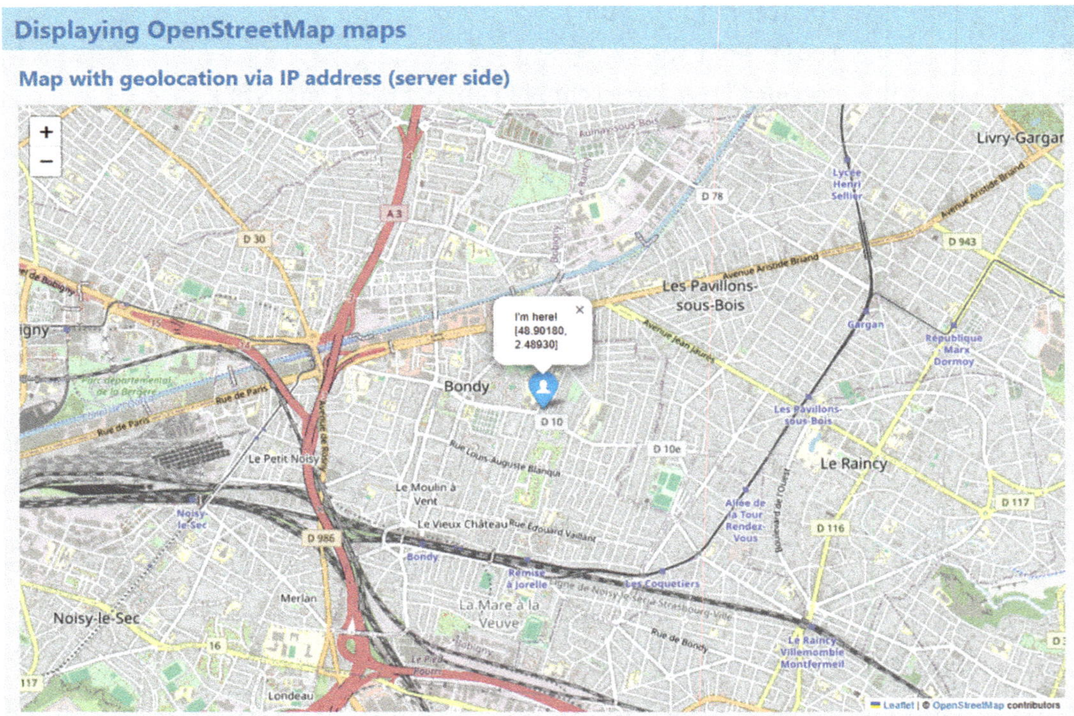

Figure 10-1. *Map created with Folium using OpenStreetMap tiles. Data © OpenStreetMap contributors, licensed under the ODbL*

The position indicated is about 4.5 miles away from the user's actual position, as well as from the web server running the Python code. This approach thus demonstrates limited precision. Remember that:

- On the one hand, as the Python code is executed on the server, it is the server's location that we can at best hope to find, whereas it is the browser's coordinates that we are interested in.

- In many cases (as above), this method does not even find the geographic coordinates of the server executing the Python code. The main reasons for that are the following:
  - `geocoder.ip('me')` usually asks an external geolocation service for the coordinates associated with the IP address it sees.
  - The server may be behind a NAT, firewall, or proxy.
  - The IP that the geolocation service sees might be the public IP of the network, rather than the server's physical location.
  - If the server is in a cloud provider (AWS, Azure, GCP, etc.), the IP may be registered to the provider's data center location, which can be far from the actual physical server.

Using Geolocation by IP Address in the Browser and Creating a Map with Leaflet

In this approach, we will still search for the location via the IP address, but this time via JavaScript code executed in the browser. The position found will then be presented on a map built with Leaflet in the browser.

To perform geolocation based on the client IP address from JavaScript code, an external web service must be invoked through an HTTP request. Several providers offer such IP-based geolocation services:

- Some are open source (such as **IPWho**, **GeoLite2**, **FreeGeoIP**, and others), while others are proprietary and offered as commercial products (such as **IPinfo**, **ipapi**, **ipstack**, and others).
- Some of these services are available over **HTTP** (unsecured), while others use **HTTPS** (secured).

It is considered best practice to always use an **HTTPS** endpoint. In particular, if your web application itself is served over HTTPS, attempting to access a geolocation service via HTTP will cause modern browsers to block the request due to a **mixed-content violation**.

When selecting a geolocation service, it is also important to review its **usage policies** and ensure that they meet your requirements. For example, some providers limit the number of requests per day or throttle requests that occur too frequently.

In the following example, we will be using **IPWho**, an open source service which provides geolocation data through a secured endpoint (`https://ipwho.is/`) and returns the results in JSON format.

The code to be added to the `advanced_maps.py` blueprint to implement this map is very simple:

```
@bp.route('/am/map_gl2')
def map_gl2():
    return render_template('maps/page_with_leaflet_map_for_geoloc2.html',
        name="Map with geolocation via IP address (client side)")
```

It is limited to rendering an HTML page with the `maps/page_with_leaflet_map_for_geoloc2.html` template, passing a name as an argument.

The template `page_with_leaflet_map_for_geoloc2.html` must be created in the `prototypes/templates/maps` directory with the following contents:

```
{% extends 'base_ll.html' %}

{% block title %}{{ name }}{% endblock %}

{% block header %}
<h2>{{ name }}</h2>
{% endblock %}

{% block script %}
<script type="text/javascript">
let map;
fetch("https://ipwho.is/")
.then(response => response.json())
.then(data => {
    if (!data || !data.latitude || !data.longitude) throw new Error("Coordinates not found");
    const [lat, lon] = [data.latitude, data.longitude];

    map = L.map('map').setView([lat, lon], 14);
```

```
19
20      L.tileLayer('https://tile.openstreetmap.org/{z}/{x}/{y}.png', {
21          maxZoom: 19,
22          attribution: '&copy; <a href="http://www.openstreetmap.org/
            copyright">OpenStreetMap</a>'
23        }).addTo(map);
24
25      L.marker([lat, lon]).addTo(map).bindPopup(`<b>I'm here!</b>
        <br>${data.city}, ${data.country}`).openPopup();
26  })
27  .catch(error => {
28      if (typeof map !== 'undefined' && map) {map.remove();}
29
30      const div = document.getElementById("map")
31      if (div) {div.remove();}
32
33      const elemDiv = document.createElement('div');
34      elemDiv.innerHTML = `<p>${error}</p>`;
35      document.body.appendChild(elemDiv);
36  });
37  </script>
38  {% endblock %}
```

- **Line 12**: The IPWho site is queried via a secured HTTP request to obtain geolocation data via the client's IP address.

  The `fetch()` function submits the request asynchronously, which means we cannot use a standard `try/catch` block for synchronous error handling.

  Depending on the environment, invoking a geolocation service directly from browser-based JavaScript may trigger a CORS (cross-origin resource sharing) error. In such cases, a common solution is to use a server-side proxy that adds a contact User-Agent (typically of the form: `MyApp/1.0 (me@example.com)`) and appropriate headers. Appendix C provides details on how to implement this approach. Another possible solution is to route the request through a public CORS proxy service (such as **corsproxy.io**).

- **Lines 13 and 14**: The response is parsed and placed in a JSON-structured object.
- **Line 15**: This line verifies whether the request was successful and whether coordinates were returned. If not, it raises an error.
- **Line 16**: This line extracts the latitude and longitude from the geolocation response.
- **Line 18**: A map is built and centered on the coordinates returned by the geolocation.
- **Lines 20 to 23**: A tile layer is associated with the map.
- **Line 25**: A marker is created at the position found by the geolocation, and it is added to the map. Then a popup is added to the marker with the message "I'm here", as well as the city and country identified.
- **Lines 27 to 36**: These lines handle any errors that may occur.

Launching the server and opening the address `http://127.0.0.1:5000/am/map_gl2` in the browser produces the following map:

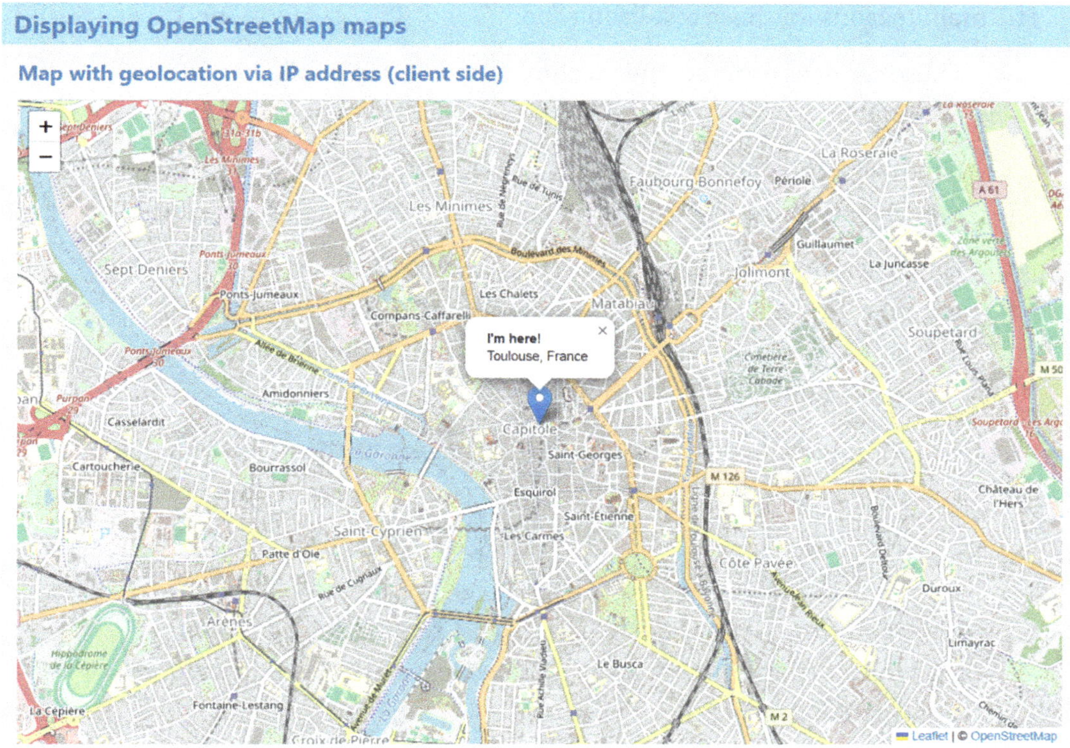

Figure 10-2. *Map created with Leaflet using OpenStreetMap tiles. Data © OpenStreetMap contributors, licensed under the ODbL. Geolocation data found by IPWho*

If the map fails to display, it may be caused by a CORS error (as explained above). Verify this possibility and, if necessary, implement a server-side proxy as described in Appendix C or use a public CORS proxy service (such as **corsproxy.io**). However, if you do so, the coordinates obtained will be those of the server where the proxy runs, not the client's. It is possible to modify both the HTML template and the Python proxy to pass the client's IP address to the proxy and then forward it to IPWho, so that the client's coordinates are obtained (we will not prototype this solution).

Regarding the position obtained, the country is correct, but the position is more than 350 miles from the actual one.

Another test performed with **IPinfo** (a proprietary/commercial service) instead of IPWho produced better results, although they were not fully accurate (about 4.5 miles away from the actual position).

The main reasons for these results may be:

- These services estimate the location based on the public IP address of the client.
- If the client is behind a VPN, proxy, or corporate network, the IP seen by the service may be that of the VPN server or proxy, not the actual device.
- Databases mapping IP addresses to locations are not perfectly accurate, especially for mobile networks or dynamic IPs. Often, the location is only accurate at the city level, and sometimes only at the region or country level. Moreover:
    - Commercial services such as IPinfo aggregate their data from multiple proprietary databases and registries, giving them richer and more frequently updated mappings between IP address blocks and real-world locations.
    - Open source services, such as IPWho, rely on open datasets which are less granular and updated less often, leading to larger positional errors.
- Free tiers often limit accuracy or throttle requests. Some services may only provide approximate location unless you have a paid plan or token.
- Some services may have incomplete mapping for IPv6 addresses, leading to inaccurate results.

Using the Browser's Geolocation API and Creating a Map with Leaflet

The two previous attempts were unsuccessful and demonstrated that geolocation via IP address should be avoided. Our new approach will be to determine the workstation's geographic position using the browser's Geolocation API. This API can rely on several sources, depending on the device and its environment:

- **GPS satellites**: If the device is equipped with a GPS chip (e.g., a smartphone or tablet), the position is determined via GPS and is highly accurate.

- **Nearby Wi-Fi access points**: If GPS is unavailable, the browser can use the identifiers of nearby Wi-Fi access points, matched against a database of known locations, to estimate the position with reasonable accuracy.

- **IP address-based geolocation**: If neither GPS nor Wi-Fi data can be used, the browser falls back to geolocation via the device's IP address. This method is the least accurate and can sometimes place the device hundreds of kilometers from its actual location.

Our next prototype will use JavaScript code running in the browser to invoke the Geolocation API, determine the workstation's coordinates, and then build a Leaflet map to display the detected position.

The JavaScript statements used to retrieve the user's coordinates via the browser's Geolocation API are as follows:

```
1   function setPositionOnMap(position) {
2       const latitude = position.coords.latitude
3       const longitude = position.coords.longitude;

        (Map construction statements)

13  }
14
15  if (navigator.geolocation) {
16       navigator.geolocation.getCurrentPosition(setPositionOnMap);
17  }
```

- **Lines 1 to 13**: A function named `setPositionOnMap()` is defined:
    - **Lines 2 and 3**: The latitude and longitude values obtained by the browser's Geolocation API (and passed as arguments to this function) are retrieved.
    - This should be followed by Leaflet statements for building a map showing the position found.
- **Line 15**: This line checks whether the browser supports geolocation.

- **Line 16**: If it does, the browser's Geolocation API is invoked to request the coordinates at which the user is located. The `navigator.geolocation.getCurrentPosition()` function, used for that purpose, is asynchronous. The coordinates found are passed to the `setPositionOnMap()` function, which is then invoked in callback mode.

Executing the `navigator.geolocation.getCurrentPosition()` function may trigger a permission prompt, asking the user whether to allow access to their location. If the user denies permission, geolocation will not succeed. This permission request appears only the first time the site attempts to access the user's location.

The code added to the `advanced_maps.py` blueprint to obtain this map is very simple:

```
@bp.route('/am/map_gl3')
def map_gl3():
    return render_template('maps/page_with_leaflet_map_for_geoloc3.html',
        name="Map with geolocation via the browser's geolocation API")
```

It is limited to rendering an HTML page with the `maps/page_with_leaflet_map_for_geoloc3.html` template, passing a name as an argument.

The template `page_with_leaflet_map_for_geoloc3.html` must be created in the `prototypes/templates/maps` directory with the following contents:

```
{% extends 'base_ll.html' %}

{% block title %}{{ name }}{% endblock %}

{% block header %}
<h2>{{ name }}</h2>
{% endblock %}

{% block script %}
<script type="text/javascript">
let map;

```

```
function setPositionOnMap(position) {
    const lat = position.coords.latitude;
    const lon = position.coords.longitude

    map = L.map('map').setView([lat, lon], 10);

    L.tileLayer('https://tile.openstreetmap.org/{z}/{x}/{y}.png', {
        maxZoom: 19,
        attribution: '&copy; <a href="http://www.openstreetmap.org/
        copyright">OpenStreetMap</a>'
    }).addTo(map);

    const userIcon = L.icon({
        iconUrl: '/static/silhouette.jpg',
        iconSize: [20, 40],
        iconAnchor: [10, 20],
        popupAnchor: [0, -20]
    });

    L.marker([lat, lon], { icon: userIcon })
    .addTo(map)
    .bindPopup(`<b>I'm here</b><br>[${lat.toFixed(5)}, ${lon.
    toFixed(5)}]`)
    .openPopup();

    map.setView([lat, lon], 14, { animate: true, duration: 1 });
}

function handleGeolocationError(error) {
    const errorMsg = {
        1: "Permission denied by user",
        2: "Position not available",
        3: "Timeout exceeded",
    }[error.code] || "Unknown error";

```

```
46          const elemDiv = document.createElement('div');
47          elemDiv.innerHTML = `<p>Geolocation error: ${errorMsg}</p>`;
48          document.body.appendChild(elemDiv);
49      }
50
51      if (navigator.geolocation) {
52          navigator.geolocation.getCurrentPosition(setPositionOnMap,
            handleGeolocationError);
53      } else {
54          const elemDiv = document.createElement('div');
55          elemDiv.innerHTML = "<p>Geolocation is not supported by this
            browser</p>";
56          document.body.appendChild(elemDiv);
57      }
58      </script>
59        {% endblock %}
```

- **Lines 13 to 37**: A function named `setPositionOnMap()` is defined. It is invoked by the Geolocation API in callback mode on line 52, passing the coordinates found as arguments.
    - **Lines 14 and 15**: The latitude and longitude found by geolocation are retrieved.
    - **Line 17**: A map is created and centered on this position, with an initial zoom level of 12.
    - **Lines 19 to 22**: A tile layer is associated with the map.
    - **Lines 24 to 29**: An icon to be associated with a marker is defined. This is an image in JPEG format, located in the `static` directory of the workspace.

      This image can be found at `https://github.com/Apress/Modern-Web-Cartography` (file: `carto_osm/prototypes/static/silhouette.jpg`).

      Retrieve it and download it into the `prototypes/static` directory of your workspace.

The `iconSize`, `iconAnchor`, and `popupAnchor` options define the dimensions of the icon, the point of the icon that corresponds to the marker's geographic coordinates, and the relative position where the popup will open.

- **Lines 31 to 34**: A marker is created to indicate the position found, and then it is added to the map. This marker is defined with the previous icon, and a popup containing the message "I'm here!", along with the coordinates found, is added to the marker.
- **Line 36**: The map is zoomed in on the located position with a one-second animation, gradually bringing the user's viewpoint closer.

- **Lines 39 to 49**: A function named `handleGeolocationError()` is also defined. It may be invoked by geolocation in "callback" mode on line 52: if an error occurs while searching for the workstation coordinates, this function will be triggered.
    - **Lines 40 to 44**: The cause of the error is investigated, and an error message prepared.
    - **Lines 46 to 48**: A new division is added to the page, and a message inserted in this division.

  This error handling function is necessary because the geolocation function invoked on line 52 is asynchronous, and a usual `try/catch` block would not be appropriate.

- **Line 51**: This statement verifies whether the browser supports geolocation.
    - **Line 52**: If it does, the browser's Geolocation API is invoked, and then the `setPositionOnMap()` function is called and receives the coordinates found. If the geolocation fails for some reason, the `handleGeolocationError()` function is invoked.
    - **Lines 54 to 56**: Otherwise, a new division is added to the page, and a message informing the user that the browser does not support geolocation is inserted in this division.

Launching the server and opening the address `http://127.0.0.1:5000/am/map_gl3` in the browser produces the following map. If prompted to allow access to your location, please accept; otherwise, geolocation will fail, and an error will be shown.

Figure 10-3. *Map created with Leaflet using OpenStreetMap tiles. Data © OpenStreetMap contributors, licensed under the ODbL*

This location found is correct.

The preceding code works well when the user's position is stationary, but not when the user is on the move. We will now improve it so that the map automatically tracks and updates the user's position in real time. The following update of the template `page_with_leaflet_map_for_geoloc3.html` fulfills this requirement (modifications and additions are shown in **bold**):

```
{% extends 'base_ll.html' %}

{% block title %}{{ name }}{% endblock %}

```

```
{% block header %}
<h2>{{ name }}</h2>
{% endblock %}

{% block script %}
<script type="text/javascript">
let map;

function updatePositionOnMap(position) {
    const lat = position.coords.latitude;
    const lon = position.coords.longitude;

    if (!map) {
        map = L.map('map').setView([lat, lon], 10);

        L.tileLayer('https://tile.openstreetmap.org/{z}/{x}/{y}.png', {
            maxZoom: 19,
            attribution: '&copy; <a href="http://www.openstreetmap.org/copyright">OpenStreetMap</a>'
        }).addTo(map);

        const userIcon = L.icon({
            iconUrl: '/static/silhouette.jpg',
            iconSize: [20, 40],
            iconAnchor: [10, 20],
            popupAnchor: [0, -20]
        });

        userMarker = L.marker([lat, lon], { icon: userIcon }).addTo(map);
        userMarker.bindPopup(`<b>I am here!</b><br>[${lat.toFixed(5)}, ${lon.toFixed(5)}]`).openPopup();

        map.setView([lat, lon], 14, { animate: true, duration: 1 });
```

```
    } else {
        userMarker.setLatLng([lat, lon]);
        userMarker.setPopupContent(`<b>I am here now!</b><br>[${lat.toFixed(5)}, ${lon.toFixed(5)}]`);

     map.setView([lat, lon], map.getZoom(), { animate: true });
    }
}

function handleGeolocationError(error) {
    const errorMsg = {
        1: "Permission denied by user",
        2: "Position not available",
        3: "Timeout exceeded",
    }[error.code] || "Unknown error";

    const elemDiv = document.createElement('div');
    elemDiv.innerHTML = `<p>Geolocation error: ${errorMsg}</p>`;
    document.body.appendChild(elemDiv);
}

if (navigator.geolocation) {
    navigator.geolocation.watchPosition(updatePositionOnMap, handleGeolocationError, {
        enableHighAccuracy: true,
        maximumAge: 5000,
        timeout: 10000
    });
} else {
    const elemDiv = document.createElement('div');
    elemDiv.innerHTML = "<p>Geolocation is not supported by this browser</p>";
    document.body.appendChild(elemDiv);
}
</script>
{% endblock %}
```

The changes are the following:

- **Lines 57 to 61**: The `navigator.geolocation.watchPosition()` function is now used instead of the `navigator.geolocation.getCurrentPosition()` function to continuously monitor the user's position. This allows the geolocation to update automatically whenever the position changes.
    - The `enableHighAccuracy: true` option requests the device to use the most precise location sources available.
    - The `maximumAge: 5000` option specifies that a position found by the Geolocation API will be regarded as valid for five seconds. After this delay, it will be regarded as stale, and the Geolocation API will try to get a new one.
    - The `timeout: 10000` option specifies the maximum amount of time (ten seconds here) that the browser will wait to obtain a location before throwing a timeout error.
- **Lines 13 to 42**: The `updatePositionOnMap()` function (replacing `setPositionOnMap()`) checks whether the map has already been created.
    - If the map does not exist:
        - It is initialized and centered on the current location.
        - A marker is added at the initial position with a popup.
    - If the map already exists:
        - The existing marker is repositioned, and its popup content is updated.
        - The map is not recreated, but smoothly moved (thanks to the `animate: true` option) to the new position.

Finally, note that the positions returned by a browser's geolocation feature may differ slightly between browsers (such as Firefox, Chrome, Edge, or Safari). This variation occurs because each provider maintains its own database of Wi-Fi access points, cell towers, and IP address ranges, so their reported coordinates may differ - sometimes by only a few meters, and sometimes by hundreds.

Alternative Solutions with Leaflet and Folium Plugins

In the previous section, we demonstrated how to write JavaScript code for geolocating the user with the browser's Geolocation API. However, Leaflet and Folium plugins are also available for geolocation.

Among the numerous Leaflet plugins designed for geolocation are:

- **Geolet** (developed by Ruben Holthuijsen): This plugin provides a new control for searching the user's position using the device's geolocation and displaying it on the map.
- **Leaflet.AccuratePosition** (developed by Michael Schmidt-Voigt): This plugin provides a new control for searching the user's position using the device's geolocation and displaying it on the map. It uses the `navigator.geolocation.watchPosition()` function, which may take more time but offers better accuracy.
- **Leaflet.Locate** (developed by Dominik Moritz): Provides a new control for searching the user's position using the device's geolocation and displaying it on the map.
- **Leaflet.SimpleLocate** (developed by "mfhsieh"): Provides a new control for locating the user's position using the device's geolocation and displaying it on the map. The control also takes the form of a compass to indicate the map's orientation.

A Folium plugin also supports geolocation:

- **LocateControl** (developed by Dominik Moritz): This plugin adds a control button that, when clicked, uses the device's geolocation to display the user's position as a circle marker on the map.

CHAPTER 11

Measuring Distances and Areas

In this chapter, we will learn how to **measure distances** and **areas** by drawing polylines and polygons on a map.

This functionality differs from what was discussed in Chapter 9, "Route Import, Calculation, and Visualization," where a route was computed between two locations and its length was determined based on existing carriageways or pathways. Here, the objective is simply to allow the user to draw a line on the map and measure its length, or to draw a polygon and measure its area, without relying on an existing road network or factoring in elevations.

Measuring Distances and Areas with Leaflet

Since this new functionality requires a high level of interactivity, **Leaflet** appears to be the most suitable and natural choice for implementing it.

Measuring Distances

We will start by allowing the user to draw a polyline by clicking several points on the map. Each time a point is clicked, the application will add a segment to the polyline and recalculate and display its total length.

To compute the distance between two points, we need to calculate the length of an arc on the surface of a sphere (remember, the Earth is not flat!). We will use the **Haversine formula** (for more details, see `https://en.wikipedia.org/wiki/Haversine_formula`) for that purpose.

P. Marie, *Modern Web Cartography*, https://doi.org/10.1007/979-8-8688-2335-0_11

In JavaScript, this formula can be implemented as follows:

```
1   function haversine(latlng1, latlng2) {
2       //const R = 6371; // Earth's radius in kilometers
3       const R = 3959; // Earth's radius in miles
4
5       const toRad = angle => angle * Math.PI / 180;
6
7       const dLat = toRad(latlng2.lat - latlng1.lat);
8       const dLon = toRad(latlng2.lng - latlng1.lng);
9
10      const phi1 = toRad(latlng1.lat);
11      const phi2 = toRad(latlng2.lat);
12
13      const a = Math.sin(dLat / 2) ** 2 +
14          Math.cos(phi1) * Math.cos(phi2) *
15          Math.sin(dLon / 2) ** 2;
16
17      const c = 2 * Math.atan2(Math.sqrt(a), Math.sqrt(1 - a));
18
19      return R * c;
20  }
```

This function takes two points (each with a latitude and a longitude) as arguments and returns the distance between them in miles. To obtain the distance in kilometers, simply uncomment line 2, and comment out line 3.

The code to be added to the advanced_maps.py blueprint to build this new map is straightforward:

```
@bp.route('/am/map_measure1')
def map_measure1():
    return render_template('maps/page_with_leaflet_map_for_measure1.html',
        name="Map for measuring distances and areas")
```

It is limited to rendering an HTML page with the maps/page_with_leaflet_map_for_measure1.html template, passing a name as an argument.

The template page_with_leaflet_map_for_measure1.html must be created in the prototypes/templates/maps directory with the following contents:

```
{% extends 'base_ll.html' %}

{% block title %}{{ name }}{% endblock %}

{% block header %}
<h2>{{ name }}</h2>
{% endblock %}

{% block script %}
<script type="text/javascript">
let map;
try {
    map = L.map('map').setView([38, -100], 5);

    L.tileLayer('https://tile.openstreetmap.org/{z}/{x}/{y}.png', {
        maxZoom: 19,
        attribution: '&copy; <a href="https://www.openstreetmap.org/copyright">OpenStreetMap</a> contributors'
    }).addTo(map);

    let points = [];
    let circleMarkers = [];
    let polyline = null;
    let dist = 0;

    map.on('click', function(e) {
        const latlng = e.latlng;

        points.push(latlng);

        const newCircleMarker = L.circleMarker(latlng, {radius: 5}).addTo(map);
        circleMarkers.push(newCircleMarker);
```

```
          if (points.length === 1) {
              polyline = L.polyline(points).addTo(map);
              dist = 0;
          } else {
              polyline.addLatLng(latlng);
              dist = dist + haversine(points[points.length - 2],
              points[points.length - 1])
              newCircleMarker.bindPopup(`Length: ${dist.
              toFixed(3)} miles`)
              .openPopup();
          }
      });
      document.addEventListener('keydown', function(e) {
          const key = (e.key || '').toLowerCase();
          if (key === 'r') {
              if (polyline) polyline.remove();
              for (let marker of circleMarkers) marker.remove();
              points = [];
              circleMarkers = [];
          }
      });
  } catch(e) {
      if (typeof map !== 'undefined' && map) {map.remove();}
      const div = document.getElementById("map")
      if (div) {div.remove();}
      const elemDiv = document.createElement('div');
```

```
65      elemDiv.innerHTML = `<p>${e}</p>`;
66      document.body.appendChild(elemDiv);
67  }
68
69  function haversine(latlng1, latlng2) {
70      //const R = 6371; // Earth's radius in kilometers
71      const R = 3959; // Earth's radius in miles
72
73      const toRad = angle => angle * Math.PI / 180;
74
75      const dLat = toRad(latlng2.lat - latlng1.lat);
76      const dLon = toRad(latlng2.lng - latlng1.lng);
77
78      const phi1 = toRad(latlng1.lat);
79      const phi2 = toRad(latlng2.lat);
80
81      const a = Math.sin(dLat / 2) ** 2 +
82          Math.cos(phi1) * Math.cos(phi2) *
83          Math.sin(dLon / 2) ** 2;
84
85      const c = 2 * Math.atan2(Math.sqrt(a), Math.sqrt(1 - a));
86
87      return R * c;
88  }
89  </script>
90    {% endblock %}
```

- **Lines 13 to 18**: A map is created with a tile set.
- **Lines 20 to 23**: Several variables are initialized.
- **Lines 25 to 45**: These lines handle the event when a user clicks on the map:
    - **Line 26**: The geographic coordinates of the point the user clicked are extracted from the event.

- **Line 28**: This point is added to the array of points already clicked by the user.
- **Line 30**: A circle marker is added to the map at this point.
- **Line 31**: This circle marker is added to the array of circle markers already added to the map.
- **Lines 33 to 36**: If this point is the first for the current measurement:
    - **Line 34**: A polyline is created with this single point and added to the map.
    - **Line 36**: The measured length is initialized to zero.
- **Lines 37 to 43**: If this point is not the first:
    - **Line 38**: A new segment is added to the polyline, connecting the last-but-one point to the newly added point.
    - **Line 40**: The measured length is updated by adding the length of the new segment (connecting the last-but-one point to the new point), calculated using the `haversine()` function.
    - **Lines 42 and 43**: A popup is added to the circle marker with the current value of the measured length.

- **Lines 47 to 56**: These lines handle the event when the user presses a key.

  A `document.addEventListener('keydown', function(e) {...})` event listener is preferred over a `map.on('keypress', function(e) {...})` listener (provided by Leaflet) because it will trigger **regardless of whether the map has focus**, whereas the latter only works when the map is focused.

    - **Line 48**: This line retrieves the key pressed and converts it to lowercase.
    - **Line 49**: This line checks which key has been pressed. The next statements are executed only if the "r" or "R" key (for "reset") has

been pressed, meaning that the user wants to cancel the current measurement in order to start a new one.

- **Line 50**: The polyline is removed from the map (if it exists).
- **Line 51**: All the circle markers are removed from the map.
- **Lines 53 and 54**: The arrays of points and circle markers are reset.

- **Lines 58 to 67**: These lines handle any error occurring in the `try` part above.
- **Lines 69 to 88**: The code of the `haversine()` function, used to calculate the distance between two points (taking into account the curvature of the earth).

Launching the server and opening the address `http://127.0.0.1:5000/am/map_measure1` in the browser produces the following map (after clicking on a few points):

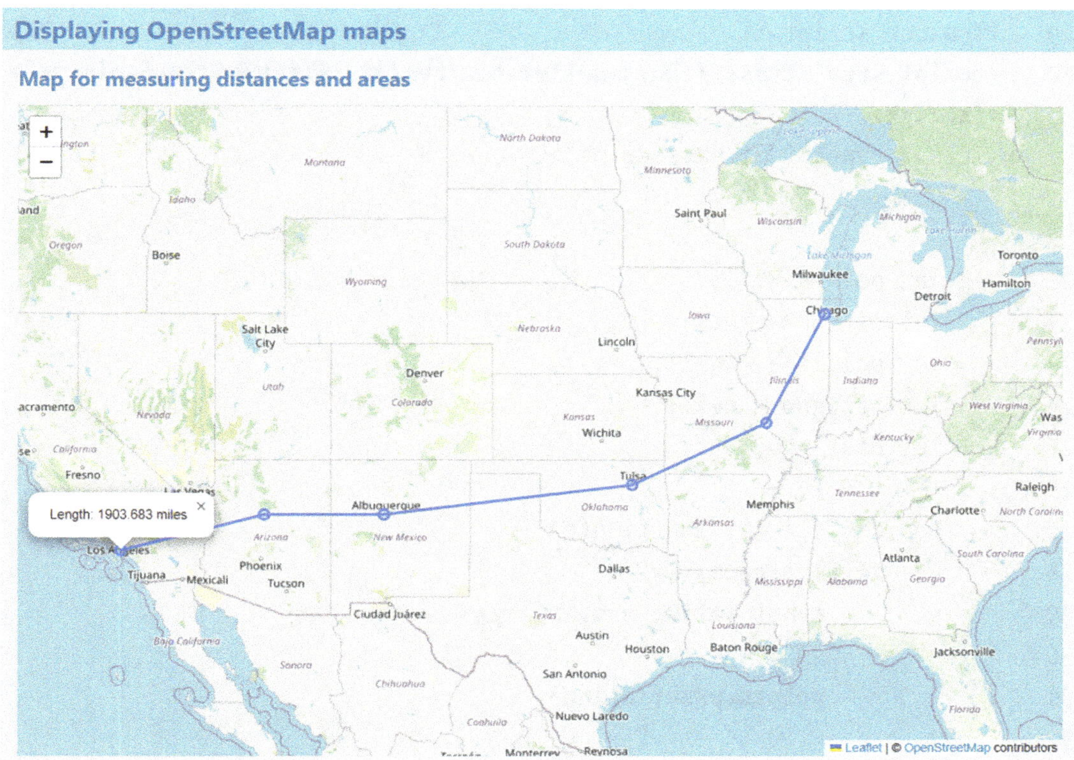

Figure 11-1. *Map created with Leaflet using OpenStreetMap tiles. Data © OpenStreetMap contributors, licensed under the ODbL*

Each time a point is clicked on the map, a new segment is added and a popup displays the distance from the origin to that point.

The map can be reset for a new measurement by pressing the “r” or “R” key.

Measuring Areas

We will now enhance the preceding code by adding the ability to close the current polyline to form a **polygon** and to display both its **perimeter** and **area**.

To calculate the area of a polygon, we will use the `turf.js` library, a JavaScript library dedicated to advanced geospatial analysis. Its documentation is available at `https://turfjs.org/`.

The following changes (in **bold**) must be made to the `page_with_leaflet_map_for_measure1.html` template:

```
...

{% block script %}
<script src="https://cdn.jsdelivr.net/npm/@turf/turf@6/turf.min.js"></script>
<script type="text/javascript">

...

    let points = [];
    let circleMarkers = [];
    let polyline = null;
    let polygon = null;
    let dist = 0;

    map.on('click', function(e) {
        if (!polygon) {
            const latlng = e.latlng;

            points.push(latlng);

            const newCircleMarker = L.circleMarker(latlng, {radius: 5}).addTo(map);
```

```
                circleMarkers.push(newCircleMarker);

                if (points.length === 1) {
                    polyline = L.polyline(points).addTo(map);

                    dist = 0;
                } else {
                    polyline.addLatLng(latlng);

                    dist = dist + haversine(points[points.length - 2],
                    points[points.length - 1])

                    newCircleMarker.bindPopup(`Distance: ${dist.
                    toFixed(3)} miles`)
                    .openPopup();
                }
            }
        });

        document.addEventListener('keydown', function(e) {
            const key = (e.key || '').toLowerCase();
            if (key === 'r') {
                if (polyline) polyline.remove();
                if (polygon) polygon.remove();
                polygon = null;
                for (let marker of circleMarkers) marker.remove();

                points = [];
                circleMarkers = [];
            } else if (key === 'c' && !polygon && points.length > 2) {
                dist = dist + haversine(points[points.length - 1],
                points[0]);

                polygon = L.polygon(points).addTo(map);
                polyline.remove();
```

```
66
67                  const geojson = polygon.toGeoJSON();
68                  const area = turf.area(geojson) / (1609.344 * 1609.344);
69
70                  polygon.bindPopup(`<strong>Perimeter: </strong> ${dist.toFixed(3)} miles<br/><strong>Area:</strong> ${area.toFixed(3)} square miles`)
71                  .openPopup();
72              }
73          });
74
75      } catch(e) {
        ...
84      }
85
86      function haversine(latlng1, latlng2) {
        ...
105     }
106     </script>
107     {% endblock %}
```

- **Line 10**: This line specifies where the code of the `turf.js` library can be found.
- **Line 24**: An additional variable named `polygon` is initialized.
- **Line 28**: Once a measurement has been completed by closing the polyline into a polygon, no additional points may be added.
- **Lines 55 and 56**: When resetting the current measurement to start a new one, the polygon is removed from the map (if it exists), and the `polygon` variable is reset.
- **Lines 61 to 71**: These lines handle the event when the user presses the "c" or "C" key (for "close"), after having clicked at least three points, to close the current polyline into a polygon and display its perimeter and area.

- **Line 61**: Closing the polyline into a polygon is allowed only once and when at least three points have been created.
- **Line 62**: The length is updated with a closing segment (using the `haversine()` function).
- **Line 64**: A polygon is built with the array of points and added to the map.
- **Line 65**: The polyline is removed from the map.
- **Lines 67 and 68**: The area of the polygon is calculated by first converting it into a GeoJSON object, which is then passed to the `turf.area()` function.
- **Lines 70 and 71**: A popup is added to the polygon with its perimeter and area.

Above, the perimeter is calculated using the `haversine()` function, as was done previously for calculating polyline length. Alternatively, the `turf.js` library could also be used. The following code demonstrates this approach:

```
const line = turf.lineString(geojson.geometry.coordinates[0]);
const perimeter = turf.length(line, { units: 'miles' });
```

- The first statement extracts the outer ring coordinates of the polygon (`coordinates[0]`) and builds a `LineString` with it.
- The second measures the length of the `LineString`.

Launching the server and opening the address `http://127.0.0.1:5000/am/map_measure1` in the browser displays the following map, after clicking on a few points to draw a polyline and then pressing the “c” key to close it into a polygon.

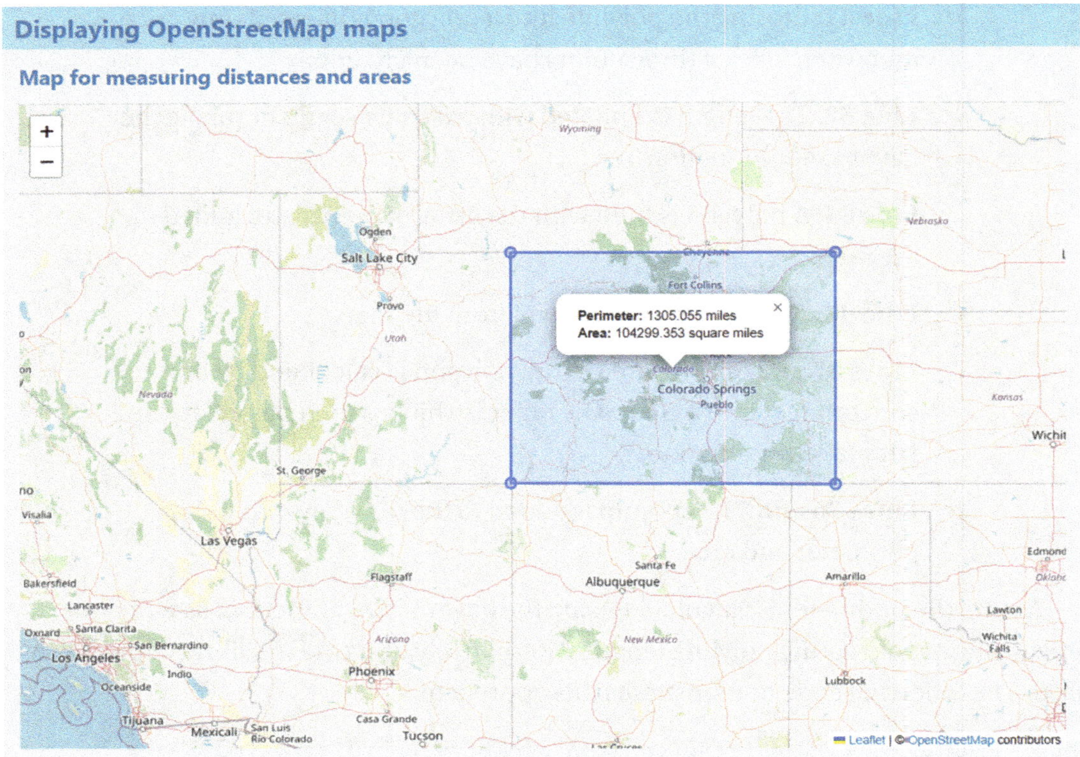

Figure 11-2. Map created with Leaflet using OpenStreetMap tiles. Data © OpenStreetMap contributors, licensed under the ODbL

We have thus obtained rough estimates of the perimeter and area of Colorado.

To complete this prototype, we will add a legend that explains how to use the map. This can be done by inserting the following statements (in **bold**) into the code of `page_with_leaflet_map_for_measure1.html`:

```
    ...
        }).addTo(map);

        const legend = L.control({position: 'bottomleft'});
        legend.onAdd = function(map) {
            const div = L.DomUtil.create('div', 'info legend');

            div.innerHTML += `<strong>Distance and area measurement:
            </strong><br>`;
```

```
            div.innerHTML += `- To measure a distance, draw a polyline by clicking two or more points on the map<br>`;
            div.innerHTML += `- To close a polyline into a polygon and display its perimeter and area, press 'C'<br>`;
            div.innerHTML += `- To finish the current measurement and start a new one, press 'R'`;

            return div;
        };

        legend.addTo(map);

          let points = [];
...
```

This produces the following result.

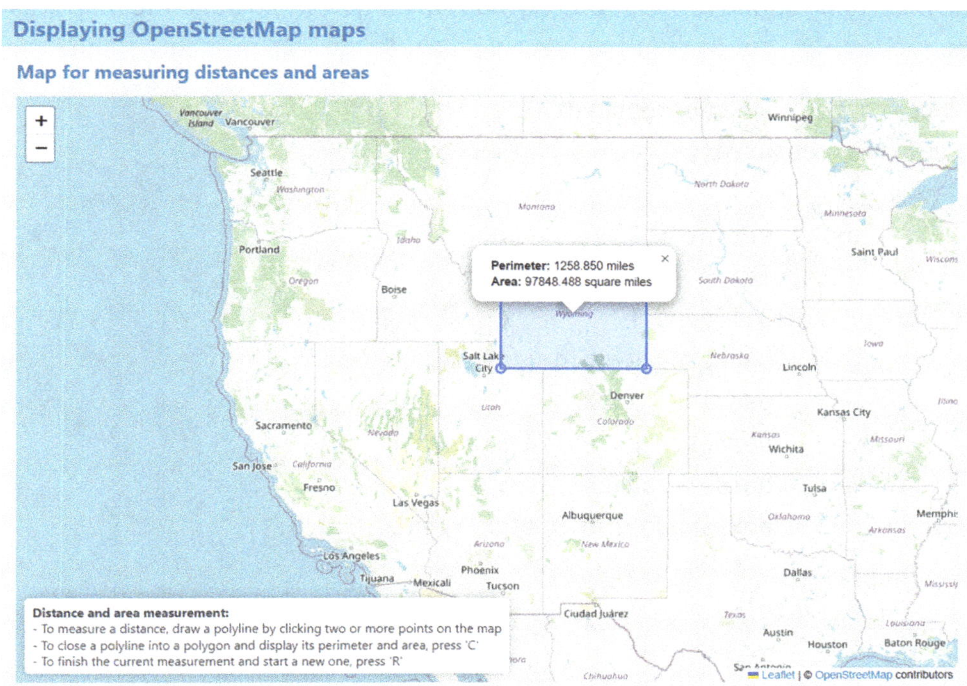

***Figure 11-3.** Map created with Leaflet using OpenStreetMap tiles. Data © OpenStreetMap contributors, licensed under the ODbL*

Measuring Distances and Areas with Folium

Although the choice of **Folium** for implementing this functionality is questionable, as it can hardly provide the high level of interactivity required, we will proceed with developing this option.

Our implementation will be largely based on the preceding one that used Leaflet:

- The map itself will be created using Folium.
- The interaction logic, responsible for drawing points, polylines, and polygons and for measuring distances and areas, will be written in JavaScript using Leaflet statements. This code will be injected into the HTML structure of the Folium map using the same technique that we explained at the end of Chapter 5 ("Injecting Leaflet Statements into Folium Maps to Add Interactivity" section) and used in Chapter 7 ("Searching for an Address by Clicking on a Map with Folium" section).

The code to be added to the advanced_maps.py blueprint to build the map is as follows:

```
@bp.route('/am/map_measure2')
def map_measure2():

    leaflet_script = """
        <style>
        .info.legend {
            background-color: white;
            padding: 10px;
            border: 2px solid black;
            border-radius: 5px;
            box-shadow: 0 0 15px rgba(0,0,0,0.2);
        }
        </style>

        <script src="https://cdn.jsdelivr.net/npm/@turf/turf@6/turf.min.js"></script>
        <script type="text/javascript">
```

```
        window.onload = function() {
            var map;
            for (var key in window) {
                if (window[key] instanceof L.Map) {
                    map = window[key];
                    break;
                }
            }
            try {
                const legend = L.control({position: 'bottomleft'});
                legend.onAdd = function(map) {
                    const div = L.DomUtil.create('div', 'info legend');

                    div.innerHTML += `<strong>Distance and area measurement:</strong><br>`;
                    div.innerHTML += `- To measure a distance, draw a polyline by clicking two or more points on the map<br>`;
                    div.innerHTML += `- To close a polyline into a polygon and display its perimeter and area, press 'C'<br>`;
                    div.innerHTML += `- To finish the current measurement and start a new one, press 'R'`;

                    return div;
                };

                legend.addTo(map);

                let points = [];
                let circleMarkers = [];
                let polyline = null;
                let polygon = null;
                let dist = 0;
```

```

                map.on('click', function(e) {
                    if (!polygon) {
                        const latlng = e.latlng;

                        points.push(latlng);

                        const newCircleMarker = L.circleMarker(latlng, {radius: 5}).addTo(map);
                        circleMarkers.push(newCircleMarker);

                        if (points.length === 1) {
                            polyline = L.polyline(points).addTo(map);

                            dist = 0;
                        } else {
                            polyline.addLatLng(latlng);

                            dist = dist + haversine(points[points.length - 2], points[points.length - 1])

                            newCircleMarker.bindPopup(`Length: ${dist.toFixed(3)} miles`)
                            .openPopup();
                        }
                    }
                });

                document.addEventListener('keydown', function(e) {
                    const key = (e.key || '').toLowerCase();
                    if (key === 'r') {
                        if (polyline) polyline.remove();
                        if (polygon) polygon.remove();
```

```
                for (let marker of circleMarkers) marker.remove();

                points = [];
                circleMarkers = [];
                polygon = null;
            } else if (key === 'c' && !polygon && points.length > 2) {
                dist = dist + haversine(points[points.length - 1], points[0]);

                polygon = L.polygon(points).addTo(map);
                polyline.remove();

                const geojson = polygon.toGeoJSON();
                 //const perimeter = turf.length(geojson, { units: "miles" });
                const area = turf.area(geojson) / (1609.344 * 1609.344);

                polygon.bindPopup(`<strong>Perimeter:</strong> ${dist.toFixed(3)} miles<br/><strong>Area:</strong> ${area.toFixed(3)} square miles`)
                .openPopup();
            }
        });

    } catch(e) {
        if (typeof map !== 'undefined' && map) {map.remove();}

        const div = document.getElementById("map")
        if (div) {div.remove();}

```

```
                const elemDiv = document.createElement('div');
                elemDiv.innerHTML = `<p>${e}</p>`;
                document.body.appendChild(elemDiv);
            }

            function haversine(latlng1, latlng2) {
                //const R = 6371; // Earth's radius in kilometers
                const R = 3959; // Earth's radius in miles

                const toRad = angle => angle * Math.PI / 180;

                const dLat = toRad(latlng2.lat - latlng1.lat);
                const dLon = toRad(latlng2.lng - latlng1.lng);

                const phi1 = toRad(latlng1.lat);
                const phi2 = toRad(latlng2.lat);

                const a = Math.sin(dLat / 2) ** 2 +
                    Math.cos(phi1) * Math.cos(phi2) *
                    Math.sin(dLon / 2) ** 2;

                const c = 2 * Math.atan2(Math.sqrt(a), Math.
                sqrt(1 - a));

                return R * c; // distance in miles
            }
        };
        </script>
    """

    try:
        m = folium.Map(location=[38, -100], zoom_start=5)

        m.get_root().html.add_child(folium.Element(leaflet_script))

```

```
135             return render_template('maps/page_with_map.html',
136                 name="Map for measuring distances and areas (built with
                    Folium)",
137                 map_html=m._repr_html_())
138
139         except Exception as e:
140             flash("An error has been encountered:<br/>" + str(e),
            category="error")
141                return render_template('maps/page_with_map.html', name =
               "No map to display")
```

- **Lines 4 to 128**: The JavaScript and Leaflet code to be injected into the Folium map is defined in a variable. This code is almost the same as the one used in the previous section, with the following differences:
  - **Lines 5 to 13**: A specific style must be defined for the legend so that it appears inside a white frame.
  - **Lines 18 to 24**: Since the map is defined with Folium outside the script, it is not known as a JavaScript variable and must first be retrieved before it can be used.
- **Line 131**: A map is created using Folium.
- **Line 133**: The map's HTML structure is accessed and augmented with the JavaScript code, including Leaflet statements, defined earlier on lines 4 to 128.

Launching the server and opening the address `http://127.0.0.1:5000/am/map_measure2` in the browser displays the following map, after clicking on a few points to draw a polyline and then pressing the "c" or "C" key to close it into a polygon:

Figure 11-4. *Map created with Folium using OpenStreetMap tiles. Data © OpenStreetMap contributors, licensed under the ODbL*

Alternative Solutions with Leaflet and Folium Plugins

In the previous section, we demonstrated how to write JavaScript and Python code to allow users to measure distances and areas on a map. However, some Leaflet plugins are also available for this purpose, including:

- **Leaflet.MeasureControl** (developed by Makina Corpus): This plugin allows users to measure distances on a map by drawing a polyline.
- **Leaflet.MeasurePolygon** (developed by Erick S. Escalante Olano): This plugin enables users to draw polygons on the map and calculate their area and perimeter.

- **leaflet-measure** (developed by "ljagis"): This plugin provides a new control that enables users to draw polylines and measure their length or to draw polygons and measure their perimeter and area.

There is also a Folium plugin available for measuring distances and areas:

- **MeasureControl** (developed by "ocefpaf"): This plugin allows users to measure distances and areas by drawing polylines on the map. It wraps a Leaflet plugin (the **leaflet-measure** plugin mentioned above) into an easy-to-use control.

CHAPTER 12

Conclusion

Throughout this book, we explored the rich ecosystem of open source tools that support the creation, manipulation, and display of digital maps. We started with an understanding of OpenStreetMap, the foundation technology. We then went on to learning how to query and extract geographic data for general or specific needs using OSM-associated APIs – Overpass and Nominatim.

Then, practical skills were developed through hands-on projects using Leaflet and Folium, showing how to create interactive web maps both on the client side (with JavaScript) and the server side (with Python). We discovered how to incorporate markers, layers, geometric shapes, and images, how to create choropleth maps, how to integrate external data, and how to leverage the interactive capabilities of Leaflet.

We also tackled advanced topics, such as combining multiple tools within a single workflow – for example: retrieving OSM data on the server and displaying the elements on a map, interacting with maps to show geocoded information, building advanced choropleth maps, querying external services like OSRM to find and plot routes, geolocating the user, or measuring distances and areas on maps. As a result, we gained the ability to create robust, customized, and interactive mapping applications using free and open source technologies.

To conclude our exploration of digital cartography, it is worth considering the future directions and emerging trends that may shape this rapidly evolving field.

If society continues to evolve toward increasingly ubiquitous and sophisticated technology, cartography will also undergo major transformations. However, disasters can never be ruled out, and survivors are likely to revert to much more rudimentary and ancestral techniques.

Among the probable developments, 3D maps, which already exist, will play an increasingly important role, although they will not dethrone 2D maps. As our planet is vast enough that some may believe it to be flat, two-dimensional representations will remain relevant and more practical in many situations.

P. Marie, *Modern Web Cartography*, https://doi.org/10.1007/979-8-8688-2335-0_12

Certain 3D maps will become more realistic and offer immersive experiences. They will enable virtual navigation enriched by augmented reality, displaying dynamic information about the places visited.

Interactive maps will become more widespread: instead of being simply frozen images, they will offer users the possibility of interacting, querying their content, and accessing precise information about places or objects. Some will incorporate artificial intelligence capabilities, answering questions and providing tailored recommendations.

Moreover, short-term, if not real-time, updating will become the norm. Rather than reflecting a state frozen in time, maps will be continually fed by data streams, offering the most up-to-date possible view of the world around us. Artificial intelligence techniques will contribute to this goal by analyzing aerial images, detecting objects in them, and updating geographic databases very rapidly.

Finally, the role of cartography will extend far beyond simple consultation. All systems requiring spatial management, whether for travel or other purposes, will rely on these services. However, the display of maps will become less frequent:

- Many devices will use mapping services internally, without explicit visual representation.
- Users, meanwhile, will increasingly delegate the search and analysis of geographic data to intelligent assistants. Rather than searching for information or interpreting a route on a map themselves, they will simply express their need - for example, asking an autonomous vehicle to drive them to a destination - leaving artificial intelligence and cartography, hidden in the background, to do the rest.

As a result, cartography will increasingly become an essential but unobtrusive back-end IT service, providing spatial data and geographic services to a wide range of applications and devices.

APPENDIX A

Development and Deployment Tools Used for the Prototypes

All the prototypes prepared for this book were developed on a **Windows** workstation using **Visual Studio Code**, a freeware development environment from Microsoft that lets you develop applications in a wide variety of programming languages: JavaScript, C/C++/C#, Java, Python, PHP, etc.

For this project, VS Code was configured to support the Python language and the Flask framework.

The integrated debugger of VS Code was used to test and correct the applications.

The prototypes were first tested with the local test server of VS Code, then deployed on the **PythonAnywhere** web server (`https://www.pythonanywhere.com`) and further verified there.

On the **server side**, the following technologies were used to implement the prototypes:

- The **Python** development language.
- The **Flask** framework, chosen for its ease of use when implementing web applications with Python (it is considered a **lightweight framework**). However, other frameworks such as Django would also have been suitable.

  Flask allows you to define routes (URLs) that trigger the execution of the web application's functions within **blueprints**. It also uses the **Jinja** template engine to dynamically inject data into HTML pages based on **templates**.

P. Marie, *Modern Web Cartography*, https://doi.org/10.1007/979-8-8688-2335-0

- The **Folium** Python library, to build some of the maps, offering a rich set of functions for creating and enhancing maps in order to display them in a browser.

All server-side code using these technologies was developed and tested with VS Code, with final tests also performed on the PythonAnywhere server.

On the **client side**, the following technologies were used:

- **HTML** pages, most often generated dynamically on the server from templates using Jinja, but also sometimes static.

  HTML describes the content of a web page to be displayed in a browser. Tags are used to structure this content (headings, paragraphs, lists, etc.). HTML documents typically reference external CSS files to apply styling and layout rules.

- A **CSS** style sheet, to manage page appearance.

  A CSS file defines the graphic presentation of HTML tags.

- **JavaScript** code, embedded in HTML pages, to manage interactivity (defining appropriate processing to respond to various user actions) and build maps with Leaflet.js.

  Part of this JavaScript code was written manually, and part is generated dynamically (when executing the prototypes) on the server by the Folium library.

- The **Leaflet.js** JavaScript library, to build and display maps.

  Leaflet offers a rich set of functions to create and enhance maps, then to display them in a browser.

  Leaflet statements are involved in all the prototypes:

  - Sometimes these statements were written manually.
  - Sometimes they are generated dynamically (when executing the prototypes) on the server from Folium statements (written manually).

All these elements are executed in a **browser** (client-side) and have also been developed and tested with VS Code (a PythonAnywhere server was also used for the final tests). The **Firefox** and **Chrome** browsers were mainly used to check the prototypes, with **Microsoft Edge** also used occasionally.

Although the prototypes were developed on Windows, most of the tools used are also available on Linux and macOS. With some adaptation, everything presented in this book can be reproduced in these environments.

APPENDIX B

Setting Up the Prototype Development and Test Environment

The environment described below is the one that was used to build all the prototypes in this book. This environment can also enable readers to reproduce these examples with minimal effort by simply following the instructions provided in the book.

Prerequisites

All prototypes were implemented in the Windows environment, but with minimal adaptation, the same developments can be carried out on Linux and macOS.

Visual Studio Code, Microsoft's freeware development tool, must first be installed on the workstation, then the **Python** language installed in VS Code.

- Instructions on how to download and install VS Code can be found at `https://code.visualstudio.com/docs/setup/setup-overview`.
- And instructions for installing the Python language in VS Code can be found at `https://code.visualstudio.com/docs/languages/python`.

The next sections will describe how to create a new workspace, create a virtual environment, and install Flask, then implement and run a basic test application. This is also explained at `https://code.visualstudio.com/docs/python/tutorial-flask`.

P. Marie, *Modern Web Cartography*, https://doi.org/10.1007/979-8-8688-2335-0

Creating a New Workspace

This workspace will be used to develop and test prototypes that build OpenStreetMap-based maps using Leaflet or Folium as web applications.

1. Using the operating system's File Explorer, create a directory, named, for example, `carto_osm` (any other name can be used), in a convenient location on your file system (e.g., in the directory `C:/modern_web_cartography`).
2. Then start VS Code and, with its menu bar, select "File ➤ Open Folder", retrieve and select the newly created directory, and click "Select Folder". This folder will be the workspace containing the source code for the new applications. You will get a popup window with the message "Do you trust the authors of the files in this folder?"; click the `Yes, I trust the authors` button (meaning that you trust the files in this directory, which is still empty).
3. Finally, close the "Welcome" panel (in the middle) and the "Chat" panel (on the right). The VS Code Explorer (on the left) is positioned on the `carto_osm` directory (the base directory of the new workspace), which is currently empty.

Figure B-1. *Screenshot of Visual Studio Code. © Microsoft Corporation (`http://www.microsoft.com`). All rights reserved. Used with permission from Microsoft*

Creating a Virtual Environment and Installing Flask

The next step is to install Flask in the workspace and check that it works properly. To do this, we must start by creating a virtual environment in which Flask will be installed. These operations are explained below and will be performed using the command palette.

1. In VS Code, open the command palette (use "View ➤ Command Palette" or the cogwheel at bottom left), and then select the "Python: Create Environment" command to create a virtual environment in the `carto_osm` workspace.

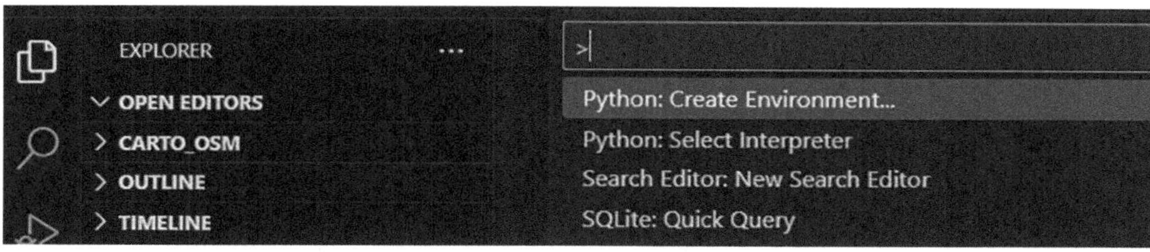

Figure B-2. *Screenshot of Visual Studio Code. © Microsoft Corporation (`http://www.microsoft.com`). All rights reserved. Used with permission from Microsoft*

Then select "Venv Create a '.venv' virtual environment in the current workspace".

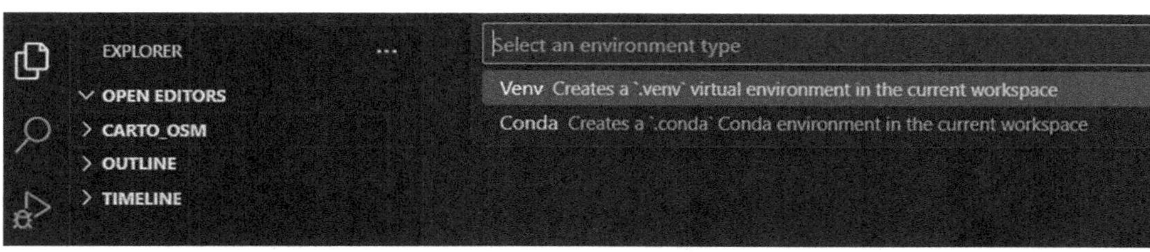

Figure B-3. *Screenshot of Visual Studio Code. © Microsoft Corporation (`http://www.microsoft.com`). All rights reserved. Used with permission from Microsoft*

Then select Python 3.11.2 (at the time of writing this book) or later.

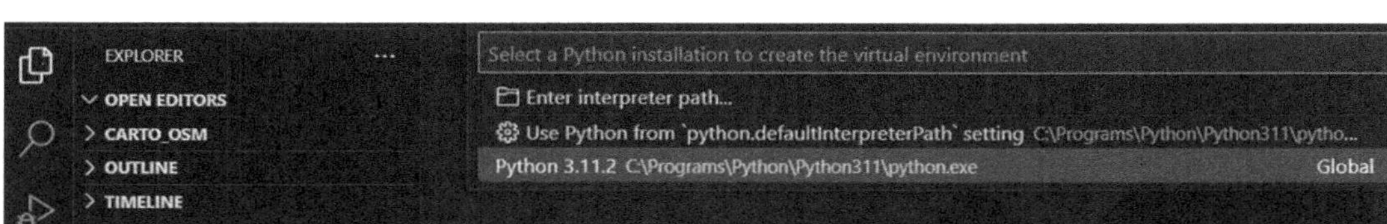

Figure B-4. *Screenshot of Visual Studio Code. © Microsoft Corporation (`http://www.microsoft.com`). All rights reserved. Used with permission from Microsoft*

2. Still in VS Code, select "View ➤ Terminal". This opens a terminal positioned on the `carto_osm` directory and displayed at the bottom of the VS Code window. At the same time, this operation activates the virtual environment just created.

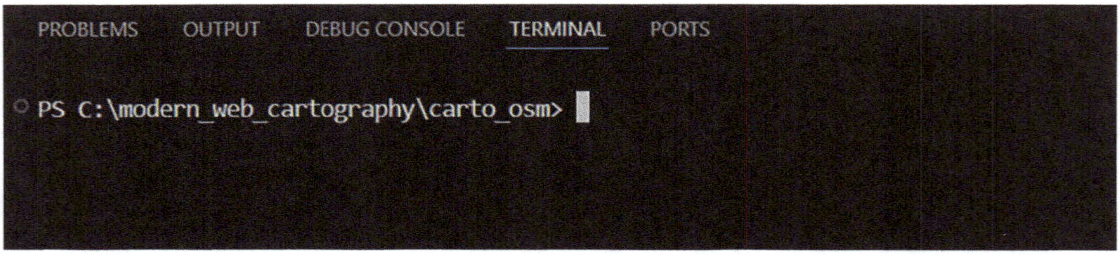

Figure B-5. *Screenshot of Visual Studio Code. © Microsoft Corporation (`http://www.microsoft.com`). All rights reserved. Used with permission from Microsoft*

Then, on the terminal command line, enter and execute the following command to install Flask in the new virtual environment:

```
python -m pip install flask
```

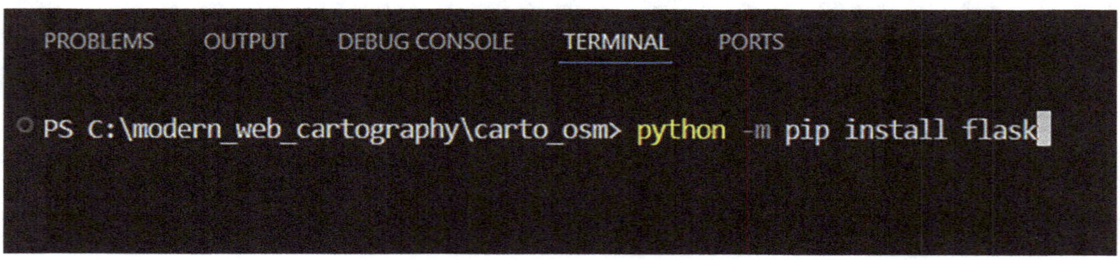

Figure B-6. *Screenshot of Visual Studio Code. © Microsoft Corporation (`http://www.microsoft.com`). All rights reserved. Used with permission from Microsoft*

Flask 3.1.2 (at the time of writing this book) and a number of additional packages are installed.

To check whether a higher level of Flask is available, run:

```
python -m pip install --upgrade flask
```

If a more recent package exists, it is installed in place of the previous one.

At this point, the `carto_osm` directory contains the `.venv` directory with a number of Python libraries in it.

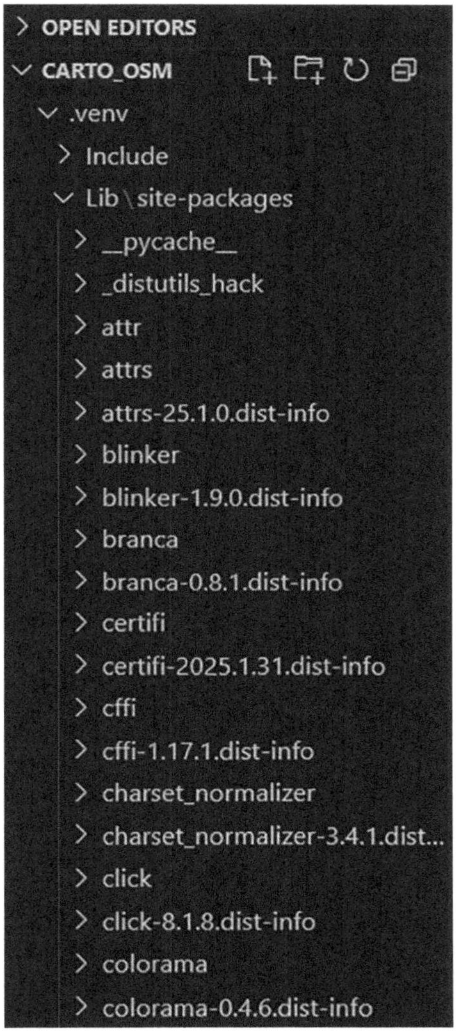

Figure B-7. *Screenshot of Visual Studio Code. © Microsoft Corporation (`http://www.microsoft.com`). All rights reserved. Used with permission from Microsoft*

The versions of Python and Flask just installed can be checked using the following commands:

```
python --version
```

```
python -m pip show flask
```

And this command displays all the Python packages installed:

```
pip list
```

Creating and Running a Basic Application

1. Using the VS Code Explorer, create a `prototypes` directory in the workspace (use the "New Folder" icon), with two sub-directories, `static` and `templates`.

Figure B-8. *Screenshot of Visual Studio Code. © Microsoft Corporation (`http://www.microsoft.com`). All rights reserved. Used with permission from Microsoft*

2. In the `prototypes` directory, create an `__init__.py` file (use the "New File" icon) with the following contents (the `__init__.py` file will mark the directory as a regular Python package and will contain initialization statements):

```
import os
from flask import Flask

def create_app(test_config=None):
    # Create and configure the application
    app = Flask(__name__, instance_relative_config=True)
    app.config.from_mapping(
        SECRET_KEY='dev'
    )
```

```
10
11      if test_config is None:
12          # Load instance configuration, if any
13          app.config.from_pyfile('config.py', silent=True)
14      else:
15          # Load test configuration
16          app.config.from_mapping(test_config)
17
18      # Check that the instance directory exists
19      try:
20          os.makedirs(app.instance_path)
21      except OSError:
22          pass
23
24      # A simple page that says Hello
25      @app.route('/hello')
26      def hello():
27          return 'Hello, World!'
28
29      return app
```

This code defines a minimal Flask application:

- The first statement (`import os`) must not be indented (otherwise, a Python syntax error will occur). For the following lines, the column alignments and indentations shown above must be respected (Python is strict about column alignment and indentation).
- **Line 25**: The route to be used to call the application is defined (resulting in the URL: `http://127.0.0.1:5000/hello`).
- **Lines 26 and 27**: The associated view function (i.e., the code to be executed for the application when the route is called) is defined: the application will simply return the message "Hello, World!".

Do not forget to save the file after having created it.

3. Next, we are going to test this simple application. First, click "Run and debug" (icon with an arrow head and a bug in the left margin just below the Explorer icon), and then select "create a launch.json file".

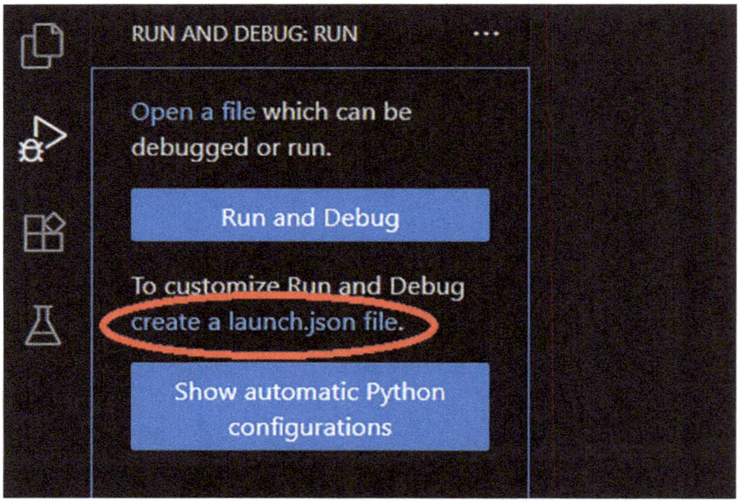

Figure B-9. *Screenshot of Visual Studio Code. © Microsoft Corporation (`http://www.microsoft.com`). All rights reserved. Used with permission from Microsoft*

Then, when prompted, select "More Python Debugger options...".

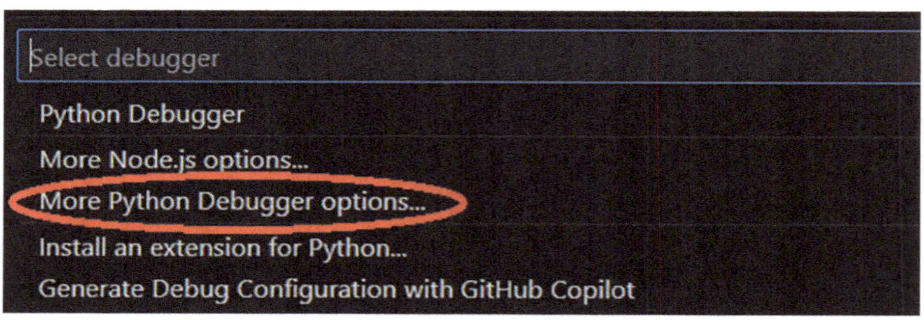

Figure B-10. *Screenshot of Visual Studio Code. © Microsoft Corporation (`http://www.microsoft.com`). All rights reserved. Used with permission from Microsoft*

Then select "Python Debugger: Flask".

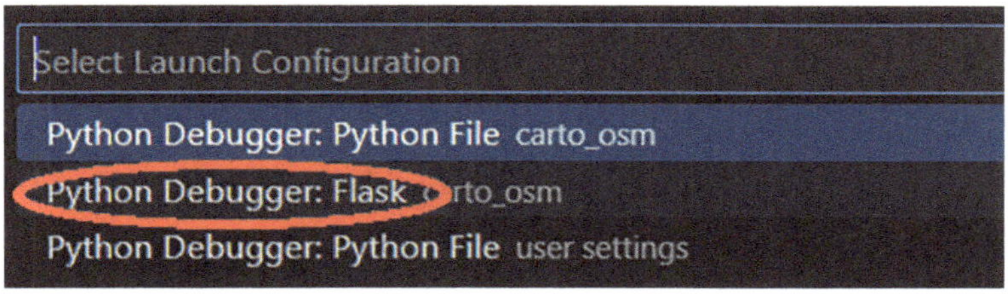

Figure B-11. *Screenshot of Visual Studio Code. © Microsoft Corporation (`http://www.microsoft.com`). All rights reserved. Used with permission from Microsoft*

Finally, select "__init__.py". VS Code then creates and opens a file named `launch.json`, which contains options to be used when starting the Python debugger.

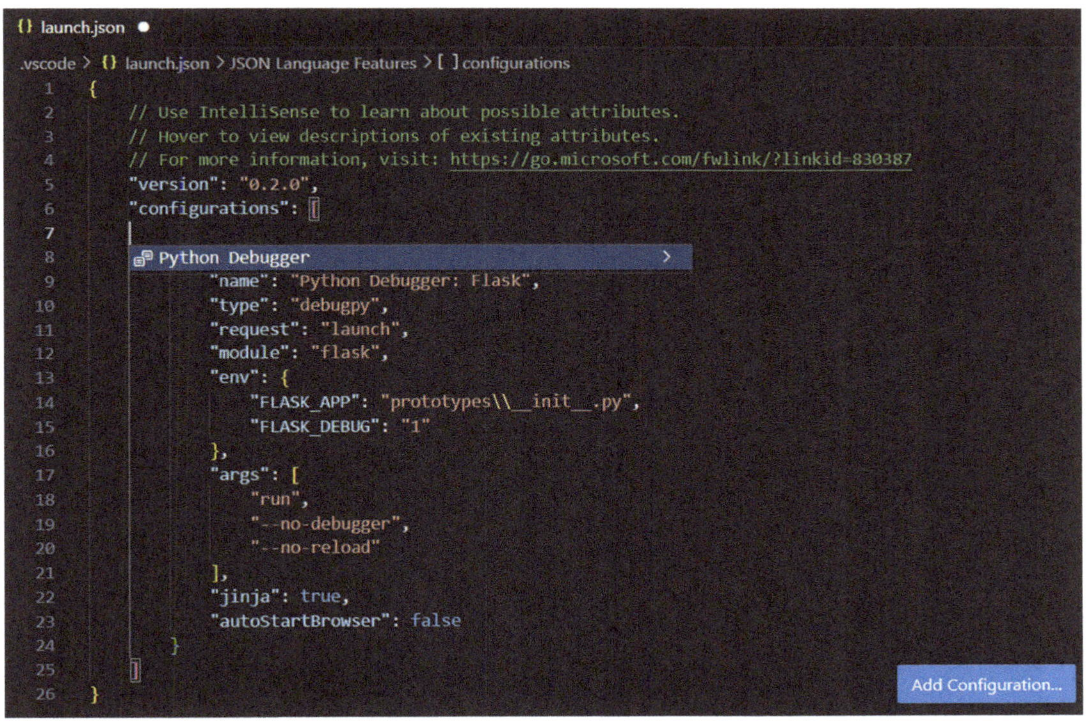

Figure B-12. *Screenshot of Visual Studio Code. © Microsoft Corporation (`http://www.microsoft.com`). All rights reserved. Used with permission from Microsoft*

Simply save and close this file without modifying it. This file only needs to be created once; it will be reused when you start the server and debugger again.

4. Now start the server and debugger using "Run ➤ Start Debugging" in the top menu bar. The VS Code terminal displays the following messages:

```
 * Serving Flask app 'prototypes\__init__.py'
 * Debug mode: on
WARNING: This is a development server. Do not use it in a
production deployment. Use a production WSGI server instead.
 * Running on http://127.0.0.1:5000
Press CTRL+C to quit
```

   This means that the development server is ready to run the application and is waiting for HTTP requests on the URL `http://127.0.0.1:5000`.

5. Finally, using a browser, open the URL `http://127.0.0.1:5000/hello`.

***Figure B-13.** Execution of the "Hello World!" simple Flask application*

The "Hello, World!" message confirms that the application is working correctly.

Then stop the server and debugger, either by selecting "Run ➤ Stop Debugging" or clicking the "Stop" icon (an orange square) in the middle of the top bar.

Detailed information about how to use the VS Code debugger (setting breakpoints, inspecting variables, etc.) can be found at `https://code.visualstudio.com/docs/debugtest/debugging`.

Installing Folium

Folium is a Python library for building and displaying OpenStreetMap maps using the Python language. These maps can be of various types and can be enriched with additional information (highlighting particular points, plotting lines and polygons, surrounding and filling areas, etc.). Folium is a Python wrapper for the `Leaflet.js` JavaScript library, designed for the same usage.

1. To install Folium, use the following command in the virtual environment created for the application (use the terminal and go to the `carto_osm` workspace directory, if not already there):

   ```
   pip install folium
   ```

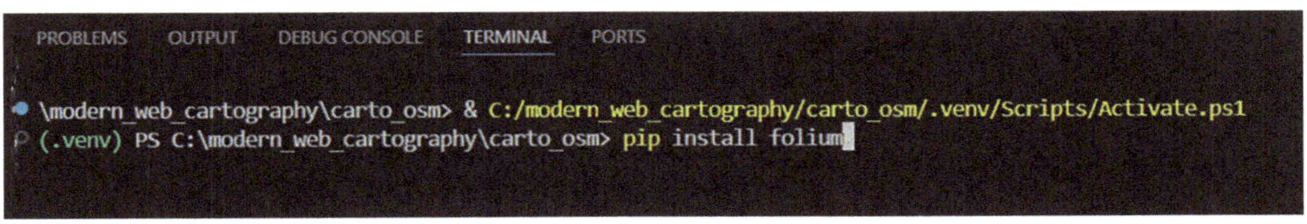

Figure B-14. *Screenshot of Visual Studio Code. © Microsoft Corporation (`http://www.microsoft.com`). All rights reserved. Used with permission from Microsoft*

   Folium 0.20.0 (at the time of writing this book) and a number of accompanying packages are installed.

   The version of Folium just installed can be checked using the following command:

   ```
   python -m pip show folium
   ```

2. To test Folium's correct operation, add the following code (in **bold**) at the end of the `prototypes/__init__.py` file created earlier, just before the `return app` statement:

```
...
29      @app.route('/map')
30      def map():
31          import folium
32          m = folium.Map(location=[45.5236, -122.6750])
33          return m._repr_html_()
34
35      return app
```

- Be careful with column alignment and indentation. Python is very strict about both.
- **Line 29**: This statement defines the application route, resulting in the URL: `http://127.0.0.1:5000/map`. It enables the application to respond to HTTP requests at this endpoint.
- **Line 30**: This line begins the definition of the `map()` function - the route's view function - which contains the logic for generating the response at this route.
- **Line 31**: The `folium` library is imported.
- **Line 32**: A map is created (using the Folium library) and centered on a geographic position whose latitude and longitude are set via the location parameter (in this case, the city of Portland, Oregon, in the USA).
- Line 33 returns the map as a response to the HTTP request and terminates the function. Before being returned, the map is converted to HTML and JavaScript code (containing Leaflet statements) by the `_repr_html_()` method, so that it can be displayed by the browser.

3. Run this simple application by selecting "Run ➤ Start debugging" (or "Restart Debugging" if the server is already running), then opening the URL `http://127.0.0.1:5000/map` in a browser. You will get the following map of Portland and the surrounding area:

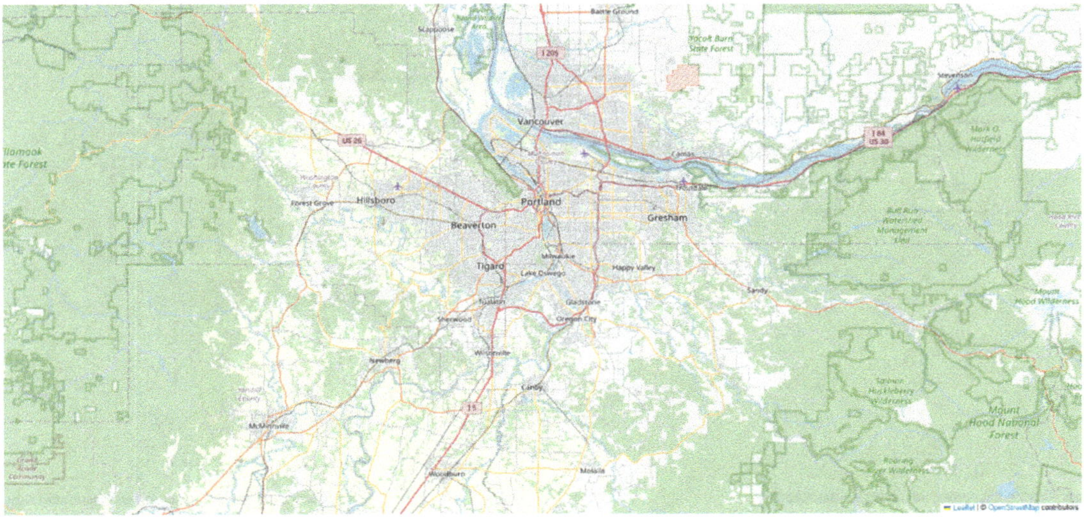

Figure B-15. *Map created with Folium using OpenStreetMap tiles. Data © OpenStreetMap contributors, licensed under the ODbL*

The architecture used by this basic application is the following:

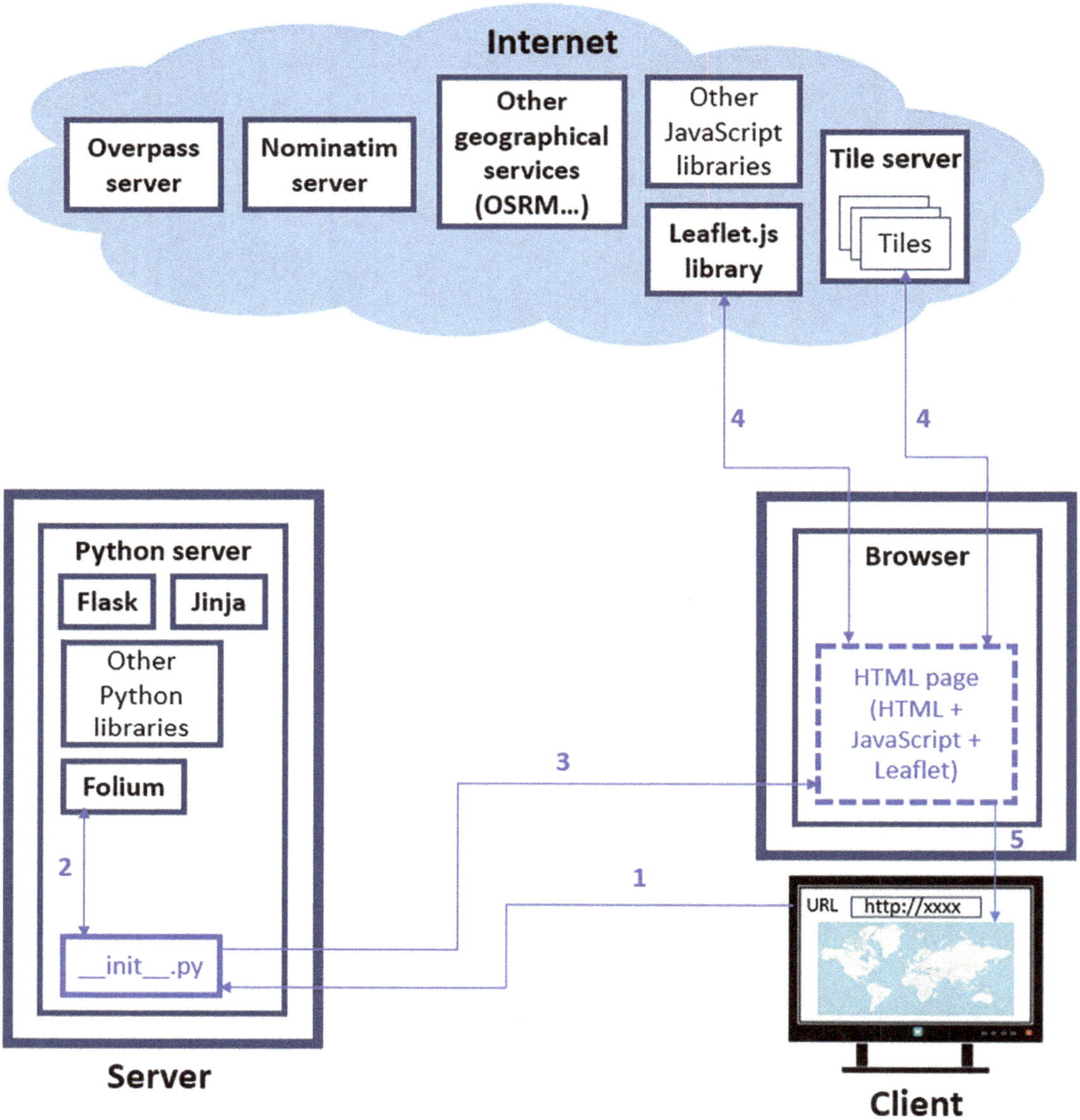

***Figure B-16.** Architecture used to test a simple Folium application*

1. The user invokes a URL from the browser. This URL points to a route on the Python server.
2. The Python server retrieves the corresponding Python code for this route in the `__init__.py` file and executes it. With this code containing Folium statements, a map is generated as HTML, JavaScript code, and Leaflet statements.

3. The resulting HTML, JavaScript code, and Leaflet statements are then returned to the browser as a response.
4. The browser receives the response, executes the embedded JavaScript code and Leaflet statements, and downloads the tiles necessary to render the map (from a tile server).
5. The browser finally displays the resulting web page, i.e., a map, on the screen.

Setting Up a Flask Package Structure

To prototype OSM maps with Folium and Leaflet, a "Flask package" structure will be used (in preference to the simpler but overly monolithic "Flask application" structure). This structure will be made up of "blueprints" and "templates".

- A "blueprint" is a Python file executable on the server, which is capable of receiving HTTP requests and performing processing to construct and return appropriate responses to these requests. Most often, these responses consist of HTML code that can be interpreted and presented as a web page by a browser.
- A "template" is a model of an HTML page designed to be displayed in a browser. Blueprints generally use templates to construct the web pages that they return to browsers (in response to requests). Templates contain mostly, but not exclusively, HTML code. They can also contain:
    - **Jinja code**: The Jinja code included in a template is executed **on the server** when a blueprint requests that the template be used to generate an HTML page in response to a client request. This happens **before** the page is sent to the browser. Jinja is primarily used to inject **dynamic data** into otherwise **static HTML**. Its syntax is inspired by the Python language.

- **JavaScript code**: Adding JavaScript code to HTML pages makes them more dynamic and interactive for the user. This code is executed **in the browser**, either automatically when the page is loaded or in response to user actions (typically, clicking on an element on the page).

For more information on blueprints and templates, see `https://flask.palletsprojects.com/en/stable/`.

Let's implement this structure in our workspace (in VS Code). Our blueprints will be located in the `prototypes` directory and our templates in the `prototypes/templates` directory. Other complementary files will be placed in the `prototypes/static` directory: Cascading Style Sheets (CSS), static data files (i.e., which are not modified) used by blueprints and templates, etc.

1. Open the `prototypes/__init__.py` file created earlier, and, just before the final `return app` statement, add the declaration of a blueprint named `simple_maps` (the two statements in **bold** below):

```
35        from prototypes import simple_maps
36        app.register_blueprint(simple_maps.bp)
37
38        return app
```

- **Line 35**: Imports the `simple_maps` module from the `prototypes` package. This module, which will be created next, will contain the blueprint definition and the routes and associated view functions.

- **Line 36**: Registers the `bp` blueprint from the `simple_maps` module with the Flask application. This integrates the routes and view functions defined in `prototypes/simple_maps.py` into the main app.

2. Create the blueprint as `simple_maps.py` file in the `prototypes` directory, with the following contents (to start with):

```
import folium
from flask import Blueprint, render_template, request, flash
```

```
3
4   bp = Blueprint('simple_maps', __name__)
5
6   @bp.route('/sm/map0', methods=('GET', 'POST'))
7   def map0():
8       m = folium.Map()
9       return render_template('maps/page_with_map.html',
10          name="Hello World!",
11          map_html=m._repr_html_())
```

- As usual with Python, be careful with column alignment and indentation. The first statement (`import folium`) must not be indented.
- **Lines 1 and 2**: The `folium` library, used for generating interactive maps, is imported. The necessary Flask functions for defining routes, rendering templates, handling requests, and flashing messages are also imported. These libraries will be used multiple times in the examples of this appendix and of Chapter 5 (and others). Importing them globally here avoids having to import them again in each example.
- Line 4: A Flask blueprint named simple_maps is defined. This blueprint will group related routes and view functions for modular application structure.
- Lines 6 to 1**1**: A first test route and its view function are defined to verify the blueprint and template integration:
  - **Line 6**: The `/sm/map0` route, accessible via HTTP GET and POST requests, is defined. When accessed locally on the VS Code test server, the full URL will be `http://127.0.0.1:5000/sm/map0`.
  - **Line 7**: This statement begins the definition of the `map0()` view, i.e., the function associated with the `/sm/map0` route, which will be executed each time this route is accessed via an HTTP request.

- **Line 8**: A basic Folium map is created using default arguments, which will result in a map of the world.
- **Lines 9 to 11**: A web page is rendered using the `page_with_map.html` template (created next). A name ("Hello World!") and the map object, converted to HTML and JavaScript code using the `_repr_html_()` method, are passed as arguments.

3. Then create a `base.html` file with the following contents in the `prototypes/templates` directory. This file will be referenced by several templates that will be created later. It defines a basic HTML structure in which elements common to several templates are positioned. This only needs to be done once, as this technique centralizes these common elements and avoids repeating them multiple times.

```
<!doctype html>
<html lang="en">
<head>
  <meta charset="utf-8">
  <title>{% block title %}{% endblock %} - Prototypes</title>
  <link rel="stylesheet" href="{{ url_for('static', filename='style.css') }}">
  {% block links %}{% endblock %}
</head>
<body>
  <nav>
    <h1>Displaying OpenStreetMap maps</h1>
  </nav>
  <section class="content">
    <header>
      {% block header %}{% endblock %}
    </header>
    {% for message in get_flashed_messages() %}
    <div class="flash">{{ message | safe }}</div>
    {% endfor %}
    {% block content %}{% endblock %}
```

```
    </section>
    {% block script %}{% endblock %}
</body>
</html>
```

The various {% block xxxx %}{% endblock %} elements will be replaced by specific content defined in the templates extending base.html.

4. In the prototypes/templates directory, create a maps subdirectory, and, in this new directory, create a page_with_map.html file with the following contents:

```
1  {% extends 'base.html' %}
2
3  {% block title %}{{ name }}{% endblock %}
4
5  {% block header %}
6  <h2>{{ name }}</h2>
7  {% endblock %}
8
9  {% block content %}
10 {{ map_html | safe }}
11 {% endblock %}
```

- **Line 1**: This template extends the base.html base template created just before and therefore inherits its structure and contents.
- **Line 3**: A specific content is added to the <title> tag (defined in base.html). Its text is received as an argument when the blueprint triggers the construction of an HTML page from this template. The expression {{ name }} is a Jinja variable that will be replaced by its value (set in the Python blueprint) by the Jinja engine. This substitution will be done on the server before the page is sent to the browser.

  This content is the web page title: it will not appear in the web page itself, but in the top bar of the browser.

- **Line 6**: A specific content is added to the <header> tag (defined in base.html). Its text is received as an argument when the blueprint triggers the construction of an HTML page from this template. The Jinja expression {{ name }} is a variable, and the Jinja engine will replace it with a value set in the blueprint.

  This content is a level 2 header (<h2>) that will appear at the top of the web page.

- **Line 10**: The HTML and JavaScript code for the Folium map is inserted into the page. This line is processed by the Jinja engine:

  - The map_html variable is replaced by the actual HTML and JavaScript code generated by Folium to render the map.

  - The safe filter is necessary to prevent Jinja from escaping the HTML/JavaScript content. Without it, Jinja would convert characters such as "<" and ">" into "<" and ">", causing the browser to display raw code instead of rendering the map.

This page_with_map.html template will be used by many prototypes in this book.

5. Then, in the prototypes/static directory, create a style.css file with the following contents:

```
/* ==========
   Global variables
   ========== */
:root {
  --color-primary: #377ba8;
  --color-bg-body: #f0f0f0;
  --color-bg-nav: lightblue;
  --color-bg-flash: #faa5a5;
  --color-border-flash: #fa5b5b;
  --font-main: system-ui, -apple-system, BlinkMacSystemFont,
"Segoe UI",
               Roboto, Helvetica, Arial, sans-serif;
  --max-width: 1060px;
  --padding-page: 1rem;
```

```
}

/* ==========
   Base styles
   ========== */
html {
  font-family: var(--font-main);
  background: #fff;
  padding: var(--padding-page);
  color: #000;
}

body {
  max-width: var(--max-width);
  margin: 0 auto;
  background: var(--color-bg-body);
  line-height: 1.5;
}

h1, h2 {
  font-family: var(--font-main);
  color: var(--color-primary);
  margin: 1rem 0;
}

h1 { font-size: 1.56rem; } /* ≈ 25px */
h2 {
  line-height: 1.1;
  font-size: 1.31rem;
} /* ≈ 21px */

a {
  color: var(--color-primary);
  text-decoration-thickness: 1px;
}

hr {
  border: none;
```

```
  border-top: 1px solid lightgray;
}

/* ==========
   Navigation
   ========== */
nav {
  background: var(--color-bg-nav);
  display: flex;
  align-items: center;
  justify-content: space-between;
  padding: 0.5rem 0.75rem;
  flex-wrap: wrap; /* responsive */
}

nav h1 {
  margin: 0;
  line-height: 1.1;
  flex: 1 1 auto;
}

nav h1 a {
  text-decoration: none;
  padding: 0.25rem 0.5rem;
}

nav ul {
  display: flex;
  list-style: none;
  margin: 0;
  padding: 0;
  flex-wrap: wrap; /* better for small screens */
}

nav ul li a,
nav ul li span,
header .action {
  display: block;
```

```
  padding: 0.5rem;
  font-weight: 400;
}

/* ==========
   Content area
   ========== */
.content {
  padding: 0 1rem 1rem;
}

.content > header {
  border-bottom: 1px solid lightgray;
  display: flex;
  align-items: flex-end;
  flex-wrap: wrap; /* ensures text doesn't overflow on small
  screens */
}

.content > header h1 {
  flex: 1 1 auto;
  margin: 1rem 0 0.25rem;
}

/* ==========
   Flash messages
   ========== */
.flash {
  margin: 1em 0;
  padding: 1em;
  background: var(--color-bg-flash);
  border: 1px solid var(--color-border-flash);
  border-radius: 6px;
}

/* ==========
   Posts
   ========== */
```

```
.post > header {
  display: flex;
  align-items: flex-end;
  font-size: 0.85em;
  flex-wrap: wrap;
}

.post > header > div:first-of-type {
  flex: 1 1 auto;
}

.post > header h1 {
  font-size: 1.5em;
  margin-bottom: 0;
}

.post .about {
  color: slategray;
  font-style: italic;
}

.post .body {
  white-space: pre-line;
}

.content:last-child {
  margin-bottom: 0;
}

/* ==========
   Forms
   ========== */
.content form {
  margin: 1em 0;
  display: flex;
  flex-direction: column;
  gap: 0.75rem;
}
```

```
.content label {
  font-weight: 600;
}

.content input,
.content textarea {
  font-family: var(--font-main);
  padding: 0.5rem;
  border: 1px solid #ccc;
  border-radius: 4px;
}

.content textarea {
  min-height: 12em;
  resize: vertical;
}

input.danger {
  color: #cc2f2e;
}

input[type="submit"] {
  align-self: start;
  min-width: 10em;
  cursor: pointer;
}

/* ==========
   Legend (map control)
   ========== */
.legend {
  background: #fff;
  padding: 6px 8px;
  font: 14px/16px var(--font-main);
  box-shadow: 0 0 15px rgba(0, 0, 0, 0.2);
  border-radius: 5px;
  line-height: 18px;
  color: #555;
```

```
}

.legend i {
  width: 18px;
  height: 18px;
  float: left;
  margin-right: 8px;
  opacity: 0.7;
}

/* ==========
   Responsiveness
   ========== */
@media (max-width: 768px) {
  body {
    background: #fff;
  }

  nav {
    flex-direction: column;
    align-items: stretch;
  }

  nav ul {
    justify-content: center;
  }

  .content {
    padding: 0.5rem;
  }
}
```

This file defines style elements for the rendering of various HTML tags by the browser. It is referenced by the template `page_with_map.html`, via the base template `base.html` (`<link>` tag on line 6 of this file), and determines the presentation style of the web pages to be displayed. It will be used by other HTML templates later on, always via a reference from these templates to a base template. Further details on the syntax of CSS style sheets are available at `https://www.w3schools.com/css/default.asp`.

6. Finally, to test the files just set up, launch (or relaunch) the VS Code test server and open the URL `http://127.0.0.1:5000/sm/map0` in a browser. You will get the following map:

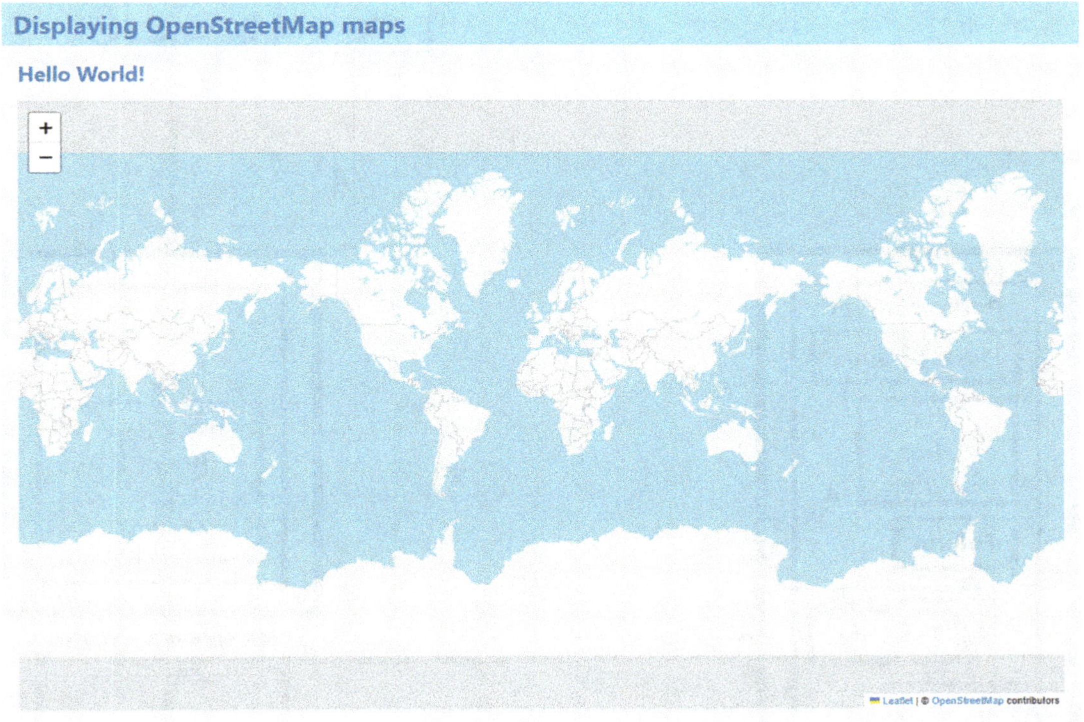

Figure B-17. *Map created with Folium using OpenStreetMap tiles. Data © OpenStreetMap contributors, licensed under the ODbL*

The architecture used by this simple application is the following:

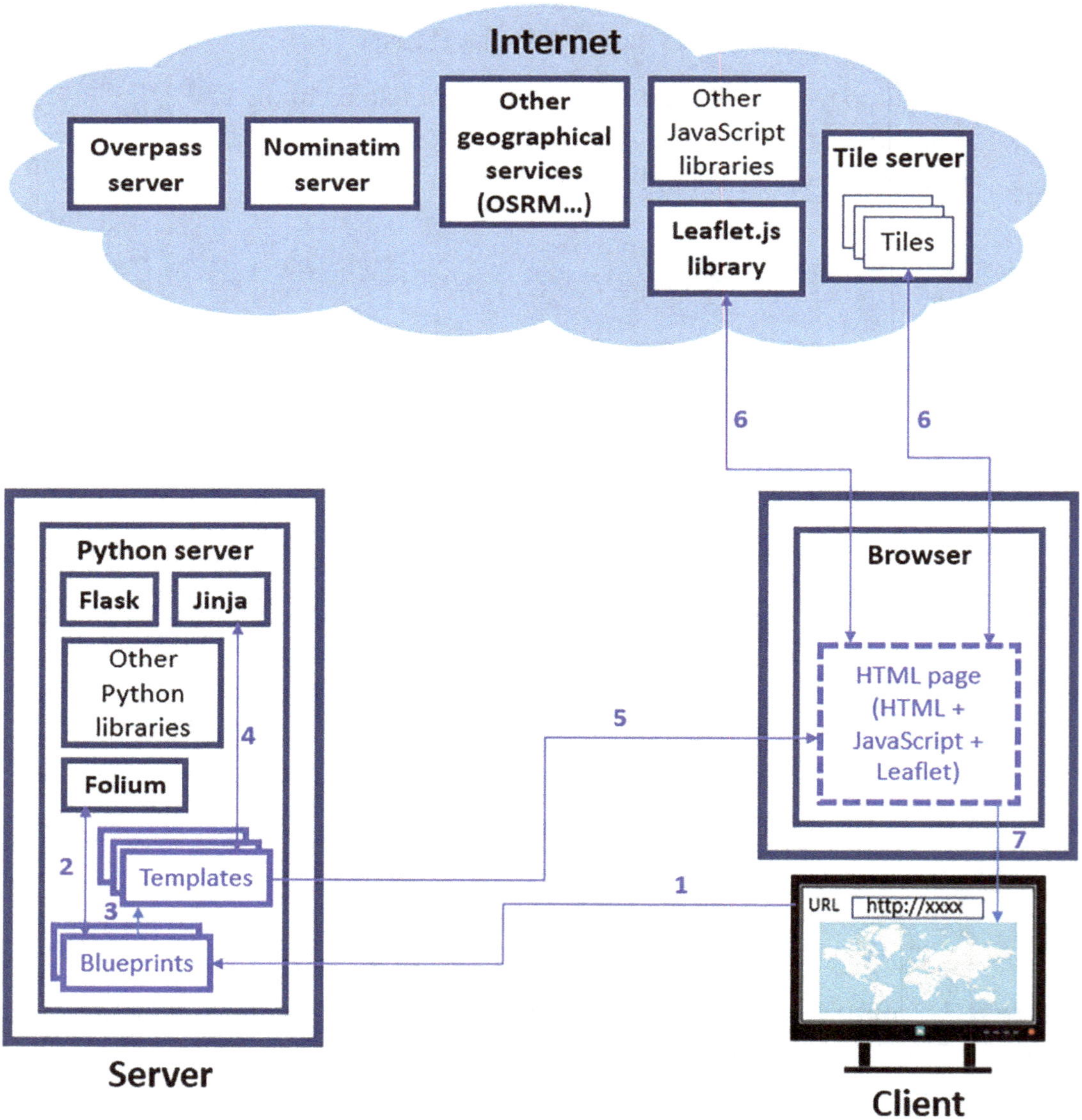

***Figure B-18.** Architecture used to test a simple Flask and Folium "Hello World!" application*

1. The user invokes a URL from the browser. This URL points to a route on the Python server.
2. The Python server retrieves the corresponding Python code for this route in a blueprint and executes it. With this Python code containing Folium statements, a map is generated as HTML, JavaScript code, and Leaflet statements.
3. The Python server then invokes a template, passing it the Folium-generated map (currently consisting of HTML, JavaScript code, and Leaflet statements) and a name.
4. The Jinja engine injects dynamic data passed by the blueprint into the template: the map (currently consisting of HTML, JavaScript code, and Leaflet statements) and a name.
5. The resulting HTML page is then returned to the browser as a response.
6. The browser receives the HTML page, executes the embedded JavaScript code and Leaflet statements, and downloads the tiles necessary to render the map.
7. The browser finally displays the resulting web page, including a map, on the screen.

APPENDIX C

Implementing a Server-Side Proxy to Avoid CORS Errors

CORS (Cross-Origin Resource Sharing) errors occur when a web page running in a browser tries to request a resource from a different origin (domain, protocol, or port) than the server that served the page, and the remote server does not explicitly allow that request. Depending on your environment and the service endpoints you use, you may experience this issue with some examples included in this book, when Overpass, Nominatim, or other remote services are invoked from JavaScript in HTML pages. In such cases, a solution is to use a server-side proxy that adds appropriate headers (that may be needed for the target servers) and a contact User-Agent (typically of the form: `MyApp/1.0 (me@example.com)`). Here is how to implement it using Python and Flask.

First, add the following proxy code to your Python blueprint as an additional route:

```
@bp.route('/proxy', methods=['GET', 'POST'])
def proxy():
    from flask import jsonify, Response
    import requests

    try:
        target_url = request.args.get('url')
        if not target_url:
            return jsonify({'error': 'Missing URL parameter'}), 400

        # Forward headers (optional: include your app's User-Agent)
```


P. Marie, *Modern Web Cartography*, https://doi.org/10.1007/979-8-8688-2335-0

```
        headers = {
            "User-Agent": "MyApp/1.0 (myemail@example.com)"
        }

        # Forward GET or POST data
        if request.method == 'POST':
            resp = requests.post(target_url, data=request.get_data(), headers=headers)
        else:
            resp = requests.get(target_url, headers=headers, params=request.args.to_dict(flat=True))

        # Return the response content and status, preserving its content type
        return Response(
            resp.content,
            status=resp.status_code,
            content_type=resp.headers.get('Content-Type', 'application/json')
        )

    except Exception as e:
        return jsonify({'error': str(e)}), 500
```

To make it more secure and avoid exposing it as an "open proxy", you can also include the following statements:

```
allowed_domains = ["overpass-api.de", "nominatim.openstreetmap.org"]
if not any(target_url.startswith(f"https://{d}") for d in allowed_domains):
    return jsonify({'error': 'Target not allowed'}), 403
```

which restrict access to the specified domains (you may need to add other domains).

Then modify the URL that you invoke in your HTML template to call the proxy instead of the target service. For example, replace:

```
...
fetch("https://overpass-api.de/api/interpreter", {
    method: "POST",
    body: "data=" + encodeURIComponent(overpassQuery)})
...
```

with:

```
...
fetch(`/proxy?url=${encodeURIComponent("https://overpass-api.de/api/
interpreter")}`, {
    method: "POST",
    body: "data=" + encodeURIComponent(overpassQuery)})
...
```

The preceding proxy code works fine when hosted by the same server that delivers the HTML pages (where CORS errors would otherwise occur). If the proxy is hosted on a different Python/Flask server, some modifications are required (highlighted in **bold**):

```
@bp.route('/proxy', methods=['GET', 'POST'])
def proxy():

    ...
            resp = requests.get(target_url, headers=headers,
            params=request.args.to_dict(flat=True))

        # Create Flask response
        flask_resp = Response(
            resp.content,
            status=resp.status_code,
            content_type=resp.headers.get('Content-Type',
            'application/json')
        )

        # Add CORS headers
```

```
        flask_resp.headers['Access-Control-Allow-Origin'] = '*'  # allow all origins
        flask_resp.headers['Access-Control-Allow-Methods'] = 'GET, POST, OPTIONS'
        flask_resp.headers['Access-Control-Allow-Headers'] = 'Content-Type, Authorization'

        # Return Flask response
        return flask_resp

    except Exception as e:
        return jsonify({'error': str(e)}), 500
```

APPENDIX D

Additional Sites and Tools Used for Prototyping

Other online tools and resources were also used to design and implement the prototypes presented in this book:

- The Overpass Turbo website (`https://overpass-turbo.eu`), used to test and fine-tune Overpass requests.
- Nominatim's Web interface (`https://nominatim.openstreetmap.org/ui/`), used to set up direct and reverse geocoding requests.
- The TagInfo website (`https://taginfo.openstreetmap.org`), used to explore the tag keys and values present in the OpenStreetMap database and identify those relevant for Overpass searches.
- The "Leaflet-providers preview" site (`https://leaflet-extras.github.io/leaflet-providers/preview`), used to visualize the renderings of the different map tile sets available and select those best suited to our needs.
- The Font Awesome cheatsheet (`https://fontawesome.com/v4/cheatsheet`), used to identify icons that can be used as symbols in map markers.
- The OSRM route planning service, used through its web user interface (`https://map.project-osrm.org/`) and through its public API.

P. Marie, *Modern Web Cartography*, https://doi.org/10.1007/979-8-8688-2335-0

- Various GeoJSON and TopoJSON files available online, providing the contours of geographical entities such as the countries of the world, American states and counties, French departments, or Canadian provinces.
- Route planning tools such as Komoot or Google Maps, used to generate exportable routes in GPX format, or programmable routers such as OSRM, used to obtain routes in the form of structured data.
- The Chrome GMaps2GPX plugin (`https://gmaps2gpx.com`), used to export routes created with Google Maps in GPX format.
- The online GPS Visualizer tool (`https://www.gpsvisualizer.com`), used to visualize GPX files graphically.
- The online GeoJSONLint tool (`https://geojsonlint.com`), used to test and visualize the contents of GeoJSON files.
- Complementary JavaScript and Python libraries, used to enhance prototype functionality.

Glossary

API (Application Programming Interface): A set of interfaces enabling programs to communicate with each other. In the context of cartography, an API can be used, for example, to request geographic data or to calculate a route.

Blueprint: A Blueprint in Flask is a Python object used to organize a web application into modular, reusable components. Each blueprint defines related routes and view functions and can be associated with its own templates and static files. This makes it easier to organize code, especially in medium to large projects.

Browser: Software for accessing web pages by interpreting HTML, CSS, and JavaScript. Popular browsers include Firefox, Chrome, Microsoft Edge, Safari, and Opera.

Cascading Stylesheet (CSS): A style sheet language used to define the presentation (colors, fonts, margins, etc.) of HTML elements on a web page.

Choropleth map: Thematic map where geographic zones (countries, regions, departments, etc.) are colored according to the value of an economic or statistical indicator (e.g., population density or unemployment rate).

CORS errors: Cross-Origin Resource Sharing errors occur when a web page running in a browser tries to request a resource from a different origin (domain, protocol, or port) and the remote server does not explicitly allow that request. A common solution to avoid CORS errors is to use a server-side proxy that adds appropriate headers and a contact User-Agent.

Data Licensing and Attribution Best Practices: Guidelines that describe how OpenStreetMap data may be used, shared, and credited in compliance with the Open Database License (ODbL).

Flask: Micro-framework in Python that facilitates the development of Web applications. It can be used to define routes, manage forms, return dynamic HTML pages, etc.

Folium: Open source Python library for creating interactive maps with simple server-side code. Folium is based on the JavaScript library Leaflet.js.

Geocoding: Process of obtaining the geographic coordinates (latitude and longitude) of a location from its address or name.

P. Marie, *Modern Web Cartography*, https://doi.org/10.1007/979-8-8688-2335-0

GeoJSON: Open format based on the JSON standard used to encode geographic data structures. It represents geographic features such as points, lines, and polygons, along with their associated properties (names, identifiers, etc.). GeoJSON is widely used in web mapping applications, especially with libraries like Leaflet or Folium.

Geolocation: Process of determining the geographical position of an object or person using GPS coordinates or other means (Wi-Fi network, IP address, etc.).

GPX (GPS Exchange Format): An open, XML-based format used to store and exchange GPS data. It can represent waypoints, tracks, and routes and is commonly used by mapping, geolocation, and route planning applications. GPX files are widely supported by GPS devices, mapping software, and online platforms, making GPX a standard format for sharing geographic data.

HTML (HyperText Markup Language): A markup language used to structure the content of a web page (headings, paragraphs, images, links, etc.).

HTTP (HyperText Transfer Protocol): Communication protocol used to exchange data on the web between a client (browser) and a server.

HTTP request: An HTTP request is a message sent by a client (usually a web browser or application) to a server, asking for a specific resource or action.

HTTP response: An HTTP response is the message that a server sends back to a client after receiving an HTTP request. It delivers the result of the request, such as the requested data, an error message, or the confirmation of an action.

Immersive views: An advanced visualization feature that provides map users with a highly realistic, interactive, and sometimes three-dimensional experience of a geographic location. It typically combines high-resolution aerial imagery, 3D modeling, street-level imagery, and sometimes real-time data (like weather or traffic) to create a lifelike representation of a place. This view allows users to virtually explore environments as if they were physically present.

IPWho: An IP-based geolocation service and open source project that provides information about the geographic location, Internet Service Provider, and network of IP addresses via a public API.

JavaScript: Programming language used to make web pages interactive. JavaScript code can be included in HTML pages and interpreted by browsers. In the context of cartography, it can be used to create maps (with the Leaflet.js library), manage user clicks, or load data on the fly.

JavaScript can also be employed on the server side, but this setup is not used in this book.

Jinja: Template engine used with Flask to dynamically generate server-side HTML pages. It lets you insert variables and control statements into HTML templates.

JSON (JavaScript Object Notation): Lightweight data exchange format, easy for humans to read and write, and simple for machines to parse and generate. It is based on a key/value pair structure, similar to that of objects in JavaScript. JSON is widely used to transmit data between a server and a Web application or between different services via APIs.

Latitude: Geographic coordinate that specifies a location's distance north or south of the equator, measured in degrees. Latitudes are positive in the Northern Hemisphere and negative in the Southern Hemisphere.

Leaflet: Lightweight JavaScript library for creating interactive maps in a browser. Often used with OpenStreetMap.

Longitude: Geographic coordinate that specifies a location's distance east or west of the Prime Meridian (Greenwich), measured in degrees. Longitudes are positive east of the Prime Meridian and negative west of the Prime Meridian.

Map: In digital cartography, a map is the canvas where geographic data, markers, layers, and interactive elements are displayed and manipulated. It serves as an interactive spatial interface, allowing users to explore geographic information through navigation, layered content, and data visualization tools.

Marker: A marker is a visual icon placed at specific geographic coordinates on a map to represent a point of interest or location (e.g., named place, shop, hotel, restaurant, office, building, community).

Node (OSM): In OpenStreetMap, a node is the most basic element in the data model. It represents a single geographic location defined by a pair of coordinates (latitude and longitude). Points can be used independently to represent features like trees, benches, or post boxes or as building blocks for more complex structures such as ways.

Nominatim: Open source geocoding (and reverse geocoding) API based on OpenStreetMap data. It is used to convert addresses into coordinates and vice versa.

Open Source Routing Machine (OSRM): A fast, open source route planning engine that uses OpenStreetMap data to provide routes by foot, bike, or car.

OpenStreetMap (OSM): A collaborative, open source project aimed at creating a free, editable map of the world. It is built and continuously improved by contributions from thousands of volunteers worldwide.

OSM XML: An XML-based exchange format used by OpenStreetMap to represent geographic data. It encodes the fundamental OSM elements: nodes, ways, relations, and tags. OSM XML is the native format handled by OSM editors such as iD and JOSM and is commonly used for importing, exporting, and processing OSM data.

Overpass: API for querying the OpenStreetMap database with a specific query language, to extract geographic elements according to various criteria.

Panoramic views: Similar to immersive views, but with a 360-degree visual representation of a geographic location. This type of view allows users to look around in all directions - left, right, up, and down - simulating the experience of standing at a specific spot in the real world.

Parser: Program or software component that reads a character string (e.g., JSON, HTML, etc.) to extract and structure the data. Parsers are specific to formats or languages. In this book, JSON and CSV parsers are used.

Polyline: A continuous line composed of several straight-line segments (connected points). In cartography, it is used to represent roads, rivers, limits, railways, or elevation lines.

Polygon: Multi-sided shape consisting of a series of connected straight-line segments that form a closed loop. In cartography, it is used to represent buildings, parks, or boundaries of cities, counties, states, or countries.

Popup: A popup is a small information box that appears on a map, typically when a user clicks on a marker or another map element, providing information about the element.

Python: A programming language widely used in data science, artificial intelligence, and web development, notably with frameworks such as Flask and Django. In the context of cartography, it can be used to create maps using the Folium library. In this book, it is used for all server-side development.

PythonAnywhere: Online platform for developing, running, and hosting Python applications. A simple Web browser is all it takes to access and work with PythonAnywhere. It offers an integrated development environment, a terminal, a code editor, and the ability to host Web applications (notably with Flask or Django), even in its free version. PythonAnywhere was used to deploy the prototypes developed in this book and verify them in a production environment.

Relation (OSM): In OpenStreetMap, a relation is a data structure that defines relationships between multiple OSM elements - nodes, ways, and even other relations. Relations are used to represent complex or grouped features, such as public transport routes, turn restrictions, multi-polygon areas (e.g., parks with holes), or administrative boundaries. Each member of a relation has a defined role (e.g., inner, outer, stop) that indicates how it contributes to the overall structure. Relations are essential for modeling real-world complexity beyond what single points or ways can capture.

Reverse geocoding: Process of obtaining an address or location from its geographic coordinates.

Route planning: Process of calculating an optimal path between two or more geographic locations. It considers factors such as road networks, travel mode (e.g., driving, walking, cycling), traffic conditions, and user preferences to generate step-by-step directions from a starting point to a destination. Tools like OSRM, GraphHopper, and Google Maps perform route planning to assist with navigation and logistics.

TagInfo: TagInfo is a web-based tool used within the OpenStreetMap ecosystem to collect, analyze, and display statistical information about the tags used in OSM data. Tags in OSM are key/value pairs that describe the attributes of geographic features (e.g., "highway=residential", "amenity=school"). TagInfo aggregates data from the OSM database to show how often specific tags and combinations are used, which values are most common, and where they are applied. This helps contributors, developers, and data consumers understand tagging practices, ensure consistency, and improve data quality.

Template: An HTML file containing placeholders for dynamic content, often written using the Jinja templating language. When a page is requested, Flask uses the template to generate the final HTML by rendering it with the necessary data provided to the Jinja engine.

Tiles: Small, square-shaped segments (often 256×256 pixels) of a larger map, used to efficiently render and display geographic data at various zoom levels. Tiles are assembled in the browser to display an entire map. Tiles are typically generated and served individually to reduce loading times and optimize performance. Tiles can be either raster - pre-rendered image files such as PNG or JPEG - or vectorial - containing geometric features and styling information that can be rendered dynamically on the client side. This tiling approach allows for scalable, seamless navigation and interaction within digital maps.

Tile server: A server that provides map tiles - pre-rendered square images (usually 256×256 pixels) representing small sections of a map at specific zoom levels. Tile servers deliver these tiles to web mapping libraries such as Leaflet which assemble them into a seamless interactive map.

Tooltip: A tooltip is a small information box that appears on a map when a user hovers over a marker or another map element, providing immediate contextual information.

TopoJSON: A compact extension of the GeoJSON format for encoding geographic data. Unlike GeoJSON, which represents geometries independently, TopoJSON encodes topology - the relationships between geometries - allowing shared boundaries to be stored once and referenced multiple times. This significantly reduces file size and improves consistency when representing adjacent features (e.g., neighboring countries or regions). TopoJSON is particularly useful for rendering maps efficiently in web applications.

Usage policies: Rules and technical guidelines that define how public OpenStreetMap services (such as the OSM tile servers, Overpass API, Nominatim, geocoding service, and other community-hosted APIs) may be accessed and used.

Visual Studio Code: Microsoft's free development environment for editing and executing code in several languages, including Python, HTML, JavaScript, etc. VS Code was extensively used to implement the prototypes in this book.

Way (OSM): In OpenStreetMap, a way is an ordered list of nodes that define a polyline or a polygon. Ways are used to represent linear features such as roads, rivers, and fences, as well as area features like buildings, parks, or lakes, when the way forms a closed loop.

XML (eXtensible Markup Language): A flexible text-based format used to store, transport, and structure data in a way that is both human-readable and machine-readable. XML allows users to define custom tags to describe the content, enabling easy data exchange between different systems and applications.

Bibliography

- OpenStreetMap wiki: https://wiki.openstreetmap.org/.
- OpenStreetMap Usage Policies: https://operations.osmfoundation.org/policies/.
- OpenStreetMap Data Licensing and Attribution best practices: https://osmfoundation.org/wiki/Licence.
- Overpass API documentation: https://wiki.openstreetmap.org/wiki/Overpass_API, https://wiki.openstreetmap.org/wiki/Overpass_API/Overpass_QL, https://wiki.openstreetmap.org/wiki/Overpass_API/Language_Guide and https://dev.overpass-api.de/overpass-doc/en/index.html.
- Nominatim API documentation: https://nominatim.org/release-docs/latest/api/Overview/.
- Leaflet.js library documentation: https://leafletjs.com/reference.html.
- Folium library documentation: https://python-visualization.github.io/folium/latest/index.html and a tutorial: https://www.callendar.tech/en/post/tutorial-creating-an-interactive-map-with-python-and-folium.
- OSRM API documentation: https://project-osrm.org/docs/v5.24.0/api/#.
- IPWho documentation: https://ipwho.org/documentation/.
- Python tutorial: https://www.w3schools.com/python/default.asp.
- Flask documentation: https://flask.palletsprojects.com/en/2.2.x/ and tutorial: https://flask.palletsprojects.com/en/2.2.x/tutorial/.

P. Marie, *Modern Web Cartography*, https://doi.org/10.1007/979-8-8688-2335-0

- Jinja 2 documentation: `https://jinja.palletsprojects.com/en/3.1.x/templates/`.
- VS Code documentation with the Python language: `https://code.visualstudio.com/docs/languages/python`.
- Flask tutorial in VS Code: `https://code.visualstudio.com/docs/python/tutorial-flask`.
- Requests library documentation: `https://requests.readthedocs.io/en/latest/`.
- Pandas library documentation: `https://pandas.pydata.org/`.
- GeoPandas library documentation: `https://geopandas.org/en/stable/index.html`.
- Turf.js library documentation: `https://turfjs.org/`.
- HTML tutorial: `https://www.w3schools.com/html/default.asp`.
- JavaScript tutorial: `https://www.w3schools.com/js/`.
- Cascading Style Sheets tutorial: `https://www.w3schools.com/css/default.asp`.

In addition to the reference documentation listed above, artificial intelligence tools such as ChatGPT can also serve as useful sources of information. In particular, they can help explain specific functions and features, analyze and diagnose problems you encounter, and suggest possible solutions. However, be aware that, unlike official reference documentation, the information provided by AI tools is not always accurate and should not be taken for granted. If the results do not match your expectations, be prepared to verify them through other sources.

Index

P. Marie, *Modern Web Cartography*, https://doi.org/10.1007/979-8-8688-2335-0

P

Q

R

S

T

U

V, W

X

Y, Z

GPSR Compliance
The European Union's (EU) General Product Safety Regulation (GPSR) is a set of rules that requires consumer products to be safe and our obligations to ensure this.

If you have any concerns about our products, you can contact us on

ProductSafety@springernature.com

In case Publisher is established outside the EU, the EU authorized representative is:

Springer Nature Customer Service Center GmbH
Europaplatz 3
69115 Heidelberg, Germany

www.ingramcontent.com/pod-product-compliance
Lightning Source LLC
LaVergne TN
LVHW082019200926
848481LV00010B/2887
* 9 7 9 8 8 6 8 8 2 3 3 4 3 *